RSAC

D1161423

JAN 2009

United States BEER CANS

The Standard Reference of
Flat Tops and Cone Tops

Beer Can Collectors of America
747 Merus Ct.
Fenton MO 63026

Copyright© 2001 by Beer Can Collectors of America

All rights reserved
Including the right of reproduction
In whole or in part in any form.

Manufactured in the United States of America
Printed on Mead Dull 80 lb. Text.
By Jostens Inc., Topeka KS
Pre-Press by Capital Graphics, Inc, Topeka KS

Publisher's Cataloging-in-Publication
(Provided by Quality Books, Inc.)

United States beer cans: the standard reference of flat tops and cone tops. - 1st ed.
p. cm.
LCCN: 00-190728
ISBN: 0-9700210-0-3

1. Beer cans--United States--Collectors and collecting. 2. Beer cans--United States--Identification. 3. Beer cans--United States--History. I. Beer Can Collectors Of America

NK8459.B36U6 2000 681.7664'075
QBI00-442

Data Entry: Herb Schwarz
Image Coordinators: Tim Hoffman Pat Kelly

A project of the scope and complexity of *United States Beer Cans* would never have been possible without the help of an army of contributors. All of the people who generously allowed us to photograph cans from their collections are included below. The list also recognizes those who volunteered their expertise to the process of establishing can values, and who provided can lore and information that found its way into the text. Each and every contributor deserves the thanks of all of us.

A very special thank you goes to Martin Landey and Fred Wolpe, whose benchmark work in cataloguing American cans some years ago provided the foundation for this book.

Contributors:

Dick Adamowicz
Steve Adydan
Keith Ajayan
Brent Alexander
Erik Amundsen
Brian Anderson
Dan Andrews
Steve Armstrong
George Arnold
Charlie Bacon
Todd Bakemeier
Hall Baker
Buck Bannon
Frank Baranco
Mike Barden
James Beaton
Ron Berg
Vinnie Berg
Jeff Berry
Curtis Boster
Bill Boyles
Cheryl Boyles
Geo Bryja
Mark Burbrink
John Burton
John Bussey
Marcia Butterbaugh
Andy Caldwell
Bob Campbell
Joe Carey
Walt Carey
John Cartwright
Dick Caughey
Tom Chegash
Bill Christensen
Lea Colvin
Pete Cornils
Brian Coughlin
Carl Covell
Kathy Covell
Gene Crane
Mark Crane
Kevin Dahmen
Robert Della Rocchetta
John Devolder
Gene DiCicco
Mark Dietrich
Bruce Dilts
Art Distelrath
Rawley Douglas
Bill Dwyer
Chris Eib
Gerald Engh-pops
Mike England
Dave Farrah
Paul Farthing
John Feinen
Bob Filus
Don Fink
Kenn Flemmons
Bob Fondren
Ed Franco
Pat Franco
John Frederickson
Jerry Gann
Joe Germino
Wally Gilbert
Mike Gisburne
Jerry Glader
Art Goetz
John Goss
Alan Green
Bruce Gregg
Cleta Gregg
Cookie Gregg
George Gregg
Kim Gregg
Ron Gregg
Adolph Grenke
John Hagedorn
Bob Hajicek
Mike Hajicek
Dick Hancock
Larry Handy
Nancy Hardaker
Warren Hardaker
Don Hardy
Ed Harker
Will Hartlep
Ken Harootunian
Bob Haefner
Rob Havrish
John Hedges
Frank Heller
Bill Helmbold
Chuck Henry
Jay Herbein
Henry Herbst
Don Hicks
Tobi Hicks
Bob Hilderbrand
Marty Hill
Chuck Hillyer
Glen Hintz
Tim Hoffman
Tom Hoffmann
John Holmes
Tim Hornseth
Ken Horstman
Tom Hull
Jerry Hyatt
Jack Isacson
Ken James
Keith Jennerjohn
Dan Jeziorski
Nick Johnson
Ray Johnson
Howard Jones
Ken Jones
Ron Jones
Tony Joynt
Neil Kanetzke
Bob Kates
Bob Kay
Elaine Kellogg
Jack Kellogg
Pat Kelly
Don Kielack
Kevin Kious
Chuck King
Shannon King
Ray Knisley
Kent Knowles
John Kottemann
Rick Kottemann
Dave Krantz
John Kretschmer
Joe Kreuser
Art LaComb
Dave Lang
Claude Lardinois
Ted Larsen
Dave Launt
Bob Lavelle
Hal Leeker
Rich Lenhard
Tom Leo
Dave Lewandowski
Mike Lewandowski
Don Lewinski
Kevin Lilek
Jerry Lorenz
Mark MacTaggart
John Marley
Bob Martin
Jerry Matonis
H. James Maxwell
Gary McClimans
Bob McCoy
Jim McCoy
Robert McCoy
Jack McDougall
Barry McGuire
Phil McLees
Norm Meier
Matt Menghini
Jim Midkiff
John Mihm
Joe Miller
Jim Mitchell
Ron Moermond
Jeff Musser
Bob Myers
Roy Nelson
Keith Niel
Bjarne Nilssen
Henry Ninde
Keith Norton
Dave Ohlendorf
Mark Oleske
Ole Olsen
Vic Olson
Marty Otto
Brent Pace
Joe Paczkowski
John Page
Augie Parochelli
Alan Paschedag
Kent Patterson
Bill Pattie
Steve Pawlowski
Doug Perry
Mark Peterson
Reed Phillips
Bob Pirie
Jim Plant
Richard Pontinen
Mark Porambo
Phil Pospychala
Dick Prazak
Dick Priessel
Joe Radman
Glenn Raisner
Mike Reilly
Jack Reisselman
George Rendl
Steve Rhodes
Dale Rogalski
Roy Rogalski
Jim Rokaitis
Jim Romine
Chick Runge
Fred Sagebaum
Cabell Sale
Don Santora
Dan Schaefer
Carl Scheurman
Gerry Schwarz
Herb Schwarz
Chuck Schwend
Dan Scoglietti
Ed Scoglietti
Terry Scullin
Kip Sharpe
Brian Sherman
Tracy Sherman
Phil Shoaf
Ed Sipos
Lonnie Smith
Scott Solie
Dave Stark
Phil Stayman
Tony Steffen
John Stejskal
Jim Stille
Wayne Stober
Greg Stoia
Mack Strickland
Greg Stroud
Richard Svec
Glenn Switcher
Bob Taylor
Hal Taylor
Jim Thole
Russ Van Nostrand
Merle Vastine
John Vetter
Don Villers
Gordon Vogel
Tom Waggoner
Harold Waite
Gerry Weishaar
Mike Weiss
Dave Wendl
Joe Wendl
Bob Werle
Nan Wick
Don Wild
Jim Wilson
Steve Wiltshire
Jim Wolf
Rusty Wyar
Tom Zalewski
Laurie Zell
Art Zerby
Adam Zoghlin

CONTENTS

PREFACE

The Beer Can Collectors of America® welcomes you to what is simply the most comprehensive guide to U.S. cone-top and flat-top beer cans ever published.

It's dedicated to every collector who has ever snooped around in the forgotten corners of beer store coolers, crawled into steamy attics or under spider-infested porches, rescued cans from years of entombment inside the walls of old buildings or resurrected them from dump sites that are home to all manner of tiny creatures that slither, buzz and bite.

Inside *United States Beer Cans* you'll find razor-sharp full-color photographs of nearly 7,300 cans. Every attempt was made to photograph the very best available example of every known cone- and flat-top can, including variations. (If you have a can that *isn't* pictured in this book, or if you see an error, let the BCCA know.) The job was so big that it took five years, visits to just about every major specialized and general collection in the country and the combined experience of scores of the hobby's most knowledgeable collectors.

> *...you'll find razor-sharp, full color photographs of nearly 7,300 cans.*

Although unknown variations of cans still pop up, only a handful of entirely different labels have come to light in the last 20 years. As a result, it's safe to assume that pictured between these covers you'll find known varieties of the vast majority of the pre-tab-top labels ever produced.

The book's pricing information is every bit as sharply focused as the photographs. It draws on the expertise of many of today's most respected collectors and leading dealers, including specialists in local or regional brands and in breweries or types of cans. Not only will it give you a general appreciation of the relative rarity of cans in excellent shape, it will also serve as an invaluable resource as you search for new gems to add to your collection.

THE HISTORY OF BEER *in* CANS

It all started with Napoleon.

In 1809, in response to the Emperor's call for a way to preserve food for his army and navy, Frenchman Nicolas Appert invented the basic process of canning. Appert used bottles and jars, but it wasn't long before England's Peter Durant went him one better by patenting a tin-coated iron container.

The can was born. Could the *beer* can be far behind?

Yes. Fast-forward a hundred years to 1909. That's when an enterprising brewer in Montana contacted the American Can Company to ask if cans were available for packaging beer. No, responded American, but we'll take a crack at it. And so they did. But the experimental pack they tested was a failure, for reasons we'll discuss later. Which meant that for the time being, Montanans and everyone else who wanted to enjoy a few cold ones at home would still have to depend on bottles and the occasional "growler" bucket filled at the local saloon.

Can manufacturers remained mildly interested. Then, on January 20, 1920, came national Prohibition, a really, really bad idea ranking right up there with the Edsel, the Tariff of Abominations and the Inquisition. Brewing, at least legally, came to a screeching halt, as did research into canning beer.

The beer can, however, was an idea whose time was inevitably to come. As the Roaring Twenties roared to a close, three factors came into play that moved the effort to create a successful metal suds container right up to the front burner.

First, more and more consumers were buying into the idea of individual unit packaging. Coffee grinders and cracker barrels were on the way out. People were now bringing home coffee in cans, baby food in jars, and biscuits, crackers and cereals in waxed cartons. If these things, why not beer in cans?

Can companies had their own motives for aggressively developing beer canning technology. Like everyone else (with the possible exception of Eliot Ness and the Untouchables), they saw that it would only be a matter of time before Prohibition came to a richly deserved end. Beer would flow again. And if a lot of it flowed into cans, can makers, hit hard by the Depression, had the potential to exploit a huge new market. Had all the beer bottled in 1934 been canned instead, for example, can companies would have added $90 million in income—a huge amount back then—to their coffers.

The Prohibition legacy

The brewing industry in the United States had been savaged by Prohibition. Many respected brewers simply went out of business. Others hung on by marketing products as diverse as ice cream, soda and malt extract. Those who would survive would discover that, despite all those images of guys and dolls downing oceans of mob-brewed beer in speakeasies and blind pigs, beer consumption in the year after Repeal was only half of what it had been in 1919, the year before Prohibition. Brewers needed something dramatically new and exciting to boost sales.

Consumers, can makers and brewers, then, were all ready for the beer can. Unfortunately, the beer can wasn't ready for *them*.

The very first beer cans to show up on store shelves: Krueger's Finest Beer and Krueger's Cream Ale.

The G. Krueger Brewery in 1953 (courtesy of John Dikun)

Two big technological hurdles had to be cleared first. When it came to beer, with all its carbonation, conventional cans just couldn't take the internal pressure—80 pounds per square inch versus the 25-35 psi created in processing green beans and tomato soup. A much stronger can would have to be developed for beer.

The second hurdle was much more critical. Beer and metal, researchers quickly found, can't stand each other. When experimental batches were packaged in conventional cans, what poured out was a milky, yellow-brown fluid with a smell and a taste that would gag a goat. A lining was needed to protect the beer from the can.

The research race

Research to solve these problems began in earnest in 1931. Anheuser-Busch and Pabst had experimented with canned beers a couple of years earlier, but the cost seemed excessive, and they weren't convinced that taste wouldn't be compromised. (Even after canned beer became a proven success, it was years before Pabst felt comfortable enough to entrust its flagship Blue Ribbon brand to cans.)

The biggest player in the can-making business by far, and the most aggressive in its pursuit of the perfect beer can, was the American Can Company. Excessive pressure caused its flat-topped experimental beer cans to buckle at the ends and burst at the seams. American engineers turned to thicker, stronger steel, formed to give the lid maximum buckling resistance. They then designed a special seam that allowed solder to flow through all three of its folds for greater strength.

The smaller Continental Can Company took a different route. Continental researchers, worried that consumers wouldn't cotton to can openers and that brewers would balk at the cost of new filling lines for flat-topped cans, came up with the cone top, a sturdy can with a short-necked conical top that could be filled on existing bottling lines and closed with a crown.

But making beer cans strong enough was an easy task compared to finding an acceptable lining. The right lining would have to be chemically inert,

1933

▼ The beer can era began in January 1935 when Krueger's Finest Beer and Cream Ale were introduced. The first beer cans, however, were really filled more than a year earlier —before the end of Prohibition!

The Eighteenth Amendment prohibited the "manufacture, sale or transportation of intoxicating liquor." But what gave it teeth was the Volstead Act, which defined "intoxicating liquor" as anything with over 0.5% alcohol.

Prohibition ended with the ratification of the Twenty-First Amendment on December 5, 1933. The beer, however, had already been legally flowing for months.

Nine days after his inauguration on March 4, 1933, Franklin Roosevelt asked Congress to increase permissible alcohol content to 4%. Legislators quickly served up the Beer-Wine Revenue Act, and by June, some 31 breweries were making 3.2% beer.

This Act made it possible for American Can to produce, and for Krueger to fill with 3.2% beer, 2,000 Krueger Special Beer cans, which were then given to Krueger drinkers to test—a month before Prohibition became history.

A photo of two Krueger Special cans appeared in the December 28, 1933 *Brewer's News*, but no example has been positively verified to exist.

Since no example of the first Krueger test can could be found for photography, we recreated it digitally.

American Can Company used this gorgeous test can to show brewers that canned beer could be a real eye-catcher on the shelf. Examples found indicate it was actually filled.

able to resist being broken down by the alcohol in beer and elastic enough to expand and contract with a container whose temperature could vary from 140°F during pasteurization to 40°F during refrigeration.

The American, Continental and National Can companies all experimented with a number of possible lining materials; each came to the conclusion that no single coating would ever be found with all of the desired characteristics. The search began, therefore, for a double-coat system.

It looked like American Can might go into production employing a first coat of enamel and a second of the brewer's pitch used inside wooden beer kegs. On September 25, 1934, American even trademarked the name "Keglined." But at the last minute it found a synthetic plastic called Vinylite. Sprayed over the enamel base, it worked like a charm.

Continental Can, simultaneously doing its own research, looked into wax coatings. A single coat of wax, however, tended to melt during pasteurization. So Continental perfected a harder wax that, when applied over a first coat of enamel, did the trick. National Can determined that a double coat of enamel would work as well as American's Vinylite and Continental's wax.

Wanted: pioneers

The technological hurdles were history. Now all that was needed was a brewer willing to take the plunge. As the can companies set out to woo brewers, they found that although everybody was interested, nobody wanted to be the pioneer. If the experiment was a failure, it might also mean curtains for the brewery. American and Continental both

The birth announcement of the beer can, from the Richmond Examiner, *January 24, 1935. (courtesy of Chris Eib)*

pursued the big breweries first, figuring that if the giants bought into the can, the smaller ones would follow suit. Except the giants were hesitant.

Enter the Gottfried Krueger Brewing Company of Newark, New Jersey.

American Can began courting Krueger in mid-1933. An average-size brewer, Krueger was pretty much in a holding pattern. Its founder had died in 1926, the family wasn't sure just where the business was headed and its workers had greeted Repeal by going on strike. An innovation like canning could be just the thing to help Krueger take off.

The can maker eventually made an offer the brewer couldn't refuse. American would install the machinery and provide the cans at its own expense. Krueger would repay American only if the test succeeded; if it failed, American would pack up its canning line and go home.

Taking the plunge

Krueger management signed on the dotted line in November 1933. By month's end, American had installed a temporary canning line and delivered 2,000 Krueger's Special Beer cans, which were promptly filled with 3.2% Krueger beer—the highest alcohol content allowed in the first few months after Repeal. Krueger's Special Beer thus became the world's first beer can.

The 2,000 cans of beer were given to faithful Krueger drinkers; 91% gave it thumbs up, and 85% said it tasted more like draft than bottled beer. So far, so good. But how would this bold, new idea fare in the real world?

For Krueger and American, the real world was Richmond, Virginia, a market carefully chosen for the test. Per capita beer consumption was low there, Richmond was far enough from Newark that a flop wouldn't hurt Krueger in its home range and there was no hometown brewer with which to contend.

By January 1935, Krueger production lines were ready. A Richmond distributor, Cavalier Distributing Company, had been contracted and advertising was ready to run. On January 24, Krueger's Finest Beer and Krueger's Cream Ale in cans went on sale for the very first time.

Brewers and can makers around the country held their collective breaths.

Then they let out a whistle of astonishment. By week's end, demand for Krueger in cans was so great that 50% of the distributors in Richmond were selling it. By month's end, that number was up to 84%.

Krueger, smelling a winner, quickly expanded the market...

Krueger, smelling a winner, quickly expanded the market to all of Virginia, then to its entire sales area, including hometown Newark. Everywhere, canned beer sold like nobody's business. By August, Krueger was producing five-and-a-half times more beer than in January and grabbing serious market share from the Big Three nationals, Pabst, Schlitz and Anheuser-Busch. American was providing Krueger with 180,000 cans *a day*.

The bandwagon had pulled up, bells ringing and trumpets blowing, and scores of brewers were clamoring to get on.

The first to come aboard was the big brewer that American Can had hoped to corral all along. In July 1935, the striking blue and silver Pabst Export Beer "TapaCan" made its debut in Milwaukee, Peoria and the Rock Island, Illinois area. By August, the brewery was filling 24,000 cans a day. Once a doubting Thomas, Pabst was now firmly in the fold of the converted.

1935

▼ In the early 1930s, while other American Can Company researchers tested beer can linings and side seams, D.F. Sampson was working on something really important: how to get the beer out of the can.

He designed a steel strip $5^{1}/_{2}$" long, 3/4" wide and 1/8" thick with a puncturing tip at one end and a crown lifter at the other. Enter the indispensable tool that soon became known as the churchkey.

American Can licensed the right to make and distribute its opener to the Vaughan Novelty and Manufacturing Company of Chicago. The first ones said "Quick & Easy Opener" and "Pats. Pend." and bore American's Canco logo. Initially, consumers found a free opener in each case of canned beer. By the end of 1935, nearly 31 million had been produced. Soon, openers began to carry brewery advertising.

The first beer cans weighed $3^{3}/_{4}$ oz. As pasteurizing pressures were reduced through improved brewing methods, can bodies and lids got thinner and lighter. As a result, openers became shorter and thinner and punched smaller holes.

This photo is from a seven-page article in the January 1936 issue of *Fortune*. Many of the cans are revered as treasures by today's collectors.

Cross-town rival Schlitz didn't let Pabst's initiative go unnoticed. The Krueger phenomenon was a real eye-opener, but it was Pabst's success that finally convinced Schlitz execs that a major brewer could profitably market canned beer. Schlitz Lager in cans was introduced in September. Instead of the flat-top route, however, The Beer That Made Milwaukee Famous came to the party all dolled up in a snazzy little number in brown and yellow with a flat bottom and a short inverted ribbed spout sealed with a crown: the first cone top from Continental can.

On a roll

The rush was now on. P. Ballantine & Sons and Adam Scheidt signed on with American that summer, as did Globe and Wehle. Genesee, Fort Pitt, Gunther and Feigenspan brews appeared in "Keglined" flat tops in September.

Its can-making capacity stretched to the limit, American had to turn some small- to medium-sized brewers away. National Can stepped up with its "Double-Lined" flat to fill the niche. In July, Northampton Brewing's Tru Blu Ale and White Seal Beer became the first brands in cans not made by American. Red Top, Fox and Stroudsburg Brewing soon followed suit. Try as it might, though, National was never able to connect with a major brewer. As big as the pond got, National seemed destined to remain a small fish.

Continental Can was another story. Heileman Old Style Lager in "Cap-sealed" cans followed Schlitz Lager into the marketplace in October. By December Burger, Blatz, Gluek, Duquesne, Fitger and Rainier had committed to the cone top as well.

Happy Birthday!

As of January 24, 1936, the beer can's official first birthday, American had signed on 20 breweries, Continental, 12 and National, 5. All together, the three companies had sold *160,000,000* beer cans. My, how the baby had grown!

The five types of cone-top cans (left to right): inverted-rib low profile, extruded-rib low profile, "J" spout, high profile and Crowntainer.

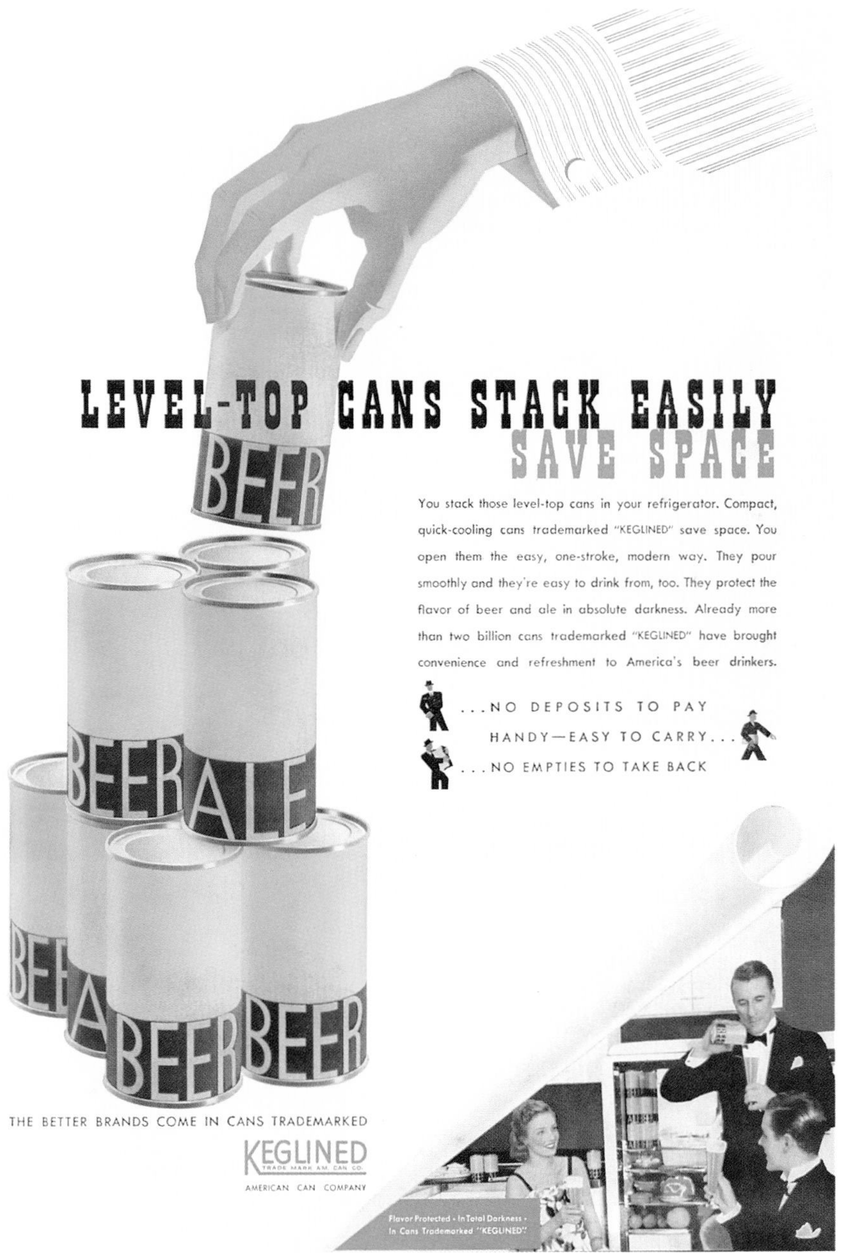

The virtues of canned beer, preached in a 1935 ad. (courtesy of Mike England)

The idea had even crossed the Atlantic; the first European can maker was Britain's Metal Box. Canned beer from Felinfoel in Wales first appeared in December 1935. Tennent's of Scotland and various English brewers followed in 1936, and in 1937 so did a few French and German brewers. Belgium, Ireland and the Netherlands came next. Virtually all European pre-war cans were cone tops.

To build on this promising beginning, brewers and can makers relentlessly extolled the virtues of beer in cans. Both spent massively to advertise the innovation. American Can awarded prizes to retail outlets that came up with the best "Keglined" store displays. Cans themselves carried promotional messages, like this one from an early Pabst Bock can:

This special container was designed to bring this beer to you with all the brewery goodness retained for your enjoyment. Here is beer as Pabst made it, in the non-refillable, brewery-filled container which protects against the harmful effects of light—tampering—anything or anybody. Now you can enjoy the true brewery flavor of this specially brewed beer, unimpaired.

Or this from one of the first cone tops: *"Cap-sealed" can for beer: Sanitary, safe. The scientific lining of this can protects the contents—insuring purity and flavor.*

Despite all this, the growth of the beer can slowed considerably after its explosive first year. Consumers still needed some convincing that beer in cans was truly as tasty as beer in bottles, especially since canned beer was more expensive. In the 1930s, you could buy

1936

▼ By the end of 1935, Krueger was on a roll. Their canned beers were selling like hotcakes and they were nabbing market share from brewing's biggest players. To build on this success, they started canning Krueger Bock and an India Pale-type stock ale called Kent Ale.

The beer can had done wonders for Krueger Beer and Cream Ale, but the magic didn't work for Kent Ale. Sales of the product were anemic from the get-go—definitely not enough to justify interrupting the filling of the popular Beer and Cream Ale on Krueger's lone canning line.

And so, sometime in mid-1936, the brand was yanked from the can lineup and relegated back to bottles. And that's how Kent Ale became the first obsolete beer can in history.

CANTASTIC MILESTONES

1944

▲ "Praise the Lord and Pass the Ammunition" was a popular WWII song. But what our troops were also passing was the Narragansett, the Rainier and the Blatz.

Beer can production had ceased in mid-1942; the steel was needed for the war effort. Then, in 1944, the U.S. military contracted with about 35 of America's largest breweries to produce beer in nonreflective olive drab-colored cans.

ODs were produced in 12 oz. flat tops, high and low profile cone tops, "J" spout cone tops and Crowntainers, all carrying the "Withdrawn Free of Internal Revenue Tax for Exportation" inscription. Although most went overseas, they were also available at stateside military bases.

Some ODs were heroes. A Continental Can Co. ad told how 960 cans of beer aboard a lifeboat adrift with no food or water kept 12 men alive for 44 days!

OD can production continued for about two years after the war's end. Over 40 different brands and variations have been discovered.

three bottles of suds for a quarter. Three *cans* cost 30¢. Canned national brands, shipped a long way, could go for 15¢ a pop.

Bottle makers initially met the threat of the can by creating a throwaway container of their own: the short, can-like bottle dubbed the "Stubby." Large regional brewers like Piels and Jacob Ruppert spurned cans for the new bottles; at the end of 1935, 28 brewers were filling Stubbies.

For many local and regional brewers, canning was simply too expensive a proposition. They produced beer in such small volume that it was difficult to justify the added cost of canning equipment, a cost they didn't dare pass on to beer drinkers.

And so, by the end of 1941, only 186 of the 507 breweries then in operation were canning beer. Although over a billion beer cans were filled that year, they represented just 10% of the packaged beer market. The can wouldn't experience another growth spurt until well after the Axis was crushed, the world was once more free, and our boys were back home. Canned beer finally caught and passed beer in bottles in 1959.

Selling the beer can by waving the flag. (courtesy of Dan Scoglietti)

Flats and cones go toe to toe

By the end of 1935, certain battle lines had been drawn. Although the bottle was the common foe, flat tops and cone tops were engaged in a struggle for dominance on the canned beer front. Each type of can had its strengths. And each would have an impact on the brewing industry that few would have expected way back then.

Larger brewers couldn't help but love the flat-top can. Beer in bottles was costly to ship. The bottles were tall, heavy, hard to stack and breakable, and the wooden cases were just heavy. The flat-top can, on the other hand, was just half the height of a 12-ounce longneck and weighed 55% less. Flat tops stacked easier and could be shipped in lightweight cardboard. More cans in every truckload brought transportation costs down tremendously, making it possible for brewers to distribute their brands to more places.

The other thing big brewers loved was that flat-top cans could be filled faster than bottles—160 cans vs. 120 bottles per minute, or nearly 20,000 more units on every eight-hour shift.

Ironically, the features that made flat-top cans so popular among brewers eventually helped put many of them out of business. Of the nearly 750 breweries that reopened after Prohibition, only a third were still around in 1960. By 1980, only 70 brewing plants survived. The fallen were victims, at least in part, of the volume efficiency of the national giants. Canning had helped to make that efficiency possible.

Smaller brewers originally tended to favor the cone-top can. But what began as the cone's major strengths turned out to be its Achilles heels. The main selling point of the cone top was that you didn't have to spend a bucket of money—money small brewers didn't have—on special canning lines. With minor adjustments, cone-top cans could be filled on bottling lines.

The bottle-like can was also consumer friendly; it was sealed with a familiar crown, and you didn't have to learn how to use a newfangled contraption, the "churchkey," to open it.

The "golden age" of the cone top was the period between 1935 and America's entry into World War II. In the fall of 1936, Continental Can introduced a redesigned cone top with a concave base and a top whose ribs were raised rather than inverted.

Continental's brief reign as the sole manufacturer of cone tops ended in 1937, when the Crown Cork & Seal Company bought the Acme Can Company of Philadelphia and started making cans of its own. Crown's "J" spout had a body shorter than a flat-top can and a concave bottom; its spout had a longer neck than Continental's can. (Crown also introduced some oddball flats that year, including an 8-ounce size and the "Tall 12," but they never caught on.)

Continental countered Crown's entry into the fray by introducing the 32-ounce quart cone top in June of 1937. Over 40 brewers filled quart-size cones during the next two years, including cones made by American Can, but the fad faded quickly after that.

The next truly significant change in can design came in 1940, when Crown Cork & Seal introduced the can it called the "Crowntainer." What made the Crowntainer so different was that, unlike other cone tops, it was made of two pieces rather than three. The body was formed by drawing the sides and top

1950

▼ In spring 1950, a Hollywood ad agency presented a novel idea to struggling Los Angeles breweries: run a campaign about a crusade to brew the perfect beer. Success would come only after 101 tries. The ultimate elixir would be called Brew 102.

The only taker was the Maier Brewing Company. Brew 102 joined Maier Select as the company's second flagship brand. And in just one year, sales soared by two hundred percent!

A light bulb went on in Maier management minds. If they could market the same beer under two labels, why not under two dozen? Or more? ABC, Amber Brau, Alpine, Bavarian and several other brands soon appeared. While some Maier brands were widely distributed, others were strictly private label. Club Special was sold by the Marines at Camp Pendleton, CA; Zodys, a pull tab, was available only at a local discount chain, in unrefrigerated cases.

Maier produced over 100 different brands, most in cans, before closing their L.A. brewery in the early '70s. Many breweries copied this sales-boosting one beer/many brands strategy. But none did it as early, or as effectively, as Maier.

1935

▼ It took a David among can makers to create the Goliath of beer cans.

The time was summer 1935, and while American and Continental were banging out 12-ounce flats and cones, the smaller Heekin Can Company of Cincinnati had a really big idea.

At $10^3/_4$ inches tall and $7^1/_2$ inches in diameter, Heekin's "Can-O-Draft" towered over the beer packaging landscape. It held 1/16 bbl., or nearly two gallons, of beer and had a unique built-in tapping system.

Sometime between July and October 1935, Heekin developed a prototype "Can-O-Draft" and started looking for takers. They didn't have to look far; a Cincinnati neighbor, the Burger Brewing Company, liked the idea enough to go into production immediately.

The king-sized cans with the camel on the front turned out to be less than a huge success. At a time when refrigerators had a lot fewer cubic feet of storage space, they were probably just too big, and the beer, once tapped, didn't stay fresh.

There's nothing to document how many big Burgers were filled or where they were distributed, but the can shown here, discovered around 1985 on the East Coast, is the only known example. "Can-O-Draft" designs exist for four brands from the Kings brewery in Brooklyn that apparently were never produced.

from a single sheet of steel. Tin-plated steel couldn't be used in this process, so although the concave bottom remained tin-plated, the body had a coating of aluminum. That's why the new can was nicknamed the "Silver Growler."

The Crowntainer proved so popular that Crown gradually phased out its "J" spout design. Crowntainers remained in production until the mid-'50s. The last brewer to use them was Louis F. Neuweiler & Sons.

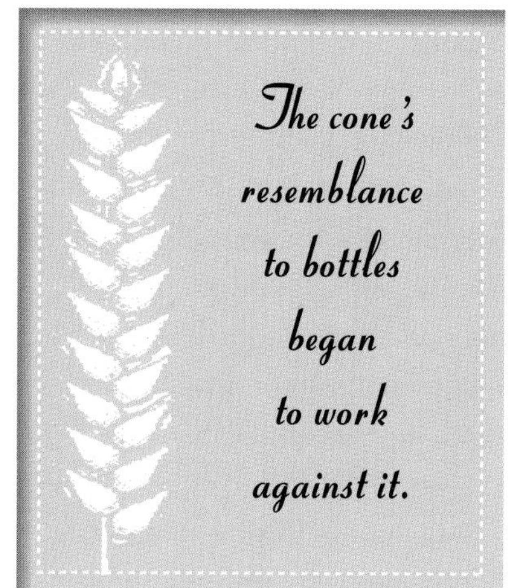

The cone-top can was an important part of beer can history for more than 20 years. By the mid-'50s, though, volume was the name of the brewing game, and cones still didn't fill any faster than bottles. This is primarily why biggies like Anheuser-Busch, Miller, Pabst and Coors never chose to market their beers in cone tops. And why more and more smaller brewers opted for faster-filling flat tops.

Even the cone's resemblance to bottles began to work against it. Cone tops were hard to stack at retail and in home refrigerators. Plus knocking back your brew straight from the flat-top can had become acceptable. The last cone tops were filled by the Rice Lake Brewing Co. in 1960.

The beer can gets drafted

Of all of the dates in the history of the beer can, one of the most important is May 31, 1942. That's the day when, in response to War Production Board Order M-81, the canning of beer for the general public came to a halt. Uncle Sam needed all that tin-plated steel to help build the military might that would save the world.

Canned beer didn't disappear, however; it simply shipped out with the troops. During the course of the war, over a billion cans of beer, painted olive drab for camouflage purposes, were distributed to our armed forces around the world. About 35 brewers canned beer for the military. The demand for flat-top cans was so great that even Continental Can started producing them.

Over There, canned beer brought out good old American ingenuity. Flight crews in the Pacific Theater, humping war supplies over the Himalayas on the India-to-China run, often

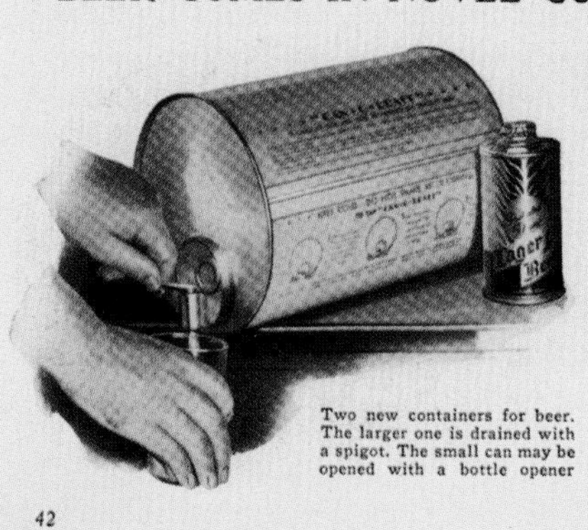

BEER COMES IN NOVEL CONTAINERS

BEER is appearing in a variety of containers, these days, and two of the newest are illustrated at the left. The small one is a tin can with a bottle-cap top; the top may be removed with an ordinary bottle opener, and the can may be thrown away when it is emptied. These small cans can be stored compactly. The larger of the two containers shown holds one sixteenth of a barrel. It is drained with a spigot, and is said to provide all the attractive features of draught beer.

Two new containers for beer. The larger one is drained with a spigot. The small can may be opened with a bottle opener

42

Big news from the January, 1936 issue of Popular Science Monthly.

stashed cases of suds in a corner of the unheated cargo hold. The frigid temperatures at high altitudes did a perfect job of cooling the beer down!

In the absence of niceties like glasses, our men and women in uniform simply sipped their beer straight from the can. When millions of them came home, they brought this habit with them. The acceptability of drinking out of a flat-top can was one of the last straws for the cone top.

In 1947, War Production Board Order M-81 went by the wayside, and beer cans, with new label designs and snappier graphics, went back on the shelves. Beer sales soared to a record 87.2 million barrels.

If the '50s were a bad time for cone tops, they were a great time for flats. While cones faded away even faster than MacArthur, brewers introduced flat-top cans in a variety of new sizes. Schlitz, often a pioneer, led the way in 1954 with the first 16-ounce flat. A couple of years later, Ballantine came out with flat-top quarts. In the belief that bigger isn't always better, Goebel and Goetz tested the waters with an 8-ouncer. And because of oddball state laws, cans started showing up in the South and West in 10-, 11-, 14- and 15-ounce sizes.

The end of an era

With the close of the '50s came the arrival of new technology that would change the beer can forever. Aluminum was coming. And it meant business. In 1958, the Hawaii Brewing Company put out the first all-aluminum flat, an 11-ounce version with a paper Primo label. Shortly thereafter, Coors became the first brewer to successfully market 7-ounce aluminum cans. These, though, were simply the first salvos in a battle that would eventually send the traditional steel can packing.

Brewers, always looking for ways to trim costs, were pressing for cheaper, lighter cans. Alcoa and Reynolds Aluminum were looking for ways to gain entry to the can business. The aluminum giants tackled the problem from the top down. Coors had experimented with aluminum tops on steel cans in the late '50s. The idea didn't catch on, though, until Schlitz trumpeted its "SofTop" can in 1960.

Popular as these aluminum-topped cans became, you still needed a churchkey to open them. The breakthrough came when Alcoa introduced the ingenious device called the tab top. Sure, it sometimes bloodied your fingers. Sure, it sometimes broke off. But you didn't need an opener!

We're reaching the end of our historical journey now; the story of the era of the tab top and the all-aluminum can belong to another book. The amazing thing is that we're finishing up just about where we started. When the Iron City Brewing Company tested the first tab top can in March 1962, they did it in Virginia—the very state where the beer can had been "born" nearly 30 years earlier.

In the words of a certain well-known Bronx Bomber, it was déjà vu all over again.

1955

▼ She was petite, lovely and noble. And she was sired by a Pittsburgh disc jockey.

In the early 1950s, KDKA radio's Rege Cordic invented a mythical brew called Olde Frothingslosh, brewed by Sir Reginald Frothingslosh IV at Upper-Crudney-On-The-Thames. Cordic made up commercials with slogans like *A whale of an ale for the pale stale male* and *Hi dittom dottom, the foam's on the bottom.*

In 1954, the Pittsburgh Brewing Company obtained rights to bottle the brew and, just for fun, printed up labels and gave 500 cases to customers for Christmas.

The brewer was deluged with requests for the coveted Pale Stale Ale. So for Christmas, 1955, they marketed not only bottled Olde Frothingslosh but also Sir Lady Frothingsloth in 8 oz. cans. Nearly a quarter of a million cans were produced, but only a small number still exist.

It was the Lady's only public appearance. When Olde Frothingslosh was eventually canned in 1968, it was not she but the voluptuous (and voluminous) Fatima Yechburgh who graced the label.

BUILDING *a* COLLECTION

Once you've been bitten by the beer can collecting bug, you'll discover that the urge to put up yards and yards of shelves and fill them with your finds is irresistible. You're not alone; there are many others who share your compulsion.

Over 4,000 worldwide are members of the Beer Can Collectors of America, founded in St. Louis on April 15, 1970.

Joining the BCCA is a great way to build a quality collection quickly. Local chapters all over the country (as well as international chapters) host hundreds of can and breweriana shows a year. And each September, BCCAers from all points of the compass flock to a huge annual *Can*vention. Recent host cities have included Atlanta, St. Louis, New Orleans and Nashville. Dates and places of all events are published in the club's colorful and informative bimonthly *Beer Cans and Brewery Collectibles* magazine, as are dozens of want ads placed by individual members.

To learn more about the BCCA, and to access a wide array of Internet links related to cans, breweriana and breweries, pay a visit to ***www.bcca.com.***

Then there's the thrill of the hunt. There are all kinds of great beer cans out there just waiting to be stumbled upon and acquired by someone like you. Finding them is a matter of knowing where to look.

Where beer cans hide

Beer cans show up in the darndest places. Abandoned structures, attics, crawl spaces, walls, barns, sheds, loading docks, the back rooms of mom-and-pop stores, dumps, hunting camps and even lovers' lanes can all be home to older cans in decent to great condition. Many a rare can has spent ages on a workbench where it has served as a container for spare screws, nuts and bolts. The more you snoop, the more you're likely to make a find.

> *There are all kinds of great beer cans out there just waiting to be stumbled upon and acquired by someone like you.*

One of the most important things you can do is make sure that absolutely everyone you know or meet is aware that you collect cans. You never know when somebody's going to remember that Great-Uncle Zeke in Paducah has this funny old beer can with a top like a gasoline additive can. It doesn't happen often, but, as almost all veteran collectors will tell you, it *does* happen.

It definitely helps to have friends who are plumbers, electricians, exterminators, rehabbers or others who frequently access places where old cans might hide. In the past, it wasn't uncommon for construction workers to knock back a couple of brews with lunch and then deposit the empty cans inside of walls, where they would lie protected from the elements.

People who collect other things can also be a valuable resource because they often come across beer cans in pursuit of their own hobbies. This kind of connection works best when it's reciprocal; be sure to clue them in when you run into the kinds of items they collect.

One person's trash

Another fun way of finding cans is to go looking for them in the great outdoors. Dumps full of rare old beer cans may be found anywhere people have needed places to dispose of empties. Recreation areas like hunting, fishing or boating resorts seem to be especially fruitful, but rural taverns, mining and logging camps and even farmhouses often have piles of old cans in the ravines or on the hillsides out back. Some cans languish forgotten in old wells, cisterns or outhouse pits. Researching old county road and topographic maps can be a big help in locating dumps.

Many of the hobby's most dedicated devotees of dumping belong to the Rusty Bunch, an at-large chapter of the Beer Can Collectors of America. Over the years Rusty Bunchers, who are known for leaving no stone—or trash pile or privy seat—unturned, have ferreted out many extremely rare cans that might otherwise have remained undiscovered. Access their home page through the BCCA's website.

Tools of the trade

When you go dumping, bring boots, a good pair of leather gloves and digging tools like a rake, hoe and shovel. It can also pay to bring a metal detector. Older dumps are often completely covered with leaves. You may not realize you've found one until you hear the crunching of cans underfoot. Other tell-tale signs may be a rusty barrel or bottles poking up through the leaves.

The best sites are those that are shaded and have good drainage. Look for beer cans that have been stuffed inside larger soup cans; these "can-in-a-can" finds are often surprisingly well preserved. Outdoor cans that have survived for many years in ideal environments can sometimes be soaked in a mild acid solution and restored to a very displayable condition.

Make sure you thoroughly rinse out any cans you find outside. You'll want to get rid of all the dirt, sticks and debris, as well as anything that may be living inside.

Now that the hobby has become so popular, antique and collectibles dealers often have beer cans and other breweriana items for sale. Cans also frequently show up at flea markets. And you can find them on an ever-increasing number of Internet sites. There are always a few sellers who are out for a quick buck, so it pays to be knowledgeable about beer cans and their relative scarcity and worth. If possible, research a can (this book will help) before you put your money down.

And, of course, there's nothing like spending a sunny summer weekend hitting every garage sale for miles around.

Running out of space

Once it dawns on you how many thousands of beer cans are really out there, you may decide, as many collectors have, to specialize. In what? Why, in whatever tickles your fancy. Some collect only U.S. cans, some only foreigns. Some covet only flat tops, some only cones, some only gallons. Some savor only ale or bock cans.

Some limit themselves to cans and breweriana from a particular region, state, city or even brewery. Some save only sports-related cans, some only cans with women on them. And some....well, you get the idea. Anything goes.

Finally, don't be surprised to discover that you take just as much pleasure in unearthing great cans as you do in owning them. To most collectors, one of the most satisfying aspects of the hobby is swapping how-I-found-it stories with fellow enthusiasts—preferably over a couple of cold ones!

There are a number of clubs for collectors of beer cans and breweriana. Here's how to contact the main ones:

Beer Can Collectors of America (BCCA)
747 Merus Ct., Fenton, MO 63026-2092
(636) 343-6486
www.bcca.com

American Breweriana Association (ABA)
P.O. Box 11157, Pueblo, CO 81001
www.a-b-a.com

Eastern Coast Breweriana Association (ECBA)
P.O. Box 349, West Point, PA 19486

National Association of Breweriana Advertising (NABA)
1380 W. Wisconsin Ave. #232, Oconomowoc, WI 53066
(262) 560-1948
www.nababrew.org

Rusty Bunch
152 Crestview Dr., Jaffrey, NH 03452

1959

▼ The story of the first all-aluminum beer can to be filled and distributed is a comedy of errors.

Hawaii achieved statehood in 1959. To coincide with this momentous event, the Hawaiian Brewing Company introduced a unique canned version of its flagship Primo brand. The whole can was aluminum, and it sported a paper label. Lightweight aluminum was cheaper to ship from the mainland than steel components, and paper labels eliminated the cost of paints and can painting machinery.

"Shiney Steineys," as they were called, were instant disasters. The can's lacquer lining didn't adequately protect the beer from the aluminum. The brewery had to dump 23,000 cases of bad beer.

It gets worse. The cans had concave lids that actually dipped down into the pressurized beer–which meant that when you pierced the lid, you were rewarded with a Primo shower! Sometimes, because the top was soft, the opener would keep going and come through the side. Thousands more cans were recalled and dumped.

The brewery said it hoped the public would forgive and forget. It didn't. The flop was so huge that Hawaii Brewing went into bankruptcy and was then purchased by Schlitz in 1963.

(ad courtesy of Adam Zoghlin)

Unfortunately, dating beer cans is far from a precise science. Brewers had no reason to put dates on cans; all the consumer needed to know was that there was beer inside. Even today's "freshness" dating has to do with the beer, not the can.

But, over the years, brewers and can makers unintentionally left a trail of clues that can help the observant collector deduce the approximate age of a can. All it takes is a little detective work. Here are some of the clues you should look for:

Exact Dating

A handful of brewers included dating information on their cans at one time or another. In 1949, for example, Schlitz decided to start putting a label design copyright date on its cans. The trouble is, they didn't change the copyright until 1954, when the design changed. Thus, all Schlitz cans filled in the years between are dated 1949!

Several Western brewers came up with a concept they called "age dated beer"; the exact date the can was filled was printed on the top or bottom lid. Examples like this, though, are few and far between. For the most part, beer cans don't have readily decipherable dating information on them.

IRTP

This one's simple. If the words "Internal Revenue Tax Paid" or something similar appear on a can, it most likely means that the can was manufactured between June 1, 1935 and March 30, 1950. So, other than a handful of cans filled between January and May 1935, all cans from this 15-year period carry the federally-mandated IRTP verbiage.

All cans, that is, except those billion or so "olive drabs" that were shipped overseas to our troops during WWII. These and other exported cans carried a designation such as "Withdrawn Free of Internal Revenue Tax for Exportation."

Can Maker Logos

When it comes to beer cans, can makers have historically believed in taking credit where credit is due. The American Can Company, for example, trademarked the "Keglined" name in September 1934. The word appears prominently on all American Can Company cans manufactured between 1935 and the early 1950s. Cans where "Keglined" appears on the face tend to be older than variations of the same can with "Keglined" on the side panel.

Continental Can, Crown Cork & Seal, National Can and others also left their marks on the cans they manufactured. Check out the accompanying chart for a summary of can maker logo information that can help you narrow down a can's age.

Can Maker Logos: Who, When, What

American Can Company
1935 - 38: "Keglined Patent/Patents Pending"
1938 - mid-40s: "Keglined Pats., 1,625,229-2,064,537 other pending"
1947 - mid-50s: "Keglined Pats., 2,064,537, 2,259,498-2,178618" printed across a full-length vertical panel
Mid - late 50s: "Keglined Pats., 2,064,537, 2,259,498-2,178618" printed in small letters near the seam
50s - early 60s: "Keglined" appears in a small oval
Mid-60s: "A" and other symbols (soldered seam)

Continental Can Company
c 1950: Flat top cans had three large "C"s
Early 50s and on: Flat top cans with three small "C"s
Late 60s: "Conoweld" (welded seams)
1935 - 38: Cone top cans had "Cap-Sealed" in block letters
1938 - c. 1950: Cone top cans had "Cap-Sealed" in script letters

Crown Cork & Seal
Pre-WWII: Flat top cans had "Crown Can"
Late 40s - 50s: Flat top cans had "Crown" in a crown-shaped box

National Can Company
1935 - early 40s: "Patents Pending National Can"
1950s and on: "National" in outline map of U.S.

Pacific Can Company (all flats; merged with National in 1955)
Mid - late 1930s: "Keglet"
Late 1930s - 1955: "Pacific Can"
1950s: "Kan Keg"

Cans Inc. (merged with National in 1953)
Early 1950s "Cans Inc." on flat top cans

Heekin Can (Cincinnati; division of Diamond International)
1960s "H" in a small circle

Opening Instructions

To us, it's something that's been around forever. But to beer drinkers in the can's earliest years, the can opener was a newfangled and unfamiliar device. That's why the majority of the beer cans produced prior to WWII feature opening instructions, complete with illustrations, on one or both side panels. Some of the earliest instructional cans from the 1930s featured either panels of instructions or a top-to-bottom picture of an opener. The more common two-panel configuration came later. Instructional panels disappeared after the war—with the exception of a few cans like Genesee 12 Horse Ale, which carried opening instructions into the early 1950s.

Can Clues

A few crow's feet under the eyes, a little gray at the temples: indicators that can tell you a lot about a person's age. In the same way, there are indicators that can tell you a lot about a beer can's age.

One of the first is the gauge of the steel from which the can was formed. If it's so heavy that the only way you could dent it is by backing over it with your car, the can was most probably made before mid-1942. Weight and gauge generally decreased over the years as manufacturing processes improved.

Next, look at which words on the label are in big type and which are in little type. When beer cans were introduced, the first thing brewers had to do was make sure consumers could easily tell what was inside. On these early cans, then, the words describing the contents—BEER, ALE and so forth—were usually a lot bigger than the brand name. As consumers became more at home with canned beer, the size relationship reversed itself. By the early '50s, it was the brand name that got top billing.

A number of the earliest beer cans sported side panels which, to increase consumer confidence in this new form of packaging, extolled the virtues of canned beer or the can itself.

Brewery Obituaries

By 1960, fully 500 of the 750 breweries which had reopened after Prohibition had bitten the dust. Before they gave up the ghost, however, many of them produced canned beer.

With a little research, it's possible to ferret out the dates of a given brewery's active operations. These dates, in turn, will give you a time frame in which that brewer's cans were produced. Two excellent sources for such information are the 1993 BCCA *Catalog of American Beer Cans*, which is available from the BCCA, and the 1995 *American Breweries II* by Dale P. Van Wieren.

Old Magazines and Newspapers

Prior to the advent of television, many brewers advertised in magazines and local newspapers. Their ads often contained pictures of cans. It's a tedious process, but by spending a few hours at the library poring over these old publications, you may find can pictures that will help you quite accurately date a particular can.

·CANTASTIC MILESTONES·

THE 1960s

▼ CinemaScope. The Edsel. The ten-gallon hat and the foot-long hot dog. Americans have a thing for bigness. But, as the history of the gallon beer can illustrates, big doesn't always turn out to be better.

The first gallon cans, filled in Hammonton, NJ, and Dunkirk, NY, made their appearance in the early 1960s. Over the next decade, 17 brewers marketed big cans. Fewer than 60 different gallons (plus one half-gallon), including variations, are known to exist, the rarest being the Wild Mustang Malt Liquor produced by Chicago's Atlantic Brewing.

So why did the gallon follow the Edsel down the highway of oblivion? By all reports, the tapping system didn't work very well. And then there was the beer. The unpopular big cans spent a lot of time on store shelves, so the beer inside was sometimes past its prime.

The size itself also turned out to be a turn-off. Gallons were heavy to lug from store to home, they took up a lot of space in the refrigerator, and once you tapped the can, you had to drink all the beer.

By the very early 1970s, the era of the gallon can was over. Sometimes the bigger they come, the harder they fall....

GUIDELINES *for* GRADING

There's a lot of difference in terms of value and desirability between a beer can that looks like it was just shipped from the brewery and one that looks like it's spent the past 40 years as cheap housing for a family of field mice.

That's why it's so important to have a system for accurately describing a can's physical condition. Numerous grading systems have been devised over the years; the Beer Can Collectors of America considers the following to be a standard and easily understandable set of grading criteria. Can prices in *United States Beer Cans* are based on original cans in Grade 1 condition.

A five-grade system

Ours is basically a five-grade system. But since can condition has such an important impact on pricing, the top two grades have plus and minus variations to take into account subtle but important distinctions.

You'll also find mention of "indoor" and "outdoor" cans. Indoor cans are generally found in places where they have been protected from light and moisture, a beer can's greatest enemies. Most indoor cans fall in Grades 1 and 2. Those below Grade 1 are often referred to as "off grade." Outdoor cans have been exposed to the elements to varying degrees and are rarely in good enough shape to earn more than a grade 2 rating.

GRADE 1+

Virtually perfect, with no noticeable imperfections of any kind. Only a tiny number of cans come off the canning line in such pristine condition.

GRADE 1/1+

Extremely minor scratches and dings around the rim, and lids may show some tarnish or discoloration, but no humidity spots, rust, dents or imperfections. (A can whose imperfections are so miniscule that it falls between 1/1+ and 1+ may be classified as About Grade 1+.)

GRADE 1

Very minor scratches, marks or humidity spots on the surface. The condition used as the pricing standard for all cans in this book.

GRADE 1-

Very small humidity spots, dents and scratches or very light color fading.

GRADE 2+

Noticeable spotting, blotches, scratches, small dents or tiny rust spots, or some color deterioration. Normally applies to indoor cans, although an occasional outdoor can falls in this grade.

GRADE 2

Clean label, but with rust and other blemishes on or near the seam, lid or bottom or with some irregular color.

GRADE 2-

A somewhat clean label, but with rust and other imperfections on the label or elsewhere on the can.

GRADE 3

Multiple scratches, spotting, rust and dents. The label is readable, and the color is still reasonably good.

GRADE 4

Rusted but easily identified by brand and label. Fading, dents, rust, holes and scratches.

GRADE 5

Very rusted and damaged and barely recognizable by brand and label.

The whole point of grading is to enable the collector to describe a can's condition to another collector clearly and honestly. In addition to assigning a grade number, many collectors will provide additional detail, including descriptions of seam holes, fading, pitting or bubbling. While the label is the most important part of the can, the entire can must be taken into account when determining the grade. Any damage on the sides or on one face should be described.

If a can is lidless, has had a lid added, or has been rolled from a flat sheet—or if restoration work has been done—it's very important to convey this information as well. These factors will affect the can's value.

Grade 1+

Think conservative

As a rule, you should be as conservative as possible when you grade cans. It's also an excellent idea to ask for or send photographs.

Every collector would like to have shelf after shelf of nothing but Grade 1 and 1+ cans. Few collectors actually achieve this goal. Others, however, may choose to collect cans in any grade just to have them represented in their collections. With less-than-perfect cans, there's always the opportunity to upgrade when a better quality can comes along.

Grade 1/1+

Grade 1

Grade 1-

Grade 2+

Grade 2

Grade 2-

Grade 3

Grade 4

Grade 5

1962

▼ Ermal Fraze was an unhappy person. On a hot, hazy day in the summer of 1959, he was at a picnic with lots of friends and lots of icy cold canned beer. Except nobody had brought an opener. Legend says he opened cans on a car bumper instead.

A lesser man might have taken this as one more banana peel strewn along the path of life. Fraze, a Dayton, Ohio tool and die designer and manufacturer, took it as a challenge. He proceeded to invent and patent the biggest innovation since the beer can itself: the pull tab.

Fraze touted his idea to a number of brewers. Eventually he sold the concept to the Pittsburgh Brewing Company and the rights to his invention to Alcoa.

Iron City cans with pull-tab tops produced by Alcoa went into test market in March 1962. A year later, both Iron City and Schlitz cans boasted pull tabs as standard equipment. In June 1963, 40 brands sported tabs; in August, 65. And by 1965, some 75% of all U.S. cans produced had an easy-open device. Ermal Fraze, needless to say, was unhappy no more.

What's a particular beer can worth?

Age, condition, desirability and scarcity all come into play. But, especially with really tough cans, we're talking the good old law of supply and demand: that point in a transaction between knowledgeable collectors where "What will you give for it?" meets "What will you take for it?"

Knowledge is the key. So here's a short course in the art of valuing beer cans. Following these guidelines will help you avoid paying too much for a can you buy—or accepting too little for a can you sell.

Some expert advice

The perfect place to start is with the values assigned to the cans pictured in this book. Each value is for an original can in Grade 1 condition. These values are based on demand and rarity, and they represent the collective buying and selling experience of some of the most respected people in the hobby. The roster of over 60 experts who collaborated on this effort includes specialists in local or regional brands, or breweries or types of cans, as well as leading dealers.

Cans worth more than $1,000 are listed at $1,000+, and for a simple reason. Being more precise is difficult because only a small percentage of the cans that are offered for sale fall into this category, and the competition for them can be intense. Ultimately, their value is usually determined by how badly potential purchasers want them.

Values change dramatically with condition; a dumper is obviously worth a whole lot less than a more pristine can. So how do you determine the value of lesser grade cans?

Use the Grade 1 values in United States Beer Cans as a base. There are no hard and fast rules, and rarity and availability are always variables, but veteran collectors would generally agree that the following scale of percentages of Grade 1 value is fair and realistic:

Grade 1- — 50% to 75%
Grade 2 — 20% to 50%
Grade 3 — 5% to 20%
Grade 4 — less than 5%

And for Grade 1+, most cans will usually bring a 25% to 50% premium over the same can in Grade 1 condition.

Cones, quarts and extremely rare cans could bring in slightly more than these percentages, and very common cans slightly less.

Restored cans: another story

Beer cans are like people; when something goes, they may need a little reconstructive surgery. This, too, affects value.

A can with a replacement lid generally has a value of 75% to 90% of the original. A can that has been rolled from a flat sheet will usually bring in only 50% to 75% as much as the original can. A touched-up can is worth much less than one in original condition, and extensive restoration can shrink its value to next to nothing.

Remember that what's presented here is meant to be taken as guideline, not gospel. One of the most enjoyable aspects of our hobby is being able to engage in the age-old pastime of haggling.

USING UNITED STATES BEER CANS

Not only is this book the most complete guide to U.S. flats and cones ever published, it's also designed to be user friendly. Photographs and information are organized in a very logical way. Once you get the hang of it, locating a specific can is as easy as falling off a beer keg.

Let's go for a quick guided tour.

Eight color-coded sections

For starters, the book is divided into eight color-coded sections by can size, as follows:

- **Flats** — 10, 11 and 12 oz. (Red Section)
- **Cones** — 10, 12 and 16 oz. (Green Section)
- **Crowntainers** — 12 oz. (Orange Section)
- **Quarts** — 32 oz. cones/flats (Purple Section)
- **Pints** — 14, 15 and 16 oz. cones/flats (Yellow Section)
- **Big Cans** — 24 and tall 32 oz. flats (Black Section)
- **Small Cans** — 6 1/2, 7 and 8 oz. flats (Gray Section)
- **Gallons** — 64 and 128 oz. (pre-1970s) (Navy Section)

Within each section, cans are presented in alphabetical order by brand name. There are exceptions, but overall nothing could be simpler. Once you find the brand, look for the type of brew. In most cases, alphabetical order is followed here as well. Ale comes first, followed by Beer, Bock, Draft, Half/Half, Malt Lager, Malt Liquor, Stout Malt Liquor, Porter and Near Beer.

Some brands—Tudor, Brown Derby, Budweiser and Kol to name a few—were filled by more than one brewer. Thus, cans that may be otherwise identical will boast different brewery and/or location information. In cases like this, you'll find a photograph of each brewery/location variation. And if more than one variation was produced at a particular location, cans will appear in order, as closely as possible, from oldest to newest. You get the picture.

A view to variations

Speaking of pictures, there's one of each variation of the face of a given can. If versions with and without opening instructions exist, both are represented. Some major side panel variations have also been photographed, but to picture most panel differences and distribution information would have been beyond the scope of this book. Two-faced cans were photographed with the seam on the left.

We attempted to include every can in every known set, with the exception of the huge Rainier Jubilee and Esslinger Parti Quiz series. Representatives of these two sets are pictured.

Loads of information

Below each photo is the pertinent information about the can, including the name of the brewer, the city and state of origin, description of any slight design, color or enamel vs. metallic color differences, a dollar value based on original Grade 1 condition and a check-off box where you can indicate if the can is in your collection. You may also find letter codes like O (opening instructions), T (Internal Revenue Tax Paid) or W (Withdrawn Free For Export). To keep the information beneath the can photos concise, we used a number of abbreviations. Take a few moments to familiarize yourself with the abbreviations and glossary.

This is where the tour ends. And the fun starts. We hope you'll enjoy using *United States Beer Cans* for years to come.

ABBREVIATIONS

GLOSSARY

ABBREVIATIONS

General

bf	Bottom face
Blk.	Black
Bot.	Bottom
Dk.	Dark
Enam.	Enamel
Frt.	Front
Lg.	Large
Lt.	Light
Med.	Medium
Met.	Metallic
O	Opening instructions
Pre.	Pre-IRTP
Sm.	Small
T	Internal Revenue Tax Paid
tf	Top face
W	Withdrawn Free For Export
Yrs.	Years

Alcohol Content

CMT	Contains more than
CNMT	Contains not more than
CNO	Contains not over
DNCMT	Does not contain more than
Exc	Exceeds
LT	Less than
MT	More than
NIEO	Not in excess of
N.O.	Not over
NU	Not under
PC	Per centum
%	Percentage of alcohol

Brewery

Amer.	America
Assoc.	Associated
Atl.	Atlantic
Bev.	Beverage
Br.	Brewing
Burg.	Burgermeister
Corp.	Corporation
DBA	Doing business as
Div.	Division
Form.	Formerly
Gett.	Gettleman
Heile.	Heileman
Ind.	Independent
Indus.	Industries
Jac.	Jacob
Ltd.	Limited
Malt.	Malting
Mass.	Massachusetts
Milw.	Milwaukee
Prod.	Products
Ref.	Refining
Sch.	Schmidt
Schuy.	Schuyler
Twp.	Township
Succ.	Successor
W.E.	West End

25 ❑ 12-1-OT
Falls City
Louisville, KY
Extra Pale **75**

Front of 33, 34

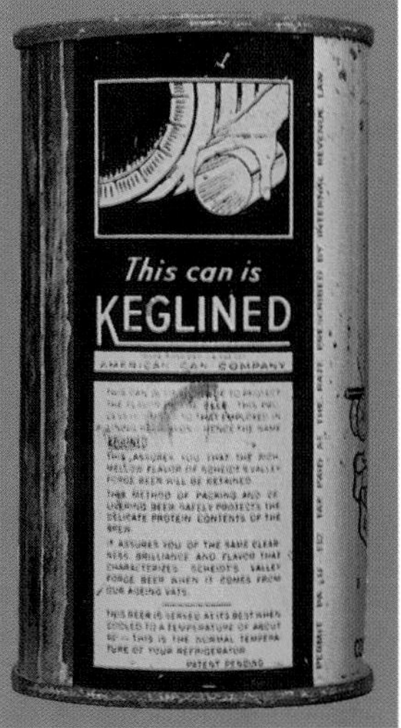

33 ❑ 12-1-OT
Adam Scheidt
Norristown, PA
200

34 ❑ 12-1-OT
Adam Scheidt
Norristown, PA
200

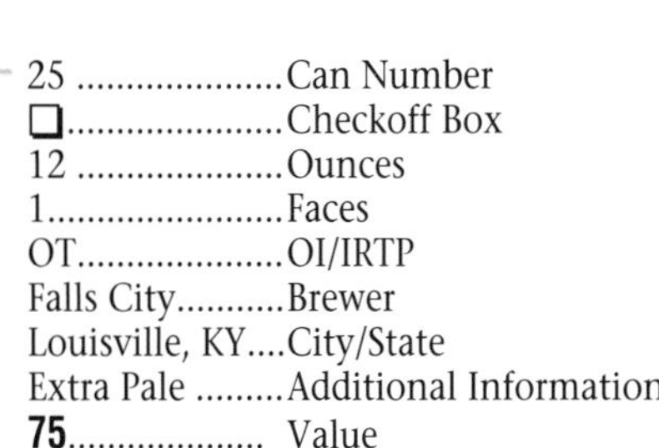

25 ....Can Number
❑......................Checkoff Box
12 ....Ounces
1........................Faces
OT.....................OI/IRTP
Falls City...........Brewer
Louisville, KY....City/State
Extra Pale ....Additional Information
75.................... Value

In some cases, a can's primary identifying feature is on its side or back rather than its front, as in numbers 33 and 34 above. In these instances, an unnumbered front face is shown for reference.

GLOSSARY

Concave Bottom — A dome-shaped bottom for cone-top cans introduced by Continental Can in September 1936. Concave bottoms resisted the pressure created by carbonation and pasteurization better than flat bottoms, so they could be made of thinner metal. Their shape also made complete filling easier by helping to create less foam.

Crowntainer — A two-piece crown-sealed can from Crown Cork & Seal. Its extruded, tinless, black steel body had a protective coating of aluminum and was attached by crimped seam to a concave bottom. Crowntainers were marketed by a number of Eastern and Midwestern brewers from the early 1940s to the mid-1950s.

Dull Gray — The dull gray enamel paint that was used during a short period toward the end of World War II as a substitute for the aluminum coating on Crowntainers. It was also used on a very small number of other types of cans during that period.

Dumper — An outdoor can in less than Grade 2 condition.

High Profile — A three-piece can with a "high" spout, produced by three different can companies. The Continental Can version, made from 1942 until the mid-1950s, the Crown Cork & Seal version, made from 1945 to the mid-1950s, and the American Can version all had spouts 1 1/8" tall. The American can was 1/8" shorter overall than the others and had a flat bottom.

Indoor Can — Found in places where it's been protected from the elements. Most fall in Grades 1 and 2.

Inverted Rib — The spouts of the first Continental Can cone tops featured a series of inverted, or indented ribs and flat bottoms. They were short-lived. When the cans were stored in coolers, condensation caused the unplated spouts to corrode. Raised-rib plated spouts, stronger and made of thinner metal, replaced the inverted-rib design early in 1936.

"J" Spout — A three-piece can with an elongated spout top. Produced by Crown Cork & Seal from 1937 to 1942, it had a body shorter than a flat top and a noticeably longer-necked spout than earlier cone tops. Concave bottoms replaced flat bottoms in 1938.

Low Profile — A three-piece can with a "low" spout 7/8" tall. Introduced in late 1935 by Continental Can Co., low-profile cones originally had flat bottoms and unplated inverted rib spouts. These were replaced by March 1936 with concave bottoms and plated raised-rib spouts. Except for some low profiles made out West in the late '40s, production ended in 1942.

Off Grade — A can in less than Grade 1 but at least Grade 2 condition and still very displayable.

Olive Drab — The color of almost everything military during World War II—including the cans in which beer was shipped to the European and Pacific theaters of battle. Today, these cans are generally referred to as Olive Drabs, ODs or Camouflage cans.

On Grade — A can in Grade 1 or better condition.

Original Can — A can whose original top and bottom were attached at the brewery or canning plant (as opposed to a can rolled from a flat sheet or with a replacement top or bottom).

Outdoor can — Having been exposed to the elements, it rarely merits more than a Grade 2 rating.

Now the tangy, English-tavern flavor of

KENT ALE

is protected in KEGLINED cans

Kent Ale is a Stock Ale of the type found on tap in taverns throughout the British Isles — the perfect brew for colder weather enjoyment. Full-bodied and mellow with age, it has a delicious, tangy quality that has delighted lovers of fine ale since we first began brewing it generations ago.

Kent Ale is the *third* fine Krueger brew to be brought to you in this wonderful new container. It follows quite naturally on the tremendous, overnight popularity of Krueger's *Finest* Beer and Krueger's *Cream* Ale in Keglined Cans, wherever they have been introduced.

We urge you to try Krueger's — first brewer to give you beer and ale in Keglined Cans.

Available at fine grocers, delicatessens, hotels, taverns, restaurants, and department stores.

G. KRUEGER BREWING COMPANY • NEWARK • NEW JERSEY

Despite its early success, the beer can was no sure thing. A massive advertising effort was going to be needed to sell consumers on the virtues of beer in these unique new metal containers, and both can makers and brewers took up the cause. Ads like the ones pictured here from the American Can Company and G. Krueger Brewing Company ran regularly during the late 1930s.

Flats Section Explanatory Notes

While many side panel and instructional differences are shown, there are still many more that are unpictured due to space limitations. For brands like Ruppert, Schaefer, Scheidt's Ram's Head and Valley Forge, side panels were pictured for one or two of the labels. These side panels represent similar sides on other cans within that brand. For example, 143-5 and 143-7 are sides that can be found on Valley Forge Beer and Ram's Head Ale. It is hoped all of these panel variations will be catalogued further in the future.

Altes Sportsman cans are listed according to the sports fact beginning in the upper left corner. It is believed that this set contains 20 cans. The missing facts are:

Detroit Baseball 1945...

First Lansing Capitol...

Michigan Ruled by French...

A-1 31-23 and 31-24. The primary difference between these cans is the can company. As a result the label has some size variation in the three stars above the eagle and some slight difference in the feathers on the eagle. There is also a difference in the black text around the oval.

The front view of **Ballantine's Bock** 34-15 and 34-16 is similar to 34-12 but with 4 line mandatory like 34-17.

Note the differences in Plain and Fancy "L" on **Black Label** cans.

Black Label 37-37 has "from Carling's" on red square. 37-38 does not.

Budweiser 43-33 with side view 43-34. There is also a side variation with a different paragraph in the circle that begins-"This is the Famous Budweiser Beer...." This also applies to 44-3.

Note that the City order for all **Budweiser** cans has a five-city can with Miami listed preceding the four-city can. The five-city can with Houston listed follows these because the brewery in Miami closed before Houston opened.

Canadian Ace 48-5 has "Brewing Company" under wings while 48-6 has "Brewing Co." This also applies to 48-11 "Brewing Co." and 48-13 "Brewing Company."

Pilgrim cans 52-15, 52-16, 52-17 are listed with Croft.

Stock Ale by Croft 52-25 listed with Croft.

Gamecock Ale by Croft 52-29 listed with Croft.

Drewrys cans 56-4 through 56-15 are shown by color of the can and then the color of the sports figures.

Edelweiss 58-38 lists Schoenhofen-Edelweiss on the front but has Atlas Brewing Co. at the seam.

English Lad 60-2 through 60-8. Two side panel variations are shown for the Ale cans. These sides apply to the Beer cans and are captioned accordingly.

Esslinger Parti Quiz. While space limitations allowed for only one representative can from each set in the Beer, all Bock variations that were found have been shown. These are listed by the fact beginning in the upper left corner.

Old Frisco 67-10 is located with Frisco to maintain brand continuity.

All **Grace Bros.** cans have been grouped together under GB.

GB 71-23 and 71-28 are shown with **Gold Bond** to show brand progression.

Gretz Fleet Car cans have been represented with only one of the front designs followed by the set of twelve different cars. It is important to note that each car has its own picture on the front of the can, making the front views on each can different.

All **Griesedieck Bros.** cans are grouped together under Griesedieck Bros. A wide variety of the GB color set has been shown with a description of the color as close as possible. This particular set of cans seems to be much larger in the shades of colors possible than other sets of cans that used various colors.

Guinness cans were shown with all lot numbers that could be located.

Gunther 78-27 and 78-28. The difference is in the mandatory statement along bottom of cans as captioned.

Heileman's Special Export cans are shown under Heileman.

Ye Old English Style Ale 83-19 is shown with Hop Gold.

India cans from Puerto Rico have been listed.

Meister Brau 97-20 through 97-24 are the same as the enamel set listed 96-40 through 97-19 except the border color is orange instead of tan. It has not been established if all twenty cans come with this border difference.

Mitchell's 100-14 has the word "Harry" in the loop of the "M" of Mitchell in the mandatory.

National Bohemian cans have the brand name in either White or Silver as captioned.

National Bohemian 102-13 is the transition can from 102-12 and has that label on the back.

Old Crown 105-8 is part of a subset of the crown's color cans that has the "lazy aged" logo on the side. It is not known how many colors of this design exist.

Old German 106-31 has the date 1901 in the lower left corner of the red ribbon.

Heileman's Old Style cans are listed under Old Style.

Old Tankard Ale and Pabst Old Tankard Ale are shown together.

Peter Hand's Reserve and Reserve are listed together.

Pfeiffer outdoor scene cans list both dull and metallic gold versions.

Pfeiffer 114-25 has "Great Lakes Country" on red label.

Pfeiffer 114-26 has "From the Water Wonderland" on red label.

Pfeiffer 114-27 has no writing on red label.

Pfeiffer 114-31 has "From the Water Wonderland" on red label.

Pfeiffer 114-32 has no writing on red label.

All **Scotch Thistle Brand Ale** cans are pictured with Rheingold to show progression of label.

Ruppert 126-39 mandatory reads "Jacob Ruppert Virginia Inc. Under Contract"

Ruppert 126-40 mandatory reads "Jacob Ruppert Virginia Inc. At Norfolk, VA"

San Miguel cans show side views of the shield with differences in the castle and lion.

Senate 132-12 also comes without 75th Year logo.

Senate 132-14 does not have 75th logo on side.

Senate 132-15 has a front like 132-14.

Senate 132-18 also comes without 75th Year logo.

Brew 66 cans are pictured with Sicks' Select cans.

Old Ox Cart and Ox Cart Dry cans are pictured with Standard cans.

Storz Triumph 137-27 "Lager Beer" in red.

Storz Triumph 137-28 "Lager Beer" in blue.

Waldorf Red Band. Captions list statements on the first line of text in red label.

FLATS

10, 11, & 12 Oz.

In General. Some can differences require more explanation than the space beneath the pictures permits. You'll find these extended explanations at the beginning of each can section.

Most cans appear in the order spelled out on page 23, but there are exceptions. Some are placed with similar cans from the same brewery (Croft). Others have been located by their common rather than their proper names (Acme Bull Dog).

Not all cans with the "Withdrawn Free" designation were produced during WWII. These cans were located as best as possible by the age of the label.

Virtually all 2-faced cans were photographed with the seam to the reader's left. This is important to note on some Bull Dog and Meister Brau Fiesta Pack cans.

Mandatory Information. To save space, only the brewer's name is listed (Pittsburgh vs. Pittsburgh Brewing Co.). In some rare cases, the difference between cans is in the brewery statement itself (Eastern Brewing Corp./Eastern Beverage Corp.).

Where no brewery information is listed on a can, the entry will read "No Mandatory." This mostly occurs on Manhattan Brewing Co. cans. On cans such as the Class (47-32) or Imperial (83-9) the mandatory is listed as it appears on the can even though it was not necessarily brewed at that location. The best resource for this information is the BCCA's *Catalog of American Beer Cans*.

Can Company Information. No attempt has been made to show the different can company variations even though some of the differences shown are a result of cans being produced by different can companies. Most of these varieties appear as slight color or typeface differences.

Alcohol Statements. Alcohol content statements appearing on the bodies (not the lids) of cans are noted in the entries in abbreviated form. You'll find a list of abbreviations on page 24. The exceptions are cans whose alcohol content appears as a numerical percentage due to the fact that the actual statement wording has not yet been verified. These percentages are noted as numeric entries only, i.e. 4%. Also, the difference between alcohol percentage "by weight" and "by volume" is not addressed at this time.

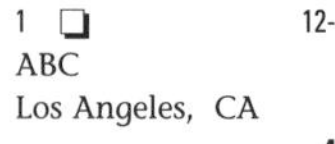

1 ❑ 12-2
ABC
Los Angeles, CA
40

2 ❑ 12-2
Maier
Los Angeles, CA
75

3 ❑ 12-2
Maier
Los Angeles, CA
12

4 ❑ 12-1
Gold Brau
Chicago, IL
35

5 ❑ 12-1
Gold Brau
Chicago, IL
60

6 ❑ 12-1
Gold Brau
Chicago, IL
25

7 ❑ 12-1-OT
ABC
St. Louis, MO
1000+

8 ❑ 12-1-OT
ABC
St. Louis, MO
325

9 ❑ 12-1-OT
ABC
St. Louis, MO
300

10 ❑ 12-1-OT
ABC
St. Louis, MO
325

11 ❑ 12-2
Sioux City
Sioux City, IA
200

12 ❑ 12-2
Ace
La Crosse, WI
60

13 ❑ 12-2
Ace
La Crosse, WI
60

14 ❑ 12-2
G. Heileman
La Crosse, WI
60

15 ❑ 12-2
G. Heileman
La Crosse, WI
40

16 ❑ 12-1
Ace
Chicago, IL
125

17 ❑ 12-1
Ace
Chicago, IL
Metallic
20

18 ❑ 12-1
Ace
Chicago, IL
Enamel
20

19 ❑ 12-1
Canadian Ace
Chicago, IL
30

20 ❑ 12-2-T
Acme
Los Angeles, CA
450

21 ❑ 12-2-T
Acme
Los Angeles, CA
125

22 ❑ 12-2-T
Acme
Los Angeles, CA
30

23 ❑ 12-2-T
Acme
Los Angeles, CA
35

24 ❑ 12-2-T
Acme
Los Angeles, CA
25

25 ❑ 12-2-T
Acme
Los Angeles, CA
12

26 ❑ 12-2
Acme
Los Angeles, CA
12

27 ❑ 12-2
Acme
Los Angeles, CA
"Vernon"
25

28 ❑ 12-1
Acme
Los Angeles, CA
15

29 ❑ 12-1
Acme
Los Angeles, CA
15

30 ❑ 12-1
Acme
Los Angeles, CA
15

31 ❑ 12-1
Acme
Los Angeles, CA
10

32 ❑ 12-1
Acme
Los Angeles, CA
25

33 ❑ 12-1
Acme
Los Angeles, CA
600

34 ❑ 12-1
Acme
Los Angeles, CA
175

35 ❑ 12-1-OT
Acme
San Francisco, CA
450

36 ❑ 12-2-T
Acme
San Francisco, CA
450

37 ❑ 12-2-T
Acme
San Francisco, CA
175

38 ❑ 12-2
Acme
San Francisco, CA
125

39 ❑ 12-1-OT
Acme
San Francisco, CA
50

40 ❑ 12-1-OT
Acme
San Francisco, CA
50

1 ❑ 12-2-T
Acme
San Francisco, CA
35

2 ❑ 12-2-T
Acme
San Francisco, CA
35

3 ❑ 12-2-T
Acme
San Francisco, CA
30

4 ❑ 12-2-W
Acme
San Francisco, CA
600

5 ❑ 12-2-T
Acme
San Francisco, CA
30

6 ❑ 12-2-T
Acme
San Francisco, CA
35

7 ❑ 12-2-W
Acme
San Francisco, CA
75

8 ❑ 12-2-T
Acme
San Francisco, CA
12

9 ❑ 12-2
Acme
San Francisco, CA
12

10 ❑ 12-1
Acme
San Francisco, CA
15

11 ❑ 12-1
Acme
San Francisco, CA
15

12 ❑ 12-1
Acme
San Francisco, CA
15

13 ❑ 12-1
Acme
San Francisco, CA
10

14 ❑ 12-1
Acme
San Francisco, CA
25

15 ❑ 12-1
Acme
San Francisco, CA
600

16 ❑ 12-1
Acme
San Francisco, CA
175

17 ❑ 12-1-OT
Cereal Prod. Ref. Corp.
San Francisco, CA
100

18 ❑ 12-2
Grace Bros.
Santa Rosa, CA
40

19 ❑ 12-2
Geo. Walter
Appleton, WI
40

20 ❑ 12-2
Geo. Walter
Appleton, WI
20

21 ❑ 12-2
Geo. Walter
Appleton, WI
15

22 ❑ 12-2
Geo. Walter
Appleton, WI
65

23 ❑ 12-2
Geo. Walter
Appleton, WI
15

24 ❑ 12-1
Southern
Los Angeles, CA
225

25 ❑ 12-1
Atlas
Chicago, IL
50

26 ❑ 12-1
Atlas
Chicago, IL
50

27 ❑ 12-1
Atlas
Chicago, IL
75

28 ❑ 12-2
Drewry's Ltd.
South Bend, IN
45

29 ❑ 12-2
Storz
Omaha, NE
8

30 ❑ 12-1-OT
Manhattan
Chicago, IL
950

31 ❑ 12-1-OT
No Mandatory
Chicago, IL
800

32 ❑ 12-1-OT
Manhattan
Chicago, IL
750

33 ❑ 12-2
All Star
Green Bay, WI
350

34 ❑ 12-1-OT
San Francisco
San Francisco, CA
450

35 ❑ 12-1-OT
San Francisco
San Francisco, CA
1000+

36 ❑ 12-1-OT
San Francisco
San Francisco, CA
Metallic
600

37 ❑ 12-1-OT
San Francisco
San Francisco, CA
Enamel
600

38 ❑ 12-2
Burgermeister
San Francisco, CA
20

39 ❑ 12-2
Burgermeister
San Francisco, CA
10

40 ❑ 12-2
Maier
Los Angeles, CA
75

1 ❑ 12-2
Maier
Los Angeles, CA
60

2 ❑ 12-1
Peter Fox
Chicago, IL
75

3 ❑ 12-1
Peter Fox
Chicago, IL
65

4 ❑ 12-1
Fox Deluxe
Grand Rapids, MI
175

5 ❑ 12-2
Alpine
Potosi, WI
10

6 ❑ 12-2
Maier
Los Angeles, CA
65

7 ❑ 12-2
Grace Bros.
Santa Rosa, CA
75

8 ❑ 12-2
Grace Bros.
Santa Rosa, CA
60

9 ❑ 12-2
Centlivre
Fort Wayne, IN
Gold Trim
20

10 ❑ 12-2
Centlivre
Fort Wayne, IN
Silver Trim
15

11 ❑ 12-2
Old Crown
Fort Wayne, IN
8

12 ❑ 12-1-OT
Columbia
Tacoma, WA
800

13 ❑ 12-1-OT
Columbia
Tacoma, WA
650

14 ❑ 12-1-OT
Columbia
Tacoma, WA
DNCMT 4%
650

15 ❑ 12-1-OT
Columbia
Tacoma, WA
900

16 ❑ 12-1-OT
Columbia
Tacoma, WA
500

17 ❑ 12-1
Columbia
Tacoma, WA
45

18 ❑ 12-1
Columbia
Tacoma, WA
30

Front of 20,21

20 ❑ 12-1-OT
Grace Bros LTD.
Los Angeles, CA
450

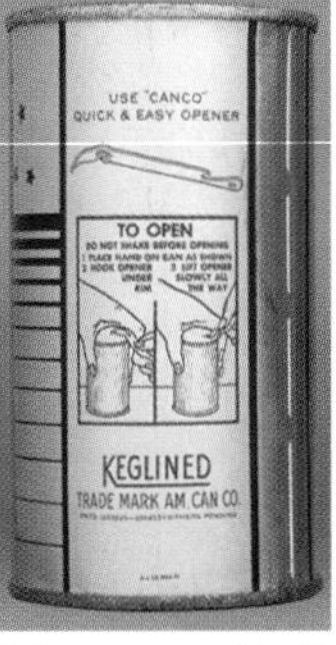

21 ❑ 12-1-OT
Grace Bros LTD.
Los Angeles, CA
325

22 ❑ 12-1-OT
Grace Bros.
Santa Rosa, CA
375

23 ❑ 12-1
Altes
Detroit, MI
30,000 Archers...
150

24 ❑ 12-1
Altes
Detroit, MI
80,000 Miles...
150

25 ❑ 12-1
Altes
Detroit, MI
Det. Baseball 1935
150

26 ❑ 12-1
Altes
Detroit, MI
First Cement Hwy.
150

27 ❑ 12-1
Altes
Detroit, MI
Mich. Ruled by Eng.
150

28 ❑ 12-1
Altes
Detroit, MI
Mich. State Flower...
150

29 ❑ 12-1
Altes
Detroit, MI
Mich. 26th State...
150

30 ❑ 12-1
Altes
Detroit, MI
Mich. Highest Point
150

31 ❑ 12-1
Altes
Detroit, MI
Mich. Land Area...
150

32 ❑ 12-1
Altes
Detroit, MI
Mich. Nickname...
150

33 ❑ 12-1
Altes
Detroit, MI
Mich. Ranks 7th...
150

34 ❑ 12-1
Altes
Detroit, MI
Muskellunge Life...
150

35 ❑ 12-1
Altes
Detroit, MI
Sault Ste. Marie...
150

36 ❑ 12-1
Altes
Detroit, MI
Spanish Flag...
150

37 ❑ 12-1
Altes
Detroit, MI
State Capitol Cost...
150

38 ❑ 12-1
Altes
Detroit, MI
Ty Cobb's Highest..
150

39 ❑ 12-1
Altes
Detroit, MI
Walk In Water...
150

40 ❑ 12-2
Altes
Detroit, MI
60

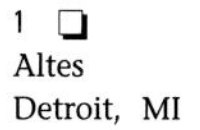
1 ❑ 12-2
Altes
Detroit, MI
35

2 ❑ 12-2
Altes
Detroit, MI
30

3 ❑ 12-1
National
Detroit, MI
40

4 ❑ 12-1
National
Detroit, MI
35

5 ❑ 12-1
National
Detroit, MI
40

6 ❑ 12-2
G. Krueger
Newark, NJ
30

7 ❑ 12-2
G. Krueger
Newark, NJ
30

8 ❑ 12-2
G. Krueger
Cranston, RI
50

9 ❑ 12-2
Maier
Los Angeles, CA
25

10 ❑ 12-2
Maier
Los Angeles, CA
20

11 ❑ 12-1-OT
Ambrosia
Chicago, IL
1000+

12 ❑ 12-1-OT
South Side
Chicago, IL
Tall Can
1000+

13 ❑ 12-1
American
Baltimore, MD
75

14 ❑ 12-2
American
Baltimore, MD
60

15 ❑ 12-1
American
Baltimore, MD
Enamel
20

16 ❑ 12-1
American
Baltimore, MD
Metallic
15

17 ❑ 12-1
American
Baltimore, MD
Enamel
300

18 ❑ 12-1
American
Baltimore, MD
Metallic
300

19 ❑ 12-2
Eastern Corp.
Hammonton, NJ
15

20 ❑ 12-2
Five Star
New York, NY
20

21 ❑ 12-2
Arizona
Phoenix, AZ
85

22 ❑ 12-2
Arizona
Phoenix, AZ
75

23 ❑ 12-1-OT
Arizona
Phoenix, AZ
Large Stars
400

24 ❑ 12-1-OT
Arizona
Phoenix, AZ
Small Stars
450

25 ❑ 12-1
Arizona
Phoenix, AZ
25

26 ❑ 10-1
Arizona
Phoenix, AZ
75

27 ❑ 12-1
Arizona
Phoenix, AZ
25

28 ❑ 12-2
Arizona
Phoenix, AZ
25

29 ❑ 12-1
Arizona
Phoenix, AZ
400

30 ❑ 12-1
Arizona
Phoenix, AZ
30

31 ❑ 12-2
Arizona
Phoenix, AZ
30

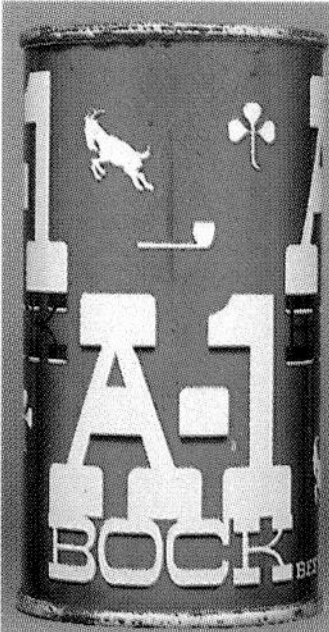

32 ❑ 12-3
Arizona
Phoenix, AZ
1000+

33 ❑ 12-2
Carling
Phoenix, AZ
40

34 ❑ 12-1-OT
American
Rochester, NY
DULL GRAY
900

35 ❑ 12-1-OT
American
Rochester, NY
900

36 ❑ 12-2
Arizona
Phoenix, AZ
500

37 ❑ 12-2
Mountain
Denver, CO
60

38 ❑ 12-2
Gold Brau
Chicago, IL
60

39 ❑ 12-2
Gold Brau
Chicago, IL
60

40 ❑ 12-1-OT
Globe
Baltimore, MD
1000+

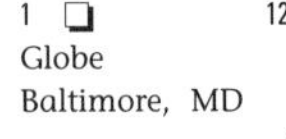

1 ❏ 12-1-OT
Globe
Baltimore, MD
1000+

2 ❏ 12-1-OT
Globe
Baltimore, MD
Enamel 110

3 ❏ 12-1-OT
Globe
Baltimore, MD
Metallic 125

4 ❏ 12-1-OT
Globe
Baltimore, MD
225

5 ❏ 12-1
Globe
Baltimore, MD
150

6 ❏ 12-1
Globe
Baltimore, MD
60

7 ❏ 12-2
Globe
Baltimore, MD
Gold Trim 25

8 ❏ 12-2
Globe
Baltimore, MD
Silver Trim 20

9 ❏ 12-1
Globe
Baltimore, MD
850

10 ❏ 12-2
Globe
Cumberland, MD
20

11 ❏ 12-2
Arrowhead
Cold Spring, MN
50

12 ❏ 12-2
Tivoli
Denver, CO
15

13 ❏ 12-1-OT
Humboldt
Eureka, CA
Paper Label 850

14 ❏ 12-1
Atlantic Co.
Charlotte, NC
225

15 ❏ 12-1
Atlantic Co.
Charlotte, NC
250

16 ❏ 12-1
Atlantic Co.
Charlotte, NC
375

17 ❏ 12-1-OT
Atlas
Chicago, IL
1000+

18 ❏ 12-1-OT
Atlas
Chicago, IL
350

19 ❏ 12-1-OT
Atlas
Chicago, IL
Enamel 110

20 ❏ 12-1-OT
Atlas
Chicago, IL
Metallic 110

21 ❏ 12-1-T
Atlas
Chicago, IL
30

22 ❏ 12-1
Atlas
Chicago, IL
20

23 ❏ 12-1
Atlas
Chicago, IL
20

24 ❏ 12-2
Atlas
Chicago, IL
15

25 ❏ 12-2
Atlas
Chicago, IL
20

26 ❏ 12-2
Atlas
Chicago, IL
20

27 ❏ 12-2
Atlas
Chicago, IL
8

28 ❏ 12-2
Atlas
Chicago, IL
85

29 ❏ 12-2
Drewry's Ltd.
South Bend, IN
5

30 ❏ 12-1
August Wagner
Columbus, OH
15

31 ❏ 12-1-OT
Manhattan
Chicago, IL
1000+

32 ❏ 12-1-OT
No Mandatory
Chicago, IL
1000+

33 ❏ 12-1-OT
Manhattan
Chicago, IL
1000+

34 ❏ 12-1-OT
Whitewater
Whitewater, WI
1000+

35 ❏ 12-1-OT
Whitewater
Whitewater, WI
1000+

36 ❏ 12-1-T
Monarch
Los Angeles, CA
400

37 ❏ 12-1-T
Monarch
Los Angeles, CA
400

38 ❏ 12-1-W
Grace Bros LTD.
Los Angeles, CA
150

39 ❏ 12-1-T
Grace Bros LTD.
Los Angeles, CA
150

40 ❏ 12-1
Southern
Los Angeles, CA
125

1 ❑ 12-1-OT
Ballantine & Son's
Newark, NJ
125

2 ❑ 12-1-OT
Ballantine & Son's
Newark, NJ
125

Front of 4,5
3 Line Mandatory

4 ❑ 12-1-T
Ballantine & Son's
Newark, NJ
30

5 ❑ 12-1-T
Ballantine & Son's
Newark, NJ
30

Front of 7-10
4 Line Mandatory

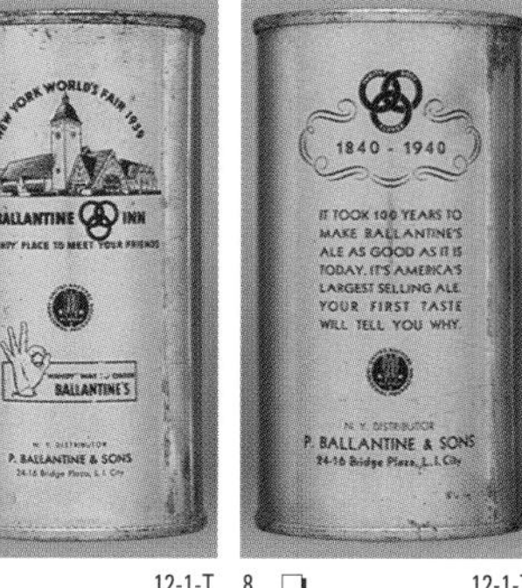
7 ❑ 12-1-T
Ballantine & Son's
Newark, NJ
30

8 ❑ 12-1-T
Ballantine & Son's
Newark, NJ
30

9 ❑ 12-1-T
Ballantine & Son's
Newark, NJ
30

10 ❑ 12-1-T
Ballantine & Son's
Newark, NJ
30

11 ❑ 12-2-T
Ballantine & Son's
Newark, NJ
25

12 ❑ 12-2
Ballantine & Son's
Newark, NJ
25

13 ❑ 12-1
Ballantine & Son's
Newark, NJ
25

14 ❑ 12-1
Ballantine & Son's
Newark, NJ
25

15 ❑ 12-2
Ballantine & Son's
Newark, NJ
8

16 ❑ 12-2
Ballantine & Son's
Newark, NJ
250

Front of 18, 19

18 ❑ 12-2
Ballantine & Son's
Newark, NJ
8

19 ❑ 12-2
Ballantine & Son's
Newark, NJ
8

20 ❑ 12-2
Ballantine & Son's
Newark, NJ
8

21 ❑ 12-1-OT
Ballantine & Son's
Newark, NJ
150

Front of 23,24
2 Line Mandatory

23 ❑ 12-1-T
Ballantine & Son's
Newark, NJ
40

24 ❑ 12-1-T
Ballantine & Son's
Newark, NJ
40

Front of 26-29
3 Line Mandatory

26 ❑ 12-1-T
Ballantine & Son's
Newark, NJ
40

27 ❑ 12-1-T
Ballantine & Son's
Newark, NJ
40

28 ❑ 12-1-T
Ballantine & Son's
Newark, NJ
40

29 ❑ 12-1-T
Ballantine & Son's
Newark, NJ
40

30 ❑ 12-1-W
Ballantine & Son's
Newark, NJ
60

31 ❑ 12-1-T
Ballantine & Son's
Newark, NJ
25

32 ❑ 12-2-T
Ballantine & Son's
Newark, NJ
25

33 ❑ 12-2
Ballantine & Son's
Newark, NJ
25

34 ❑ 12-2
Ballantine & Son's
Newark, NJ
25

35 ❑ 12-2
Ballantine & Son's
Newark, NJ
250

36 ❑ 12-2
Ballantine & Son's
Newark, NJ
250

37 ❑ 12-1
Ballantine & Son's
Newark, NJ
12

38 ❑ 12-1
Ballantine & Son's
Newark, NJ
12

39 ❑ 12-2
Ballantine & Son's
Newark, NJ
12

40 ❑ 12-2
Ballantine & Son's
Newark, NJ
12

1 ❑ 12-2
Ballantine & Son's
Newark, NJ
65

2 ❑ 12-2
Ballantine & Son's
Newark, NJ
50

3 ❑ 12-2
Ballantine & Son's
Newark, NJ
50

4 ❑ 12-2
Ballantine & Son's
Newark, NJ
Dark Brown 12

5 ❑ 12-2
Ballantine & Son's
Newark, NJ
Light Brown 12

6 ❑ 12-2
Ballantine & Son's
Newark, NJ
5

7 ❑ 12-2
Ballantine & Son's
Newark, NJ
5

8 ❑ 12-2
Ballantine & Son's
Newark, NJ
5

9 ❑ 12-2
Ballantine & Son's
Newark, NJ
White Lettering 15

10 ❑ 12-2
Ballantine & Son's
Newark, NJ
Gold Lettering 15

11 ❑ 12-1-OT
Ballantine & Son's
Newark, NJ
500

Front of 13,14

3 Line Mandatory

13 ❑ 12-1-T
Ballantine & Son's
Newark, NJ
500

14 ❑ 12-1-T
Ballantine & Son's
Newark, NJ
500

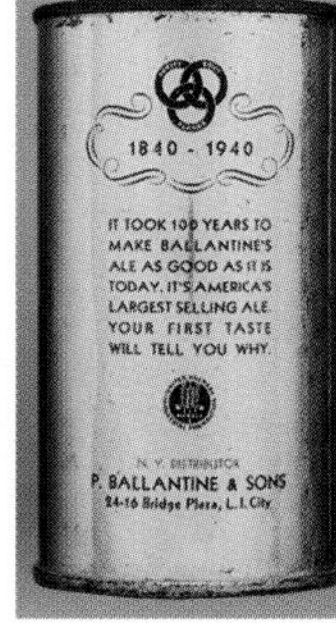

15 ❑ 12-1-T
Ballantine & Son's
Newark, NJ
4 Line Mand. 500

16 ❑ 12-1-T
Ballantine & Son's
Newark, NJ
4 Line Mand. 500

17 ❑ 12-2-T
Ballantine & Son's
Newark, NJ
200

18 ❑ 12-2
Ballantine & Son's
Newark, NJ
200

19 ❑ 12-2
Ballantine & Son's
Newark, NJ
200

20 ❑ 12-2
Ballantine & Son's
Newark, NJ
300

21 ❑ 12-2
Ballantine & Son's
Newark, NJ
250

22 ❑ 12-2
Ballantine & Son's
Newark, NJ
40

23 ❑ 12-2
Ballantine & Son's
Newark, NJ
8

24 ❑ 12-2
Falstaff
Cranston, RI
8

25 ❑ 12-1
Best
Chicago, IL
25

26 ❑ 12-1
Cumberland
Cumberland, MD
10

27 ❑ 12-1
Banner
Trenton, NJ
20

28 ❑ 12-1
Geo. F. Stein
Buffalo, NY
35

29 ❑ 12-1
Burkhardt
Akron, OH
150

30 ❑ 12-1
Burkhardt
Akron, OH
25

31 ❑ 12-1
Banner
Norfolk, VA
30

32 ❑ 12-1
Century
Norfolk, VA
25

33 ❑ 12-1
Atlantic
Chicago, IL
50

34 ❑ 12-1
Atlantic
South Bend, IN
50

35 ❑ 12-1
CV
Terre Haute, IN
55

36 ❑ 12-1
Terre Haute
Terre Haute, IN
55

37 ❑ 12-1
Red Top
Cincinnati, OH
150

38 ❑ 12-1
Wunderbrau
Cincinnati, OH
50

39 ❑ 12-2
Bartels
Edwardsville, PA
Enamel 110

40 ❑ 12-2
Bartels
Edwardsville, PA
Metallic 135

1 ❑ 12-2
Bartels
Edwardsville, PA
70

2 ❑ 12-2
Maier
Los Angeles, CA
40

3 ❑ 12-2
Maier
Los Angeles, CA
18

4 ❑ 12-2
Maier
Los Angeles, CA
18

5 ❑ 12-2
Maier
Los Angeles, CA
18

6 ❑ 12-1
Mount Carbon
Pottsville, PA
75

7 ❑ 12-2
Mount Carbon
Pottsville, PA
10

8 ❑ 12-2
Mount Carbon
Pottsville, PA
20

9 ❑ 12-2
Jos. Huber
Monroe, WI
5

10 ❑ 12-1
International
Tampa, FL
45

11 ❑ 12-1
Bavarian
Covington, KY
60

12 ❑ 12-1
Bavarian
Covington, KY
15

13 ❑ 12-1
International
Covington, KY
15

14 ❑ 12-1-OT
Commonwealth
Springfield, MA
900

15 ❑ 12-1-OT
Commonwealth
Springfield, MA
Enamel 400

16 ❑ 12-1-OT
Commonwealth
Springfield, MA
Metallic 400

17 ❑ 12-1-OT
Commonwealth
Springfield, MA
1000+

18 ❑ 12-1-OT
Commonwealth
Springfield, MA
Enamel 1000+

19 ❑ 12-1-OT
Commonwealth
Springfield, MA
Metallic 1000+

20 ❑ 12-1-T
Southern
Los Angeles, CA
N.O. 4% 375

21 ❑ 12-1
Southern
Los Angeles, CA
375

22 ❑ 12-2
North Bay
Santa Rosa, CA
1000+

23 ❑ 12-2
Tivoli
Denver, CO
18

24 ❑ 12-1-OT
Becker Products
Ogden, UT
60

25 ❑ 12-1-OT
Becker Products
Ogden, UT
60

26 ❑ 12-2
Becker Products
Ogden, UT
45

27 ❑ 12-2
Becker Products
Ogden, UT
40

28 ❑ 12-2
Becker Products
Ogden, UT
10

29 ❑ 12-2
Becker Products
Ogden, UT
10

30 ❑ 11-2
Becker Products
Ogden, UT
Dull Gold 3

31 ❑ 11-2
Becker Products
Ogden, UT
Met. Gold 5

32 ❑ 12-2
Becker Products
Ogden, UT
10

33 ❑ 12-2
Becker
Ogden, UT
15

34 ❑ 12-1-W
Jacob Ruppert
New York, NY
550

35 ❑ 12-1-T
Seattle Br. & Malt.
Seattle, WA
175

36 ❑ 12-1-W
Jos. Schlitz
Milwaukee, WI
400

37 ❑ 12-1-W
Jos. Schlitz
Milwaukee, WI
400

38 ❑ 12-1
Horlacher
Allentown, PA
45

39 ❑ 12-2
Grace Bros.
Santa Rosa, CA
400

40 ❑ 12-2
Old Reading
Reading, PA
10

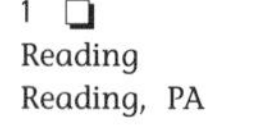

1 ❑ 12-2
Reading
Reading, PA
5

2 ❑ 12-2
Reading
Reading, PA
5

3 ❑ 12-2
Bergheim
Reading, PA
5

4 ❑ 12-1
Walter
Pueblo, CO
25

5 ❑ 12-1
Walter
Pueblo, CO
60

6 ❑ 12-1
Walter
Pueblo, CO
15

7 ❑ 12-1
Walter
Pueblo, CO
60

8 ❑ 12-1
Walter
Pueblo, CO
20

9 ❑ 12-1
Walter
Pueblo, CO
25

10 ❑ 12-2
Berghoff
Fort Wayne, IN
500

11 ❑ 12-1
Berghoff
Fort Wayne, IN
18

Front of 13, 14

13 ❑ 12-1
Berghoff
Fort Wayne, IN
60

14 ❑ 12-1
Berghoff
Fort Wayne, IN
60

15 ❑ 12-1
Berghoff
Fort Wayne, IN
325

16 ❑ 12-2
Berghoff
Fort Wayne, IN
450

17 ❑ 12-2
Berghoff
Fort Wayne, IN
600

18 ❑ 12-1
Leisy
Cleveland, OH
85

19 ❑ 12-1
Tennessee
Memphis, TN
75

20 ❑ 12-2
Spearman
Pensacola, FL
30

21 ❑ 12-2
Best
Chicago, IL
45

22 ❑ 12-1-OT
Best
Chicago, IL
600

23 ❑ 12-1
Best
Chicago, IL
50

24 ❑ 12-1
Best
Chicago, IL
35

25 ❑ 12-2
Best
Chicago, IL
8

26 ❑ 12-2
Best
Chicago, IL
10

27 ❑ 12-2
Empire
Chicago, IL
8

28 ❑ 12-2
United States
Chicago, IL
50

29 ❑ 12-2
United States
Chicago, IL
8

30 ❑ 12-2
Cumberland
Cumberland, MD
60

31 ❑ 12-2
Cumberland
Cumberland, MD
35

32 ❑ 12-2
Hornell
Hornell, NY
40

33 ❑ 12-2
Pacific
Oakland, CA
1000+

34 ❑ 12-2
Colonial
Hammonton, NJ
85

35 ❑ 12-2
Wm. Gretz
Philadelphia, PA
500

36 ❑ 12-1-T
Beverwyck
Albany, NY
90

37 ❑ 12-1
Beverwyck
Albany, NY
90

38 ❑ 12-1-T
Beverwyck
Albany, NY
125

39 ❑ 12-1
Beverwyck
Albany, NY
125

40 ❑ 12-2
Ind. Milwaukee
Milwaukee, WI
150

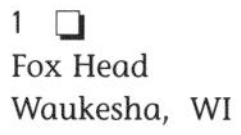

1 ❑ 12-2
Fox Head
Waukesha, WI
200

2 ❑ 12-2
Fox Head
Waukesha, WI
150

3 ❑ 12-2
Waukee
Hammonton, NJ
125

4 ❑ 12-2
Sunshine
Reading, PA
250

5 ❑ 12-2
Pabst
Peoria Heights, IL
35

6 ❑ 12-2
Colonial
Hammonton, NJ
100

7 ❑ 12-2
Menominee-Marinette
Menominee, MI
125

8 ❑ 11-2
Great Falls
Great Falls, MT
35

9 ❑ 11-2
Great Falls
Portland, OR
35

10 ❑ 12-2
Tivoli
Denver, CO
20

11 ❑ 12-2
Garden State
Hammonton, NJ
15

12 ❑ 12-1
Bismarck
Chicago, IL
30

13 ❑ 12-1
Canadian Ace
Chicago, IL
40

14 ❑ 12-1
Bismarck
Chicago, IL
30

15 ❑ 12-1-OT
Manhattan
Chicago, IL
1000+

16 ❑ 12-1-OT
Manhattan
Chicago, IL
1000+

17 ❑ 12-1-OT
Manhattan
Chicago, IL
1000+

18 ❑ 12-1-OT
ABC
St. Louis, MO
1000+

19 ❑ 12-2
Walter
Pueblo, CO
35

20 ❑ 12-2
Atlantic
Chicago, IL
60

21 ❑ 12-2
Leisy
Cleveland, OH
50

22 ❑ 12-1-OT
Manhattan
Chicago, IL
500

23 ❑ 12-1-OT
Class & Nachod
Philadelphia, PA
600

24 ❑ 12-1
Carling
Atlanta, GA
4

25 ❑ 12-1
Carling
Atlanta, GA
4

26 ❑ 12-1
Carling
Belleville, IL
3

27 ❑ 12-1
Carling
Belleville, IL
3

28 ❑ 12-1
Carling
Belleville, IL
3

Side of 28

30 ❑ 12-1
Carling
Baltimore, MD
Silver Trim
50

31 ❑ 12-1
Carling
Baltimore, MD
3

Front of 33, 34

33 ❑ 12-1
Carling
Baltimore, MD
3

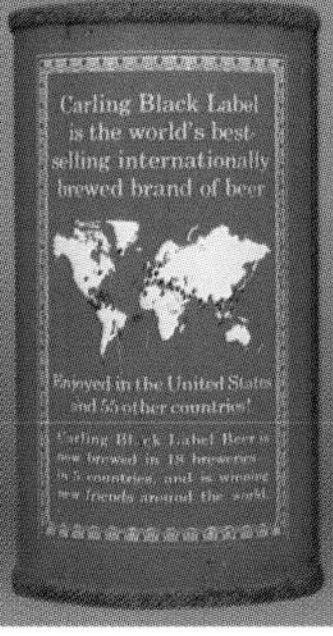

34 ❑ 12-1
Carling
Baltimore, MD
3

35 ❑ 12-2
Carling
Baltimore, MD
3

36 ❑ 12-2
Carling
Baltimore, MD
4

37 ❑ 12-2
Carling
Baltimore, MD
8

38 ❑ 12-2
Carling
Baltimore, MD
8

39 ❑ 12-1
Carling
Natick, MA
3

40 ❑ 12-1
Carling
Natick, MA
3

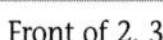
Front of 2, 3

2 ❑ 12-1
Carling
Natick, MA
3

3 ❑ 12-1
Carling
Natick, MA
3

4 ❑ 12-2
Carling
Natick, MA
4

5 ❑ 12-2
Carling
Natick, MA
4

6 ❑ 12-1
Carling
Frankenmuth, MI
3

7 ❑ 12-1
Carling
Frankenmuth, MI
3

8 ❑ 12-1
Carling
Frankenmuth, MI
5

Side of 8

10 ❑ 12-1
Carling
Frankenmuth, MI
325

11 ❑ 12-2
Carling
St. Louis, MO
8

12 ❑ 12-1
Br. Corp. of Amer.
Cleveland, OH
10

13 ❑ 12-1
Br. Corp. of Amer.
Cleveland, OH
10

14 ❑ 12-1
Carling
Cleveland, OH
10

15 ❑ 12-1
Carling
Cleveland, OH
3

16 ❑ 12-1
Carling
Cleveland, OH
3

17 ❑ 12-2
Carling
Cleveland, OH
3

18 ❑ 12-1
Carling
Cleveland, OH
325

19 ❑ 12-1
Carling
Cleveland, OH
450

20 ❑ 11-1
Carling
Tacoma, WA
3

21 ❑ 12-1
Carling
Tacoma, WA
3

22 ❑ 11-1
Carling
Tacoma, WA
3

23 ❑ 12-1
Carling
Tacoma, WA
3

24 ❑ 12-2
Carling
Tacoma, WA
8

25 ❑ 12-2
ABC
Los Angeles, CA
40

26 ❑ 12-2
Maier
Los Angeles, CA
35

27 ❑ 12-2-T
St. Claire
San Jose, CA
750

28 ❑ 12-2
Atlantic
Chicago, IL
Metallic
50

29 ❑ 12-2
Atlantic
Chicago, IL
Enamel
50

30 ❑ 12-2
Terre Haute
Terre Haute, IN
50

31 ❑ 12-2
Blackhawk
Davenport, IA
250

32 ❑ 12-2
Uchtorff
Davenport, IA
150

33 ❑ 12-2
Cumberland
Cumberland, MD
40

34 ❑ 12-2
Blackhawk
Buffalo, NY
50

35 ❑ 12-2
Blackhawk
Cleveland, OH
150

36 ❑ 12-2
Leisy
Cleveland, OH
150

37 ❑ 12-2
Waukee
Hammonton, NJ
3

38 ❑ 11-2
Pabst
Los Angeles, CA
15

39 ❑ 12-2
Pabst
Los Angeles, CA
8

40 ❑ 12-2
Pabst
Los Angeles, CA
8

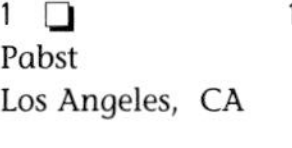
1 ❑ 11-2
Pabst
Los Angeles, CA
40

2 ❑ 12-2
Pabst
Los Angeles, CA
15

3 ❑ 12-2
Pabst
Peoria Heights, IL
5

4 ❑ 12-2
Pabst
Peoria Heights, IL
20

5 ❑ 12-2
Pabst
Newark, NJ
8

6 ❑ 12-2
Pabst
Newark, NJ
25

7 ❑ 12-1-T
Blatz
Milwaukee, WI
18

8 ❑ 12-1
Blatz
Milwaukee, WI
20

9 ❑ 12-1-T
Blatz
Milwaukee, WI
18

10 ❑ 12-1
Blatz
Milwaukee, WI
18

11 ❑ 12-1
Blatz
Milwaukee, WI
Lt. Blue 90

12 ❑ 12-1
Blatz
Milwaukee, WI
Dk. Blue 90

13 ❑ 12-1
Blatz
Milwaukee, WI
Green 90

14 ❑ 12-1
Blatz
Milwaukee, WI
Orange 90

15 ❑ 12-1
Blatz
Milwaukee, WI
Pink 90

16 ❑ 12-1
Blatz
Milwaukee, WI
Yellow 90

17 ❑ 10-1
Blatz
Milwaukee, WI
20

18 ❑ 12-1
Blatz
Milwaukee, WI
5

19 ❑ 12-2
Blatz
Milwaukee, WI
5

20 ❑ 12-1
Blatz
Milwaukee, WI
12

21 ❑ 12-1
Blatz
Milwaukee, WI
7

22 ❑ 12-2
Pabst
Milwaukee, WI
5

23 ❑ 12-2
Pabst
Milwaukee, WI
20

24 ❑ 12-1-T
Blitz-Weinhard
Portland, OR
Enamel DNCMT 4% 55

25 ❑ 12-1-T
Blitz-Weinhard
Portland, OR
Met. DNCMT 4% 55

26 ❑ 12-1
Blitz-Weinhard
Portland, OR
DNCMT 4% 55

27 ❑ 12-1-T
Blitz-Weinhard
Portland, OR
DNCMT 4% 75

28 ❑ 12-1
Blitz-Weinhard
Portland, OR
DNCMT 4% 75

29 ❑ 12-1
Blitz-Weinhard
Portland, OR
DNCMT 4% 40

30 ❑ 11-2
Blitz-Weinhard
Portland, OR
N.O. 4% 15

31 ❑ 12-2
Blitz-Weinhard
Portland, OR
N.O. 4% 15

32 ❑ 11-2
Blitz-Weinhard
Portland, OR
10

33 ❑ 12-2
Blitz-Weinhard
Portland, OR
10

34 ❑ 12-1-QT
Grace Bros.
Santa Rosa, CA
1000+

35 ❑ 12-1-QT
Regal Amber
San Francisco, CA
600

36 ❑ 12-1
North Bay
Los Angeles, CA
150

37 ❑ 12-1
Southern
Los Angeles, CA
175

38 ❑ 12-1-T
North Bay
Santa Rosa, CA
175

39 ❑ 12-1-W
North Bay
Santa Rosa, CA
250

40 ❑ 12-1
North Bay
Santa Rosa, CA
190

1 ❑ 12-1
North Bay
Santa Rosa, CA
150

2 ❑ 12-2
North Bay
Santa Rosa, CA
140

3 ❑ 12-2
Southern
Los Angeles, CA
300

4 ❑ 12-1
Metropolis
Trenton, NJ
225

5 ❑ 12-1
Hornell
Hornell, NY
275

6 ❑ 12-2
Richards
Newark, NJ
40

7 ❑ 12-2
Richards
Newark, NJ
30

8 ❑ 12-2
Enterprise
Boston, MA
25

9 ❑ 12-1
Enterprise
Fall River, MA
400

10 ❑ 12-1
Enterprise
Fall River, MA
350

11 ❑ 12-1
Enterprise
Fall River, MA
50

12 ❑ 12-2
Enterprise
Fall River, MA
40

13 ❑ 12-2
Enterprise
Fall River, MA
18

14 ❑ 12-2
Maier
Los Angeles, CA
8

15 ❑ 12-1
Southern
Los Angeles, CA
35

16 ❑ 12-2
Southern
Los Angeles, CA
Lg. Contents
30

17 ❑ 12-2
Southern
Los Angeles, CA
Sm. Contents
30

18 ❑ 12-1
Bohemian
Chicago, IL
75

19 ❑ 12-1
Bohemian
Joliet, IL
75

20 ❑ 12-2
Bohemian
Chicago, IL
25

21 ❑ 12-2
Bohemian
Chicago, IL
Enamel
15

22 ❑ 12-2
Bohemian
Chicago, IL
Metallic
15

23 ❑ 12-1
Bohemian
Joliet, IL
15

24 ❑ 12-2
Bohemian
Oconto, WI
18

25 ❑ 12-2
Bohemian
Potosi, WI
8

26 ❑ 11-2
Blitz-Weinhard
Portland, OR
20

27 ❑ 11-2
Br. Corp. of Ore.
Portland, OR
20

28 ❑ 12-2
Bohemian
Spokane, WA
Purple Vest
50

29 ❑ 12-2
Bohemian
Spokane, WA
Dk. Blue Vest
60

30 ❑ 12-2
Bohemian
Spokane, WA
50

31 ❑ 12-2
Bohemian
Spokane, WA
25

32 ❑ 11-2
Bohemian
Spokane, WA
25

33 ❑ 11-2
Bohemian Div. of Atl.
Spokane, WA
25

34 ❑ 12-1
Camden County
Camden, NJ
300

35 ❑ 12-2
Cumberland
Cumberland, MD
400

36 ❑ 12-2
Wm. Gretz
Philadelphia, PA
950

37 ❑ 12-2
Wm. Gretz
Philadelphia, PA
1000+

38 ❑ 12-1
Bosch
Houghton, MI
45

39 ❑ 12-1
Bosch
Houghton, MI
12

40 ❑ 12-1
Bosch
Houghton, MI
20

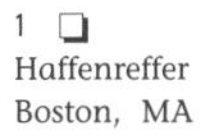
1 ❑ 12-2
Haffenreffer
Boston, MA
225

2 ❑ 12-2
Haffenreffer
Boston, MA
125

3 ❑ 12-2
Maier
Los Angeles, CA
15

4 ❑ 12-1-OT
San Francisco
San Francisco, CA
900

5 ❑ 12-2
Grace Bros.
Santa Rosa, CA
100

6 ❑ 12-2
Grace Bros.
Santa Rosa, CA
50

7 ❑ 12-2
Grace Bros.
Santa Rosa, CA
90

8 ❑ 12-1
Metropolis
Trenton, NJ
175

9 ❑ 12-1
Metropolis
Trenton, NJ
175

10 ❑ 12-1-T
Metropolis
New York, NY
225

11 ❑ 12-1-T
Old Dutch
New York, NY
1000+

12 ❑ 12-1-T
Old Dutch
New York, NY
225

13 ❑ 12-1-T
Old Dutch
New York, NY
1000+

14 ❑ 12-1
Ind. Milwaukee
Milwaukee, WI
Sm. Mandatory
10

15 ❑ 12-1
Ind. Milwaukee
Milwaukee, WI
Lg. Mandatory
10

16 ❑ 12-2
Ind. Milwaukee
Sheboygan, WI
10

17 ❑ 12-2
Ind. Milwaukee
Sheboygan, WI
5

18 ❑ 12-2
Ind. Milwaukee
Sheboygan, WI
3

19 ❑ 12-1
Kingsbury
Sheboygan, WI
20

20 ❑ 12-2
Rice Lake
Rice Lake, WI
Purple
5

21 ❑ 12-2
Rice Lake
Rice Lake, WI
Blue
5

22 ❑ 12-2
Grace Bros.
Santa Rosa, CA
50

23 ❑ 12-2
Grace Bros.
Santa Rosa, CA
50

24 ❑ 12-2
Grace Bros.
Santa Rosa, CA
55

25 ❑ 12-1
North Bay
Santa Rosa, CA
50

26 ❑ 12-2
North Bay
Santa Rosa, CA
50

27 ❑ 12-2
Uchtorff
Davenport, IA
125

28 ❑ 12-2
Brew 82
Cleveland, OH
100

29 ❑ 12-2
Leisy
Cleveland, OH
100

30 ❑ 12-1
Maier
Los Angeles, CA
45

31 ❑ 12-1
Maier
Los Angeles, CA
35

32 ❑ 12-1
Maier
Los Angeles, CA
35

33 ❑ 12-2
Maier
Los Angeles, CA
40

34 ❑ 12-2
Maier
Los Angeles, CA
Red Label
40

35 ❑ 12-2
Maier
Los Angeles, CA
Maroon Label
40

36 ❑ 12-2
Maier
Los Angeles, CA
8

37 ❑ 12-2
Maier
Los Angeles, CA
12

38 ❑ 12-2
Maier
Los Angeles, CA
12

39 ❑ 12-1
Maier
Los Angeles, CA
18

40 ❑ 12-1
Grace Bros.
Santa Rosa, CA
20

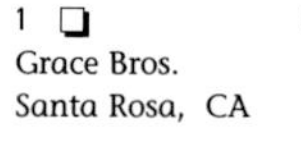

1 ❑ 12-1
Grace Bros.
Santa Rosa, CA
20

2 ❑ 12-1-T
North Bay
Santa Rosa, CA
50

3 ❑ 12-1-T
Atlantic
Chicago, IL
55

4 ❑ 12-2
Becker Products
Ogden, UT
275

5 ❑ 12-2-0T
Humboldt
Eureka, CA
1000+

6 ❑ 12-2-0T
Humboldt
Eureka, CA
50

7 ❑ 12-2-0T
Humboldt
Eureka, CA
35

8 ❑ 12-1-0T
Grace Bros LTD.
Los Angeles, CA
100

9 ❑ 12-2-0T
Los Angeles
Los Angeles, CA
40

10 ❑ 12-2-0T
Los Angeles
Los Angeles, CA
35

11 ❑ 12-2-0T
Los Angeles
Los Angeles, CA
45

12 ❑ 12-2
Maier
Los Angeles, CA
10

13 ❑ 12-2
Maier
Los Angeles, CA
Metallic 10

14 ❑ 12-2
Maier
Los Angeles, CA
Enamel 10

15 ❑ 12-2
Maier
Los Angeles, CA
12

16 ❑ 12-2
Maier
Los Angeles, CA
12

17 ❑ 12-2
Maier
Los Angeles, CA
5

18 ❑ 12-2-0T
Rainier
San Francisco, CA
35

19 ❑ 12-2-0T
Rainier
San Francisco, CA
45

20 ❑ 12-2-0T
Rainier
San Francisco, CA
45

21 ❑ 12-2-0T
Grace Bros.
Santa Rosa, CA
55

22 ❑ 12-2
Grace Bros.
Santa Rosa, CA
15

23 ❑ 12-2
Grace Bros.
Santa Rosa, CA
12

24 ❑ 12-2
Atlas
Chicago, IL
25

25 ❑ 12-2
Best
Chicago, IL
35

26 ❑ 12-2
Empire
Chicago, IL
Metallic 15

27 ❑ 12-2
Empire
Chicago, IL
Enamel 15

28 ❑ 12-2
United States
Chicago, IL
Metallic 15

29 ❑ 12-2
United States
Chicago, IL
Enamel 15

30 ❑ 12-2
Eastern Brewing Co.
Hammonton, NJ
20

31 ❑ 12-2-0T
Salem Assn.
Salem, OR
DNCMT 4% 45

32 ❑ 12-1-0T
Salem Assn.
Salem, OR
DNCMT 4% 45

33 ❑ 12-2
Century
Norfolk, VA
25

34 ❑ 11-2
Atlantic
Spokane, WA
20

35 ❑ 11-2
Atlantic
Spokane, WA
25

36 ❑ 12-2
Atlantic
Spokane, WA
20

37 ❑ 12-2
K. C. Best
Spokane, WA
30

38 ❑ 12-2-0T
Columbia
Tacoma, WA
60

39 ❑ 11-2
Silver Springs
Tacoma, WA
25

40 ❑ 12-1-T
Monarch
Los Angeles, CA
1000+

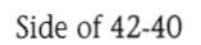
Side of 42-40

2 ❑ 12-1
Peter Bub
Winona, MN
10

3 ❑ 12-2
Pacific
Oakland, CA
500

4 ❑ 12-2
Gulf
Houston, TX
550

5 ❑ 12-2
Burgermeister
San Francisco, CA
12

6 ❑ 12-2
Meister-Brau
Chicago, IL
7

7 ❑ 12-2
Buckeye
Toledo, OH
125

8 ❑ 12-2
Buckeye
Toledo, OH
18

9 ❑ 12-2
Buckeye
Toledo, OH
12

10 ❑ 11-2
Buckhorn, Div. Hamm
San Francisco, CA
7

11 ❑ 11-2
Theo. Hamm
San Francisco, CA
7

12 ❑ 11-2
Theo. Hamm
San Francisco, CA
6

13 ❑ 11-2
Buckhorn
San Francisco, CA
6

14 ❑ 11-2
Theo. Hamm
San Francisco, CA
6

15 ❑ 11-2
Theo. Hamm
San Francisco, CA
4

16 ❑ 12-2
Buckhorn
St. Paul, MN
10

17 ❑ 12-2
Buckhorn
St. Paul, MN
8

18 ❑ 12-1-T
Wehle
West Haven, CT
1000+

19 ❑ 12-2
Anheuser-Busch
Los Angeles, CA
3 City
18

20 ❑ 12-2
Anheuser-Busch
Los Angeles, CA
3 City
20

21 ❑ 12-2
Anheuser-Busch
Los Angeles, CA
3 City
5

22 ❑ 12-2
Anheuser-Busch
Los Angeles, CA
5 City
5

23 ❑ 12-2
Anheuser-Busch
Los Angeles, CA
4 City
5

24 ❑ 12-2
Anheuser-Busch
Los Angeles, CA
5 City
5

25 ❑ 12-2
Anheuser-Busch
Los Angeles, CA
6 City
4

26 ❑ 12-2
Anheuser-Busch
Los Angeles, CA
8 City
4

27 ❑ 12-2
Anheuser-Busch
Tampa, FL
5 City
5

28 ❑ 12-2
Anheuser-Busch
Tampa, FL
4 City
5

29 ❑ 12-2
Anheuser-Busch
Tampa, FL
4 City
4

30 ❑ 12-2
Anheuser-Busch
Tampa, FL
5 City
4

31 ❑ 12-2
Anheuser-Busch
Tampa, FL
5 City
4

32 ❑ 12-2
Anheuser-Busch
Tampa, FL
6 City
4

33 ❑ 12-1-0T
Anheuser-Busch
St. Louis, MO
35

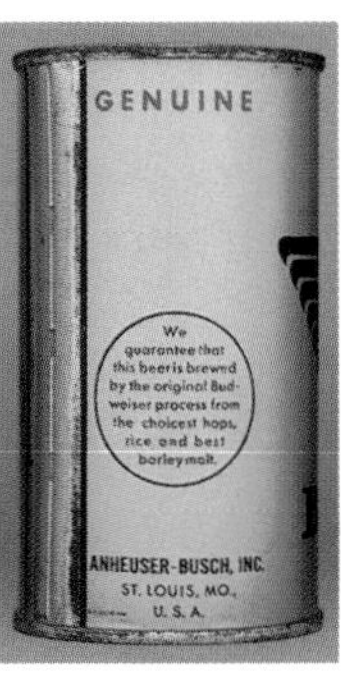

Side of 33

35 ❑ 12-1-0T
Anheuser-Busch
St. Louis, MO
35

36 ❑ 12-1-W
Anheuser-Busch
St. Louis, MO
750

37 ❑ 12-1-0T
Anheuser-Busch
St. Louis, MO
250

38 ❑ 12-1-0W
Anheuser-Busch
St. Louis, MO
100

39 ❑ 12-1-0W
Anheuser-Busch
St. Louis, MO
100

40 ❑ 12-1-0T
Anheuser-Busch
St. Louis, MO
30

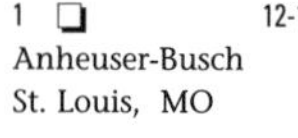
1 ❑ 12-1-W
Anheuser-Busch
St. Louis, MO
100

2 ❑ 12-1-T
Anheuser-Busch
St. Louis, MO
20

3 ❑ 12-1-T
Anheuser-Busch
St. Louis, MO
35

Side of 3

5 ❑ 12-1
Anheuser-Busch
St. Louis, MO
1 City W/Bottle
18

6 ❑ 12-1
Anheuser-Busch
St. Louis, MO
2 City W/Bottle
18

7 ❑ 12-1
Anheuser-Busch
St. Louis, MO
2 City W/Bottle
18

8 ❑ 12-2
Anheuser-Busch
St. Louis, MO
2 City
12

9 ❑ 10-2
Anheuser-Busch
St. Louis, MO
3 City
50

10 ❑ 10-2
Anheuser-Busch
St. Louis, MO
3 City
60

11 ❑ 12-2
Anheuser-Busch
St. Louis, MO
3 City
12

12 ❑ 10-2
Anheuser-Busch
St. Louis, MO
3 City
25

13 ❑ 12-2
Anheuser-Busch
St. Louis, MO
3 City
12

14 ❑ 10-2
Anheuser-Busch
St. Louis, MO
3 City
15

15 ❑ 12-2
Anheuser-Busch
St. Louis, MO
3 City
5

16 ❑ 10-2
Anheuser-Busch
St. Louis, MO
5 City
10

17 ❑ 12-2
Anheuser-Busch
St. Louis, MO
5 City
5

18 ❑ 10-2
Anheuser-Busch
St. Louis, MO
4 City
10

19 ❑ 12-2
Anheuser-Busch
St. Louis, MO
4 City
5

20 ❑ 12-2
Anheuser-Busch
St. Louis, MO
4 City
7

21 ❑ 10-2
Anheuser-Busch
St. Louis, MO
5 City
12

22 ❑ 12-2
Anheuser-Busch
St. Louis, MO
5 City
5

23 ❑ 12-2
Anheuser-Busch
St. Louis, MO
6 City
3

24 ❑ 12-2
Anheuser-Busch
St. Louis, MO
7 City
3

25 ❑ 12-2
Anheuser-Busch
St. Louis, MO
8 City
3

26 ❑ 12-1
Anheuser-Busch
St. Louis, MO
1000+

27 ❑ 12-2
Anheuser-Busch
St. Louis, MO
35

28 ❑ 12-1
Anheuser-Busch
Newark, NJ
2 City W/Bottle
20

29 ❑ 12-1
Anheuser-Busch
Newark, NJ
2 City W/Bottle
20

30 ❑ 12-2
Anheuser-Busch
Newark, NJ
2 City
12

31 ❑ 12-2
Anheuser-Busch
Newark, NJ
3 City
12

32 ❑ 12-2
Anheuser-Busch
Newark, NJ
3 City
15

33 ❑ 12-2
Anheuser-Busch
Newark, NJ
3 City
8

34 ❑ 12-2
Anheuser-Busch
Newark, NJ
5 City
6

35 ❑ 12-2
Anheuser-Busch
Newark, NJ
4 City
5

36 ❑ 12-2
Anheuser-Busch
Newark, NJ
5 City
5

37 ❑ 10-2
Anheuser-Busch
Newark, NJ
7 City
12

38 ❑ 10-2
Anheuser-Busch
Newark, NJ
8 City
12

39 ❑ 12-2
Anheuser-Busch
Columbus, OH
6 City
12

40 ❑ 12-2
Anheuser-Busch
Columbus, OH
7 City
12

1 ❑ 10-2
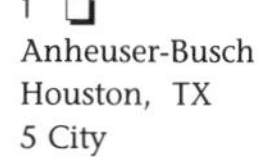
Anheuser-Busch
Houston, TX
5 City 12

2 ❑ 12-2
Anheuser-Busch
Houston, TX
5 City 10

3 ❑ 10-2
Anheuser-Busch
Houston, TX
7 City 12

4 ❑ 12-1
Buffalo
Los Angeles, CA
75

5 ❑ 12-1
Southern
Los Angeles, CA
75

Front of 7, 8, 9

7 ❑ 12-1-OT
Buffalo
Sacramento, CA
400

8 ❑ 12-1-OT
Buffalo
Sacramento, CA
400

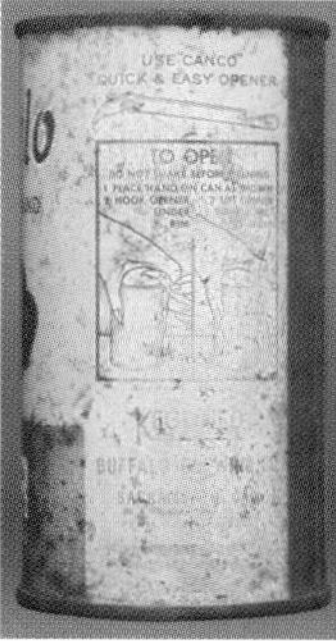

9 ❑ 12-1-OW
Buffalo
Sacramento, CA
600

10 ❑ 12-1-T
Buffalo
Sacramento, CA
175

11 ❑ 12-1-T
Buffalo
Sacramento, CA
300

12 ❑ 12-2-T
Buffalo
Sacramento, CA
200

13 ❑ 12-1-T
Buffalo
Sacramento, CA
150

14 ❑ 12-1-T
Buffalo
Sacramento, CA
150

15 ❑ 12-2
Acme
Los Angeles, CA
35

16 ❑ 12-2
Acme
Los Angeles, CA
35

17 ❑ 12-2
Acme
Los Angeles, CA
35

18 ❑ 12-2
Acme
Los Angeles, CA
30

19 ❑ 12-2
Maier
Los Angeles, CA
15

20 ❑ 12-2
Acme
San Francisco, CA
35

21 ❑ 12-2
Acme
San Francisco, CA
35

22 ❑ 12-2
Acme
San Francisco, CA
30

23 ❑ 12-2
Acme
San Francisco, CA
30

24 ❑ 12-2
California
San Francisco, CA
25

25 ❑ 12-2
California
San Francisco, CA
25

26 ❑ 12-2
California
San Francisco, CA
25

27 ❑ 12-2
California
San Francisco, CA
30

28 ❑ 12-2
California
San Francisco, CA
25

29 ❑ 12-2
California
San Francisco, CA
25

30 ❑ 12-2
Grace Bros.
Santa Rosa, CA
25

31 ❑ 12-2
Grace Bros.
Santa Rosa, CA
20

32 ❑ 12-2
Grace Bros.
Santa Rosa, CA
35

33 ❑ 12-2
Grace Bros.
Santa Rosa, CA
15

34 ❑ 12-2
Grace Bros.
Santa Rosa, CA
15

35 ❑ 12-2
Grace Bros.
Santa Rosa, CA
15

36 ❑ 12-2
Grace Bros.
Santa Rosa, CA
20

37 ❑ 12-2
Grace Bros.
Santa Rosa, CA
20

38 ❑ 12-2
Grace Bros.
Santa Rosa, CA
15

39 ❑ 12-2
Atlas
Chicago, IL
25

40 ❑ 12-2
Atlas
Chicago, IL
25

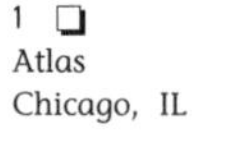
1 ❑ 12-2
Atlas
Chicago, IL
25

2 ❑ 12-2
Atlas
Chicago, IL
15

3 ❑ 12-2
Drewry's Ltd.
South Bend, IN
15

4 ❑ 12-2
Monarch
Chicago, IL
25

5 ❑ 12-2
Monarch
Chicago, IL
25

6 ❑ 12-2-OT
Golden West
Oakland, CA
150

7 ❑ 12-1
Warsaw
Warsaw, IL
15

8 ❑ 12-1
Warsaw
Warsaw, IL
Metallic
10

9 ❑ 12-1
Warsaw
Warsaw, IL
Enamel
5

10 ❑ 12-2
Burger
Akron, OH
18

11 ❑ 12-2
Burger
Akron, OH
5

12 ❑ 12-1
Burger
Cincinnati, OH
125

13 ❑ 12-2
Burger
Cincinnati, OH
70

14 ❑ 12-2
Burger
Cincinnati, OH
45

15 ❑ 12-2
Burger
Cincinnati, OH
30

16 ❑ 12-2
Burger
Cincinnati, OH
30

17 ❑ 12-1
Burger
Cincinnati, OH
25

18 ❑ 12-2
Burger
Cincinnati, OH
15

19 ❑ 12-2
Burger
Cincinnati, OH
15

20 ❑ 12-2
Burger
Cincinnati, OH
15

21 ❑ 12-2
Burger
Cincinnati, OH
8

22 ❑ 12-2
Burger
Cincinnati, OH
8

23 ❑ 12-2
Burger
Cincinnati, OH
5

24 ❑ 11-2
Burg. Div. of Schlitz
Los Angeles, CA
8

25 ❑ 12-1-OT
San Francisco
San Francisco, CA
275

26 ❑ 12-1-OT
San Francisco
San Francisco, CA
275

27 ❑ 12-1-OT
San Francisco
San Francisco, CA
650

28 ❑ 12-1-OT
San Francisco
San Francisco, CA
400

29 ❑ 12-1-T
San Francisco
San Francisco, CA
500

30 ❑ 12-1-T
San Francisco
San Francisco, CA
18

31 ❑ 12-1
San Francisco
San Francisco, CA
15

32 ❑ 12-2
San Francisco
San Francisco, CA
18

33 ❑ 12-2
San Francisco
San Francisco, CA
15

34 ❑ 12-2
San Francisco
San Francisco, CA
60

35 ❑ 12-2
San Francisco
San Francisco, CA
12

36 ❑ 12-2
San Francisco
San Francisco, CA
12

37 ❑ 12-2
Burgermeister
San Francisco, CA
8

38 ❑ 12-2
Burgermeister
San Francisco, CA
8

39 ❑ 12-2
Burgermeister
San Francisco, CA
8

40 ❑ 11-2
Burgermeister
San Francisco, CA
6

1 ❑ 11-2
Burgermeister
San Francisco, CA
8

2 ❑ 10-2
Burgermeister
San Francisco, CA
65

3 ❑ 11-2
Burgermeister
San Francisco, CA
6

4 ❑ 11-2
Burgermeister
San Francisco, CA
6

5 ❑ 12-2
Burgermeister
San Francisco, CA
6

6 ❑ 11-2
Burg. Div. of Schlitz
San Francisco, CA
5

7 ❑ 11-2
Burg. Div. of Schlitz
San Francisco, CA
5

8 ❑ 12-1
Burkhardt
Akron, OH
60

9 ❑ 12-1
Burkhardt
Akron, OH
40

10 ❑ 12-1
Burkhardt
Akron, OH
100

11 ❑ 12-2
Anheuser-Busch
Los Angeles, CA
8

12 ❑ 12-2
Anheuser-Busch
Miami, FL
15

13 ❑ 12-2
Anheuser-Busch
Miami, FL
10

14 ❑ 12-2
Anheuser-Busch
Tampa, FL
8

15 ❑ 12-2
Anheuser-Busch
Tampa, FL
22

16 ❑ 12-2
Anheuser-Busch
Tampa, FL
6

17 ❑ 12-2
Anheuser-Busch
Tampa, FL
8

18 ❑ 12-2
Anheuser-Busch
St. Louis, MO
100

19 ❑ 12-2
Anheuser-Busch
St. Louis, MO
60

20 ❑ 12-2
Anheuser-Busch
St. Louis, MO
15

21 ❑ 12-2
Anheuser-Busch
St. Louis, MO
8

22 ❑ 12-2
Anheuser-Busch
St. Louis, MO
8

23 ❑ 12-2
Anheuser-Busch
St. Louis, MO
8

24 ❑ 10-2
Anheuser-Busch
St. Louis, MO
35

25 ❑ 12-2
Anheuser-Busch
St. Louis, MO
20

26 ❑ 12-2
Anheuser-Busch
St. Louis, MO
5

27 ❑ 12-2
Anheuser-Busch
St. Louis, MO
5

28 ❑ 12-2
Anheuser-Busch
St. Louis, MO
5

29 ❑ 12-2
Anheuser-Busch
Newark, NJ
125

30 ❑ 12-2
Butte
Butte, MT
CNMT 4%
60

31 ❑ 12-2
Butte
Butte, MT
Metallic Gold
35

32 ❑ 12-2
Butte
Butte, MT
Dull Gold
35

33 ❑ 12-2
Butte
Butte, MT
CNMT 4%
35

34 ❑ 12-1-T
Wehle
West Haven, CT
1000+

35 ❑ 12-1-T
Wehle
West Haven, CT
1000+

36 ❑ 12-2
California
San Francisco, CA
Black Writing
20

37 ❑ 12-2
California
San Francisco, CA
Brown Writing
20

38 ❑ 12-1
Camden County
Camden, NJ
55

39 ❑ 12-1
Camden County
Camden, NJ
1 Park Place bf
55

40 ❑ 12-1
Camden County
Camden, NJ
251 W. 42nd St. bf
55

1 ❑ 12-1
Esslinger's
Philadelphia, PA
75

2 ❑ 12-2
Tivoli
Denver, CO
10

3 ❑ 12-1-T
Canadian Ace
Chicago, IL
20

4 ❑ 12-1
Canadian Ace
Chicago, IL
18

5 ❑ 12-2
Canadian Ace
Chicago, IL
10

6 ❑ 12-2
Canadian Ace
Chicago, IL
10

7 ❑ 12-2
Canadian Ace
Chicago, IL
10

8 ❑ 12-1-T
Canadian Ace
Chicago, IL
20

9 ❑ 12-1
Canadian Ace
Chicago, IL
10

10 ❑ 12-2
Canadian Ace
Chicago, IL
6

11 ❑ 12-2
Canadian Ace
Chicago, IL
5

12 ❑ 12-2
Canadian Ace
Chicago, IL
5

13 ❑ 12-2
Canadian Ace
Chicago, IL
Enamel
5

14 ❑ 12-2
Canadian Ace
Chicago, IL
5

15 ❑ 12-1
Canadian Ace
Chicago, IL
5

16 ❑ 12-2
Canadian Ace
Chicago, IL
45

17 ❑ 12-2
Canadian Ace
Chicago, IL
Metallic
12

18 ❑ 12-2
Canadian Ace
Chicago, IL
Enamel
12

19 ❑ 12-2
Enterprise
Fall River, MA
500

20 ❑ 12-2
E & B
Detroit, MI
125

21 ❑ 12-1
Cardinal
St. Charles, MO
3

22 ❑ 12-1
Pilsen
Chicago, IL
400

23 ❑ 11-2
Blitz-Weinhard
Portland, OR
25

24 ❑ 11-2
Blitz-Weinhard
Portland, OR
7

25 ❑ 12-2
Blitz-Weinhard
Portland, OR
15

26 ❑ 12-2
Grace Bros.
Santa Rosa, CA
200

27 ❑ 12-2
Colonial
Hammonton, NJ
10

28 ❑ 12-1-OT
Centlivre
Fort Wayne, IN
1000+

29 ❑ 12-2
Atlantic
Chicago, IL
15

30 ❑ 12-2
Atlantic
Chicago, IL
7

31 ❑ 12-2
CV
Chicago, IL
15

32 ❑ 12-1
Terre Haute
Terre Haute, IN
35

33 ❑ 12-1
Terre Haute
Terre Haute, IN
35

34 ❑ 12-1
Terre Haute
Terre Haute, IN
35

35 ❑ 12-1
Terre Haute
Terre Haute, IN
35

36 ❑ 12-1
Terre Haute
Terre Haute, IN
25

37 ❑ 12-1
Terre Haute
Terre Haute, IN
25

38 ❑ 12-2
Terre Haute
Terre Haute, IN
25

39 ❑ 12-2
Terre Haute
Terre Haute, IN
Lt. Blue
100

40 ❑ 12-2
Terre Haute
Terre Haute, IN
Dk. Blue
120

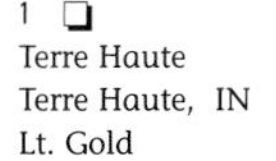
1 ❑ 12-2
Terre Haute
Terre Haute, IN
Lt. Gold 100

2 ❑ 12-2
Terre Haute
Terre Haute, IN
Lt. Green 100

3 ❑ 12-2
Terre Haute
Terre Haute, IN
Green 100

4 ❑ 12-2
Terre Haute
Terre Haute, IN
Purple 100

5 ❑ 12-2
Terre Haute
Terre Haute, IN
Red 125

6 ❑ 12-2
Terre Haute
Terre Haute, IN
20

7 ❑ 11-2
Br. Corp. of Oregon
Portland, OR
10

8 ❑ 11-2
Br. Corp. of Oregon
Portland, OR
8

9 ❑ 11-2
Bohemian
Spokane, WA
8

10 ❑ 12-2
Bohemian
Spokane, WA
25

11 ❑ 12-2
Bohemian
Spokane, WA
6

Front of 13, 14, 15

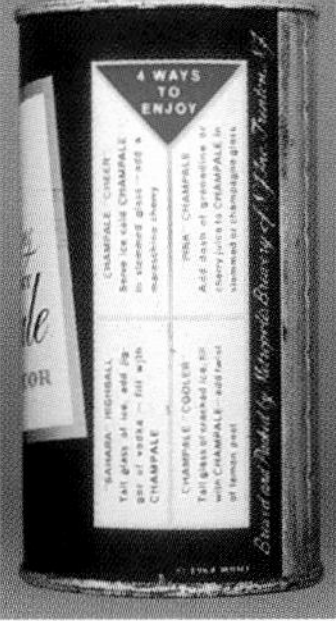
13 ❑ 12-1
Metropolis
Trenton, NJ
12

14 ❑ 12-1
Metropolis
Trenton, NJ
12

15 ❑ 12-1
Metropolis
Trenton, NJ
12

16 ❑ 12-1
Metropolis
Trenton, NJ
With Recipes 12

17 ❑ 12-1
Century
Norfolk, VA
Enamel 12

18 ❑ 12-1
Century
Norfolk, VA
Metallic 12

19 ❑ 12-1
Century
Norfolk, VA
With Recipes 12

20 ❑ 12-1
Century
Norfolk, VA
No Recipes 12

21 ❑ 12-1
Gulf
Houston, TX
550

22 ❑ 12-2
Hudepohl
Cincinnati, OH
350

23 ❑ 12-1-T
Monarch
Los Angeles, CA
1000+

24 ❑ 12-2
Oshkosh
Oshkosh, WI
35

25 ❑ 12-2
Oshkosh
Oshkosh, WI
12

26 ❑ 12-2
Oshkosh
Oshkosh, WI
12

27 ❑ 12-2
Oshkosh
Oshkosh, WI
5

28 ❑ 12-2
Oshkosh
Oshkosh, WI
20

29 ❑ 12-2
Circle
Hammonton, NJ
10

30 ❑ 12-1-OT
Class & Nachod
Philadelphia, PA
1000+

31 ❑ 12-2
Grace Bros.
Santa Rosa, CA
125

32 ❑ 12-2
Grace Bros.
Santa Rosa, CA
125

33 ❑ 12-1-T
Grace Bros LTD.
Los Angeles, CA
1000+

34 ❑ 12-1-T
Grace Bros.
Santa Rosa, CA
1000+

35 ❑ 12-2
Grace Bros.
Santa Rosa, CA
70

36 ❑ 12-1
Maier
Los Angeles, CA
35

37 ❑ 12-2
Enterprise
Fall River, MA
300

38 ❑ 12-2
Enterprise
Fall River, MA
125

39 ❑ 12-1
Horlacher
Allentown, PA
12

40 ❑ 12-1
Old Dutch
Allentown, PA
8

1 ❑ 12-2
Schoenhofen Edelweiss
Chicago, IL
25

2 ❑ 12-2
Schoenhofen Edelweiss
Chicago, IL
Enamel Red 7

3 ❑ 12-2
Schoenhofen Edelweiss
Chicago, IL
Metallic Red 7

4 ❑ 12-2
Drewry's Ltd.
South Bend, IN
7

5 ❑ 12-2
Cold Spring
Cold Spring, MN
30

6 ❑ 12-2
Cold Spring
Cold Spring, MN
12

7 ❑ 12-2
Cold Spring
Cold Spring, MN
8

8 ❑ 12-2
Cold Spring
Cold Spring, MN
8

9 ❑ 12-2
Colonial
Hammonton, NJ
40

10 ❑ 12-2
National
Miami, FL
30

11 ❑ 12-2
National
Baltimore, MD
15

12 ❑ 12-2
Lone Star
Oklahoma City, OK
30

13 ❑ 12-2
Columbia
Shenandoah, PA
40

14 ❑ 12-2
Columbia
Shenandoah, PA
40

15 ❑ 12-2
Heidelberg
Tacoma, WA
60

16 ❑ 12-2
Heidelberg
Tacoma, WA
90

17 ❑ 12-1-OT
Columbia
Tacoma, WA
1000+

18 ❑ 12-1-OT
Columbia
Tacoma, WA
1000+

19 ❑ 12-1
Haberle Congress
Syracuse, NY
Steins 150

20 ❑ 12-1
Haberle Congress
Syracuse, NY
Shrimp 150

21 ❑ 12-1
Haberle Congress
Syracuse, NY
Fish 150

22 ❑ 12-1
Haberle Congress
Syracuse, NY
Car 150

23 ❑ 12-1
Haberle Congress
Syracuse, NY
Cities 150

24 ❑ 12-1
Haberle Congress
Syracuse, NY
Summer Sports 150

25 ❑ 12-1
Haberle Congress
Syracuse, NY
Hunting 125

26 ❑ 12-1
Haberle Congress
Syracuse, NY
Hunting 125

27 ❑ 12-1
Haberle Congress
Syracuse, NY
Hunting 125

28 ❑ 12-1
Haberle Congress
Syracuse, NY
Indoor Games 125

29 ❑ 12-1
Haberle Congress
Syracuse, NY
Indoor Games 125

30 ❑ 12-1
Haberle Congress
Syracuse, NY
Indoor Games 125

31 ❑ 12-1
Haberle Congress
Syracuse, NY
Geese 125

32 ❑ 12-1
Haberle Congress
Syracuse, NY
Geese in Silver 125

33 ❑ 12-1
Haberle Congress
Syracuse, NY
Geese in Red 125

34 ❑ 12-1
Haberle Congress
Syracuse, NY
Geese 125

35 ❑ 12-1
Haberle Congress
Syracuse, NY
Table Tennis 125

36 ❑ 12-1
Haberle Congress
Syracuse, NY
Table Tennis 125

37 ❑ 12-1
Haberle Congress
Syracuse, NY
Skiing 125

38 ❑ 12-1
Haberle Congress
Syracuse, NY
Skiing Red Trim 125

39 ❑ 12-1
Haberle Congress
Syracuse, NY
Skiing Yellow Trim 125

40 ❑ 12-1
Haberle Congress
Syracuse, NY
Carnival 125

1 ❑ 12-1
Haberle Congress
Syracuse, NY
Carnival 125

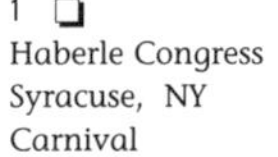

2 ❑ 12-2
Haberle Congress
Syracuse, NY
60

3 ❑ 12-2
Haberle Congress
Syracuse, NY
55

4 ❑ 12-2
Haberle Congress
Syracuse, NY
50

5 ❑ 12-1
Harvard
Lowell, MA
450

6 ❑ 12-1
Merrimack
Lowell, MA
600

7 ❑ 12-1
Harvard
Lowell, MA
350

8 ❑ 12-1
Merrimack
Lowell, MA
350

9 ❑ 12-2
F. W. Cook
Evansville, IN
225

10 ❑ 12-2
F. W. Cook
Evansville, IN
50

11 ❑ 12-2
F. W. Cook
Evansville, IN
6

12 ❑ 12-1-T
Adolph Coors
Golden, CO
225

Back of 12

Front of 15, 16

15 ❑ 12-1-T
Adolph Coors
Golden, CO
225

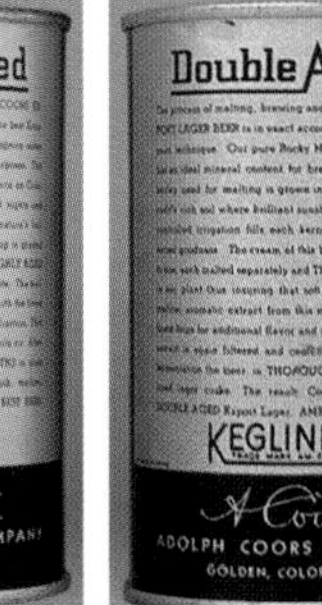

16 ❑ 12-1-T
Adolph Coors
Golden, CO
250

17 ❑ 12-1-T
Adolph Coors
Golden, CO
200

Back of 17

19 ❑ 12-1-T
Adolph Coors
Golden, CO
100

20 ❑ 12-1-T
Adolph Coors
Golden, CO
12

21 ❑ 12-1
Adolph Coors
Golden, CO
60

22 ❑ 12-1
Adolph Coors
Golden, CO
60

23 ❑ 12-2
Adolph Coors
Golden, CO
5

24 ❑ 12-2
Adolph Coors
Golden, CO
2

25 ❑ 11-2
Adolph Coors
Golden, CO
2

26 ❑ 12-1
Metropolis
Trenton, NJ
1000+

27 ❑ 12-1
Edelbrew
New York, NY
900

28 ❑ 12-1
Imperial
Baltimore, MD
85

29 ❑ 12-1
Five Star
New York, NY
65

30 ❑ 12-1-T
Monarch
Los Angeles, CA
1000+

31 ❑ 12-1
M. K. Goetz
St. Joseph, MO
90 Years 15

32 ❑ 12-1
M. K. Goetz
St. Joseph, MO
Over 90 Years 15

Front of 34, 35

34 ❑ 12-1
M. K. Goetz
St. Joseph, MO
95 Years 15

35 ❑ 12-1
M. K. Goetz
St. Joseph, MO
96 Years 15

36 ❑ 12-1
M. K. Goetz
St. Joseph, MO
125

37 ❑ 12-1
M. K. Goetz
St. Joseph, MO
12

38 ❑ 12-1
Goetz Div. of Pearl
St. Joseph, MO
12

39 ❑ 12-2
Pearl
St. Joseph, MO
12

40 ❑ 12-1
M. K. Goetz
St. Joseph, MO
75

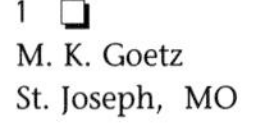

1 ❑ 12-1
M. K. Goetz
St. Joseph, MO
6

2 ❑ 12-1
Goetz Div. of Pearl
St. Joseph, MO
6

3 ❑ 12-2
Pearl
St. Joseph, MO
6

4 ❑ 12-1
M. K. Goetz
St. Joseph, MO
7

5 ❑ 12-1
Goetz Div. of Pearl
St. Joseph, MO
7

6 ❑ 12-2
Pearl
St. Joseph, MO
7

7 ❑ 12-1-OT
Manhattan
Chicago, IL
450

8 ❑ 12-1-OT
Whitewater
Whitewater, WI
550

9 ❑ 12-1-OT
Grace Bros LTD.
Los Angeles, CA
850

10 ❑ 12-1-OT
Grace Bros.
Santa Rosa, CA
850

11 ❑ 12-1-OT
Grace Bros.
Santa Rosa, CA
1000+

12 ❑ 12-2
North Bay
Santa Rosa, CA
1000+

13 ❑ 12-2
Crest
Chicago, IL
20

14 ❑ 12-1-OT
Croft
Boston, MA
200

15 ❑ 12-1-OT
Croft
Boston, MA
1000+

16 ❑ 12-1-OT
Croft
Boston, MA
1000+

17 ❑ 12-1-T
Croft
Boston, MA
Six Products 1000+

18 ❑ 12-1-T
Croft
Boston, MA
Six Products 100

19 ❑ 12-1-T
Croft
Boston, MA
Six Products 300

20 ❑ 12-1-T
Croft
Boston, MA
Six Products 750

Front of 22, 23, 24

22 ❑ 12-1-T
Croft
Boston, MA
Six Products 175

23 ❑ 12-1-T
Croft
Boston, MA
Four Products 175

24 ❑ 12-1-T
Croft
Boston, MA
Three Products 175

25 ❑ 12-1
Croft
Boston, MA
1000+

26 ❑ 12-1-T
Croft
Boston, MA
45

27 ❑ 12-1-T
Croft
Boston, MA
45

28 ❑ 12-1
Croft
Boston, MA
45

29 ❑ 12-1
Croft
Boston, MA
750

30 ❑ 12-1
Croft
Boston, MA
100

31 ❑ 12-1
Croft
Boston, MA
85

32 ❑ 12-1
Croft
Cranston, RI
65

33 ❑ 12-1
Croft
Cranston, RI
65

34 ❑ 12-1
Croft
Cranston, RI
50

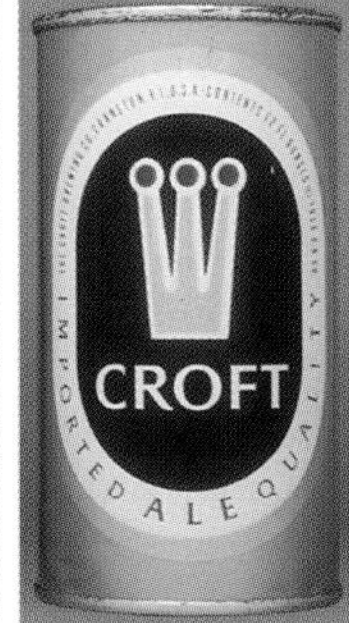

35 ❑ 12-2
Croft
Cranston, RI
30

36 ❑ 12-2
Westminster
Chicago, IL
350

37 ❑ 12-2
Walter
Pueblo, CO
12

38 ❑ 12-2
Cleveland-Sandusky
Buffalo, NY
65

39 ❑ 12-2
Cleveland-Sandusky
Cleveland, OH
600

40 ❑ 12-2
Cleveland-Sandusky
Cleveland, OH
40

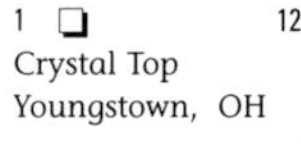

1 ❏ 12-1-T
Crystal Top
Youngstown, OH
800

2 ❏ 12-2
Grace Bros.
Santa Rosa, CA
60

3 ❏ 12-1
Dakota Malting
Bismark, ND
50

4 ❏ 12-2
Dakota Malting
Bismark, ND
30

5 ❏ 12-2
Colonial
Hammonton, NJ
60

6 ❏ 12-2-T
Dawson's
New Bedford, MA
150

7 ❏ 12-2
Dawson's
New Bedford, MA
150

8 ❏ 12-2
Dawson's
New Bedford, MA
100

9 ❏ 12-2
Dawson's
New Bedford, MA
80

10 ❏ 12-2
Dawson's
New Bedford, MA
80

11 ❏ 12-2
Dawson's
New Bedford, MA
100

12 ❏ 12-2
Dawson's
New Bedford, MA
250

13 ❏ 12-2
Dawson's
New Bedford, MA
30

14 ❏ 12-2
Dawson's
New Bedford, MA
18

15 ❏ 12-2-T
Dawson's
New Bedford, MA
250

16 ❏ 12-2
Dawson's
New Bedford, MA
350

17 ❏ 12-2
Dawson's
New Bedford, MA
150

18 ❏ 12-2
Dawson's
New Bedford, MA
150

19 ❏ 12-2
Dawson's
New Bedford, MA
80

20 ❏ 12-2
Dawson's
New Bedford, MA
125

21 ❏ 12-2
Dawson's
New Bedford, MA
250

22 ❏ 12-2
Dawson's
New Bedford, MA
25

23 ❏ 12-2
Dawson's
New Bedford, MA
20

24 ❏ 12-2
Dawson's
New Bedford, MA
1000+

25 ❏ 12-2
Tivoli
Denver, CO
50

26 ❏ 12-2
Tivoli
Denver, CO
75

27 ❏ 12-2
Tivoli
Denver, CO
50

28 ❏ 12-1
Tennessee
Memphis, TN
1000+

29 ❏ 12-2
Grace Bros.
Santa Rosa, CA
40

30 ❏ 12-1-OT
Diamond State
Wilmington, DE
450

31 ❏ 12-1-T
Diamond State
Wilmington, DE
200

32 ❏ 12-1
Diamond State
Wilmington, DE
200

33 ❏ 12-1
Wm. Gretz
Philadelphia, PA
250

34 ❏ 12-2
Genesee
Rochester, NY
175

35 ❏ 12-1-OT
Hull
New Haven, CT
1000+

36 ❏ 12-1
Diplomat
New Britain, CT
500

37 ❏ 12-1
Dixie
New Orleans, LA
100

38 ❏ 12-1
Dixie
New Orleans, LA
85

39 ❏ 10-1
Dixie
New Orleans, LA
45

40 ❏ 12-1
Dixie
New Orleans, LA
25

1 ❑ 12-1
Dixie
New Orleans, LA
8

2 ❑ 12-1
Mountain
Roanoke, VA
500

3 ❑ 12-1
Hampden-Harvard
Willimansett, MA
No Mass. Permit tf 20

4 ❑ 12-1
Hampden-Harvard
Willimansett, MA
Mass. Permit tf 20

5 ❑ 12-1
Hampden-Harvard
Willimansett, MA
18

6 ❑ 12-1
Hampden-Harvard
Willimansett, MA
No Mass. Permit tf 20

7 ❑ 12-1
Hampden-Harvard
Willimansett, MA
Mass. Permit tf 20

8 ❑ 12-1
Hampden-Harvard
Willimansett, MA
15

9 ❑ 12-1
Dobler
Albany, NY
90

10 ❑ 12-1
Dobler
Albany, NY
90

11 ❑ 12-1
Dobler
Albany, NY
80

12 ❑ 12-1
Dobler
Albany, NY
65

13 ❑ 12-1
Dobler
Albany, NY
25

14 ❑ 12-1
Dobler
Albany, NY
25

15 ❑ 12-1
Dobler
Albany, NY
1000+

16 ❑ 12-2
Maier
Los Angeles, CA
30

17 ❑ 12-2
Maier
Los Angeles, CA
25

18 ❑ 12-2
Maier
Los Angeles, CA
35

19 ❑ 12-2
Southern
Los Angeles, CA
40

20 ❑ 12-2
Drewry's Ltd.
Chicago, IL
10

21 ❑ 12-2
Great Lakes
Chicago, IL
90

22 ❑ 12-2
Schoenhofen Edelweiss
Chicago, IL
90

23 ❑ 12-2
Schoenhofen Edelweiss
Chicago, IL
15

24 ❑ 12-2
Schoenhofen Edelweiss
Chicago, IL
15

25 ❑ 12-2
Schultz
Chicago, IL
110

26 ❑ 12-2
Drewry's Ltd.
South Bend, IN
10

27 ❑ 12-2
Drewry's Ltd.
South Bend, IN
10

28 ❑ 12-2
Maier
Los Angeles, CA
70

29 ❑ 12-2
Maier
Los Angeles, CA
40

30 ❑ 12-2
Maier
Los Angeles, CA
40

Front of 32-37

32 ❑ 12-1
Drewry's Ltd.
Chicago, IL
Blue/White
40

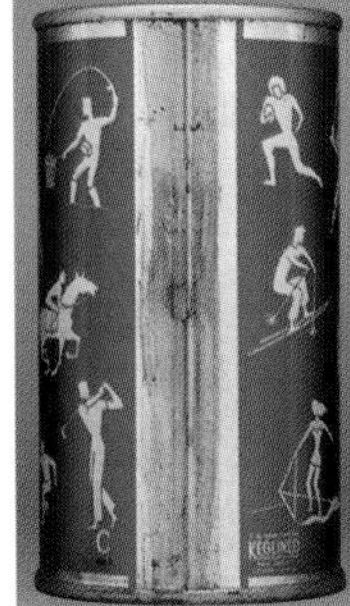
33 ❑ 12-1
Drewry's Ltd.
Chicago, IL
Bronze/White 40

34 ❑ 12-1
Drewry's Ltd.
Chicago, IL
Green/White 40

35 ❑ 12-1
Drewry's Ltd.
Chicago, IL
Purple/White 40

36 ❑ 12-1
Drewry's Ltd.
Chicago, IL
Red/White 40

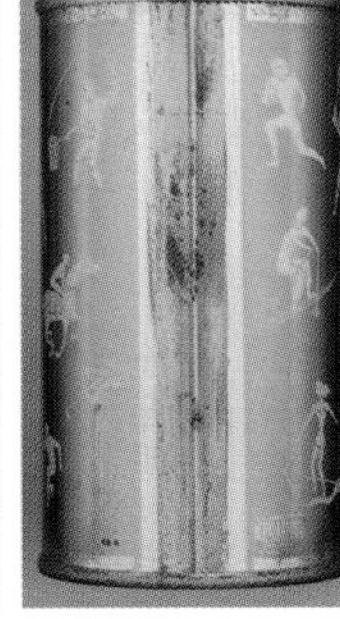
37 ❑ 12-1
Drewry's Ltd.
Chicago, IL
Yellow/White 40

Front of 39, 40
and 55-1 to 55-7

39 ❑ 12-1
Drewry's Ltd.
Chicago, IL
60

40 ❑ 12-1
Drewry's Ltd.
Chicago, IL
60

1 ❑ 12-1
Drewry's Ltd.
Chicago, IL
60

2 ❑ 12-1
Drewry's Ltd.
Chicago, IL
60

3 ❑ 12-1
Drewry's Ltd.
Chicago, IL
60

4 ❑ 12-1
Drewry's Ltd.
Chicago, IL
60

5 ❑ 12-1
Drewry's Ltd.
Chicago, IL
60

6 ❑ 12-1
Drewry's Ltd.
Chicago, IL
60

7 ❑ 12-1
Drewry's Ltd.
Chicago, IL
60

Front of 9-17

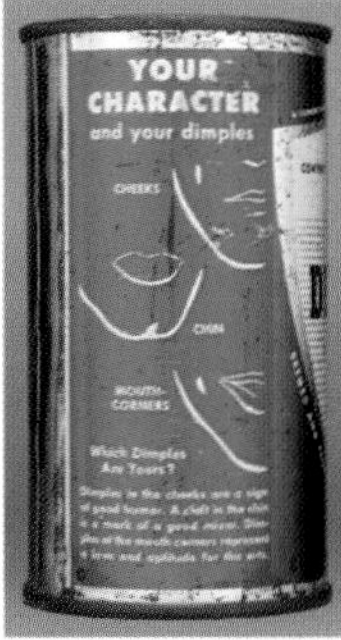

9 ❑ 12-1
Drewry's Ltd.
Chicago, IL
55

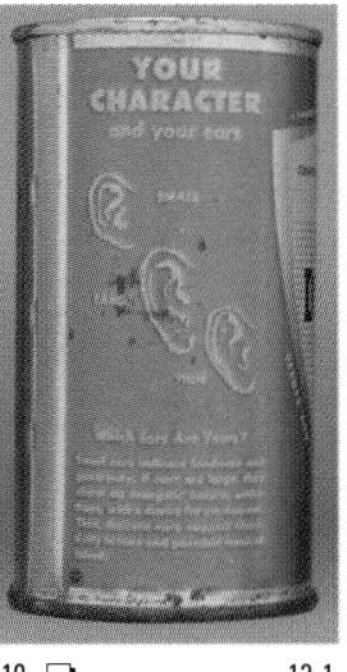

10 ❑ 12-1
Drewry's Ltd.
Chicago, IL
55

11 ❑ 12-1
Drewry's Ltd.
Chicago, IL
55

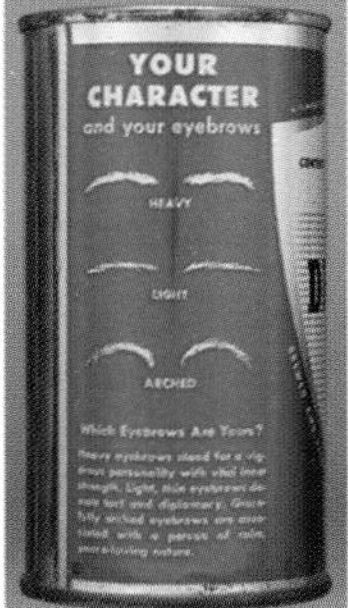

12 ❑ 12-1
Drewry's Ltd.
Chicago, IL
55

13 ❑ 12-1
Drewry's Ltd.
Chicago, IL
55

14 ❑ 12-1
Drewry's Ltd.
Chicago, IL
55

15 ❑ 12-1
Drewry's Ltd.
Chicago, IL
55

16 ❑ 12-1
Drewry's Ltd.
Chicago, IL
55

17 ❑ 12-1
Drewry's Ltd.
Chicago, IL
55

18 ❑ 12-1
Drewry's Ltd.
Chicago, IL
12

19 ❑ 12-1
Drewry's Ltd.
Chicago, IL
8

20 ❑ 12-2
Drewry's Ltd.
Chicago, IL
10

21 ❑ 12-1
Drewry's Ltd.
Chicago, IL
375

22 ❑ 12-1
Drewry's Ltd.
Chicago, IL
425

23 ❑ 12-1-OT
Drewry's Ltd.
South Bend, IN
275

24 ❑ 12-1-OT
Drewry's Ltd.
South Bend, IN
275

25 ❑ 12-1-OT
Drewry's Ltd.
South Bend, IN
150

26 ❑ 12-1-T
Drewry's Ltd.
South Bend, IN
40

27 ❑ 12-1-T
Drewry's Ltd.
South Bend, IN
40

28 ❑ 12-1
Drewry's Ltd.
South Bend, IN
35

29 ❑ 12-1
Drewry's Ltd.
South Bend, IN
25

30 ❑ 12-1
Drewry's Ltd.
South Bend, IN
25

31 ❑ 12-1-OT
Drewry's Ltd.
South Bend, IN
35

32 ❑ 12-1-OT
Drewry's Ltd.
South Bend, IN
35

33 ❑ 12-1-OT
Drewry's Ltd.
South Bend, IN
DULL GRAY
40

34 ❑ 12-1-OT
Drewry's Ltd.
South Bend, IN
30

35 ❑ 12-1-T
Drewry's Ltd.
South Bend, IN
20

36 ❑ 12-1-T
Drewry's Ltd.
South Bend, IN
20

37 ❑ 12-1-T
Drewry's Ltd.
South Bend, IN
15

38 ❑ 12-1-T
Drewry's Ltd.
South Bend, IN
15

39 ❑ 12-1-T
Drewry's Ltd.
South Bend, IN
DULL GRAY
400

Side of 39

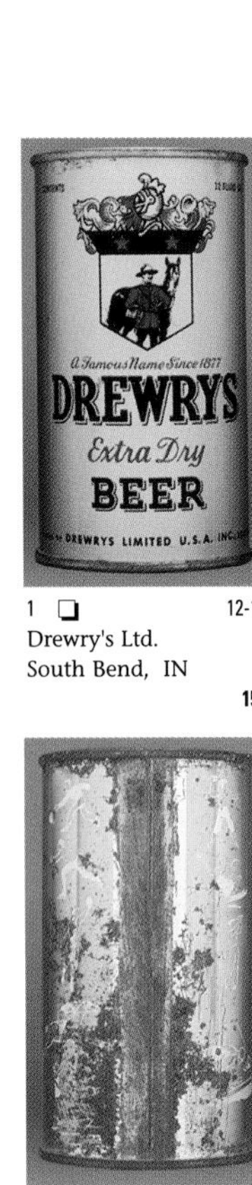

1 ❑ 12-1
Drewry's Ltd.
South Bend, IN
15

2 ❑ 12-1
Drewry's Ltd.
South Bend, IN
10

Front of 4-15

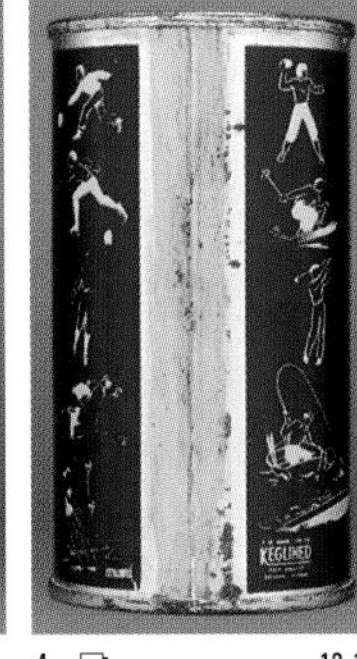
4 ❑ 12-1
Drewry's Ltd.
South Bend, IN
Blue/White 40

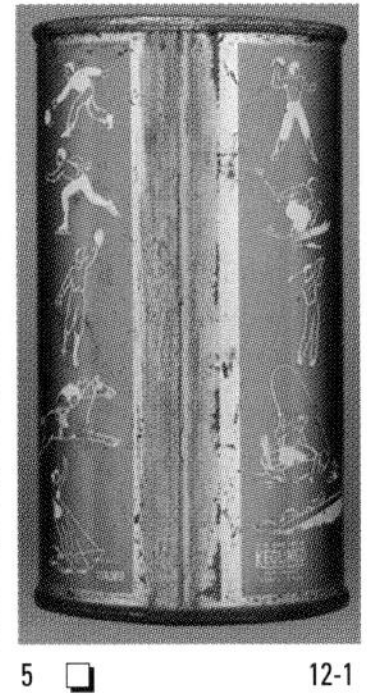
5 ❑ 12-1
Drewry's Ltd.
South Bend, IN
Bronze/White 40

6 ❑ 12-1
Drewry's Ltd.
South Bend, IN
Green/White 40

7 ❑ 12-1
Drewry's Ltd.
South Bend, IN
Purple/White 40

8 ❑ 12-1
Drewry's Ltd.
South Bend, IN
Red/White 40

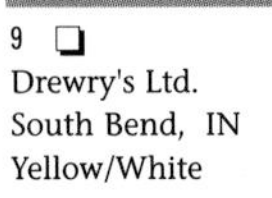
9 ❑ 12-1
Drewry's Ltd.
South Bend, IN
Yellow/White 40

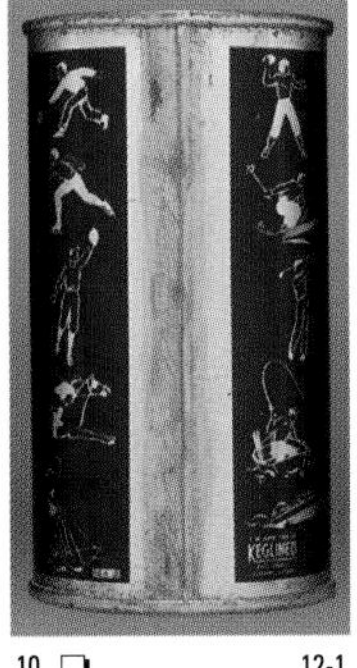
10 ❑ 12-1
Drewry's Ltd.
South Bend, IN
Blue/Silver 40

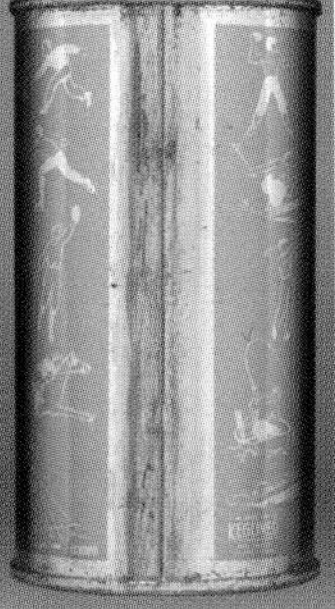
11 ❑ 12-1
Drewry's Ltd.
South Bend, IN
Bronze/Silver 40

12 ❑ 12-1
Drewry's Ltd.
South Bend, IN
Green/Silver 40

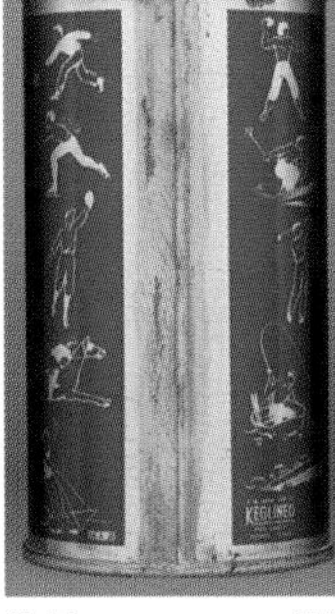
13 ❑ 12-1
Drewry's Ltd.
South Bend, IN
Purple/Silver 40

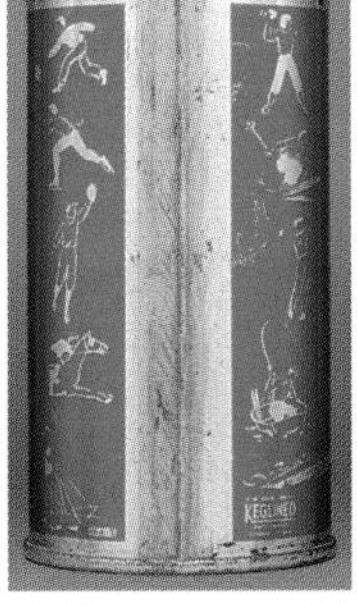
14 ❑ 12-1
Drewry's Ltd.
South Bend, IN
Red/Silver 40

15 ❑ 12-1
Drewry's Ltd.
South Bend, IN
Yellow/Silver 40

Front of 17-22

17 ❑ 12-1
Drewry's Ltd.
South Bend, IN
35

18 ❑ 12-1
Drewry's Ltd.
South Bend, IN
35

19 ❑ 12-1
Drewry's Ltd.
South Bend, IN
35

20 ❑ 12-1
Drewry's Ltd.
South Bend, IN
35

21 ❑ 12-1
Drewry's Ltd.
South Bend, IN
35

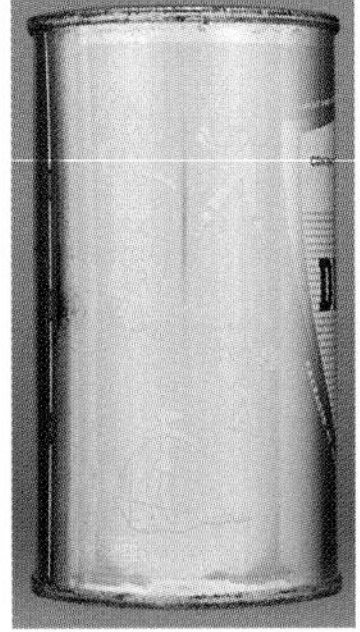
22 ❑ 12-1
Drewry's Ltd.
South Bend, IN
35

Front of 24-33

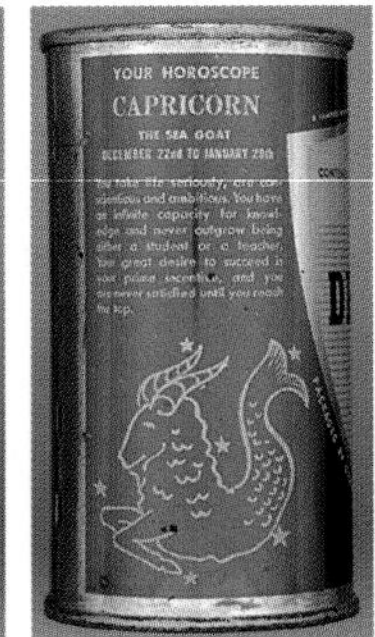

24 ❑ 12-1
Drewry's Ltd.
South Bend, IN
55

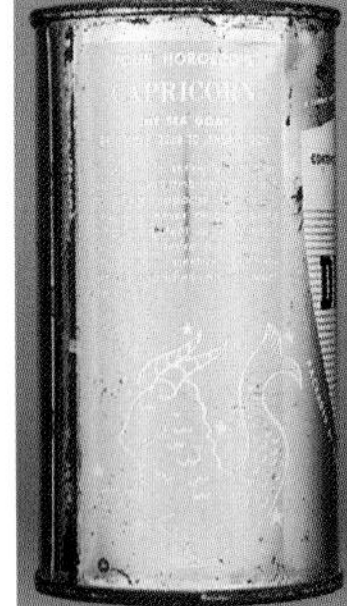
25 ❑ 12-1
Drewry's Ltd.
South Bend, IN
55

26 ❑ 12-1
Drewry's Ltd.
South Bend, IN
55

27 ❑ 12-1
Drewry's Ltd.
South Bend, IN
55

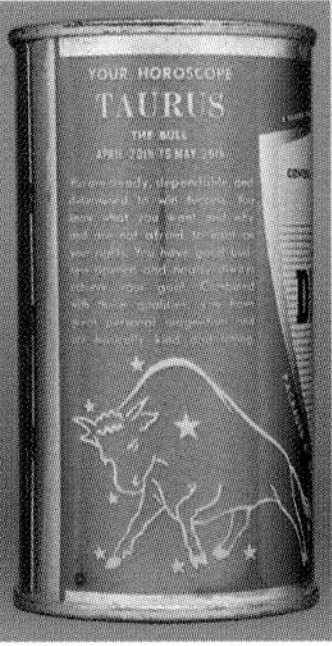

28 ❑ 12-1
Drewry's Ltd.
South Bend, IN
55

29 ❑ 12-1
Drewry's Ltd.
South Bend, IN
55

30 ❑ 12-1
Drewry's Ltd.
South Bend, IN
55

31 ❑ 12-1
Drewry's Ltd.
South Bend, IN
55

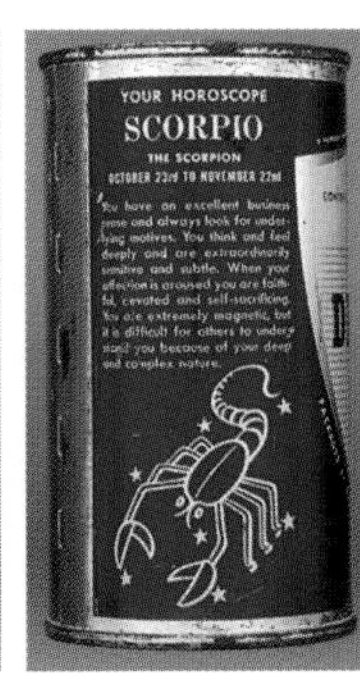

32 ❑ 12-1
Drewry's Ltd.
South Bend, IN
55

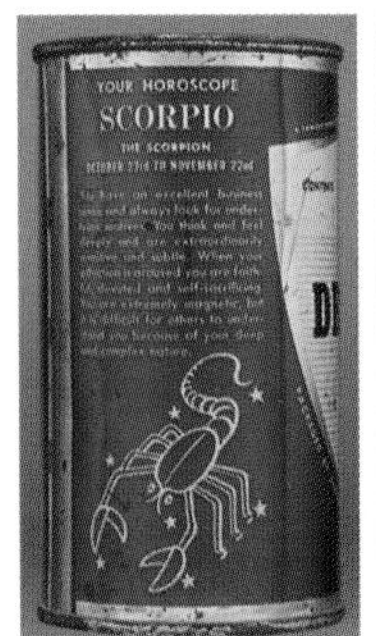

33 ❑ 12-1
Drewry's Ltd.
South Bend, IN
55

Front of 35 to 40
and 57-1 to 3

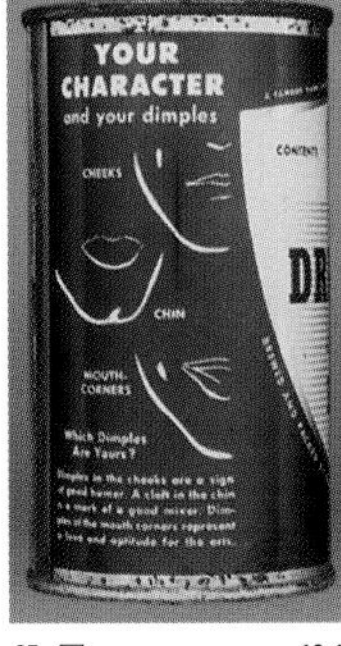

35 ❑ 12-1
Drewry's Ltd.
South Bend, IN
50

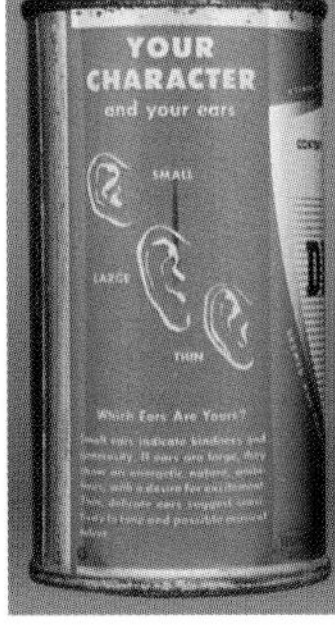

36 ❑ 12-1
Drewry's Ltd.
South Bend, IN
50

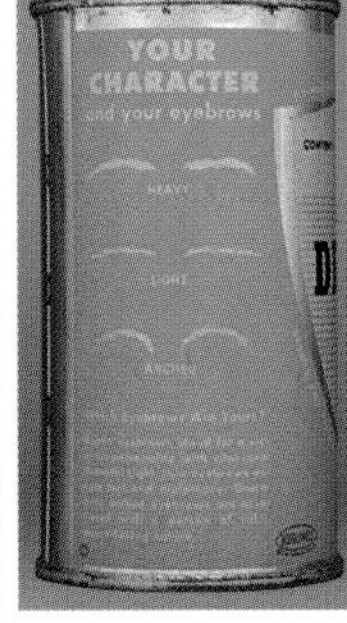

37 ❑ 12-1
Drewry's Ltd.
South Bend, IN
50

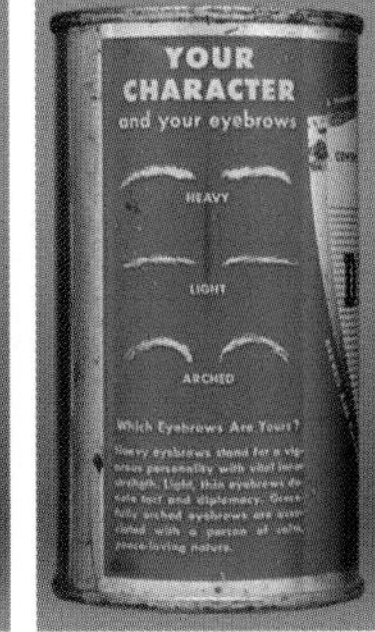

38 ❑ 12-1
Drewry's Ltd.
South Bend, IN
50

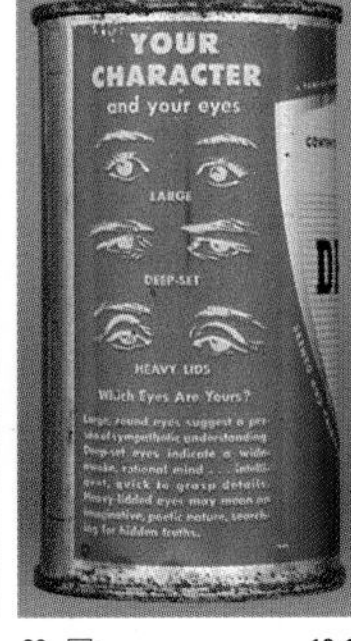

39 ❑ 12-1
Drewry's Ltd.
South Bend, IN
50

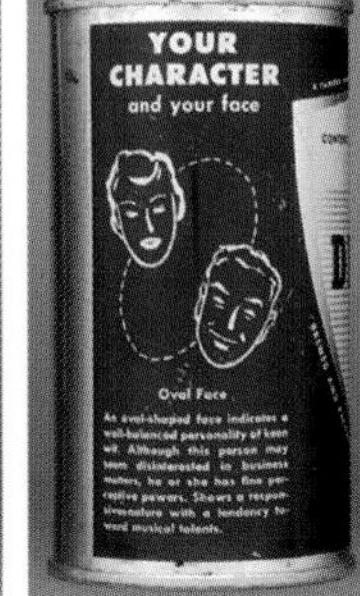

40 ❑ 12-1
Drewry's Ltd.
South Bend, IN
50

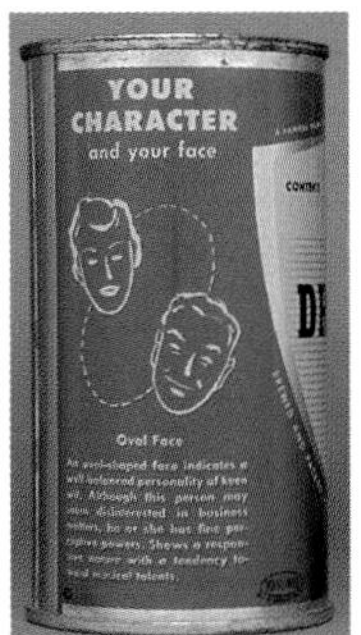

1 ❑ 12-1
Drewry's Ltd.
South Bend, IN
50

2 ❑ 12-1
Drewry's Ltd.
South Bend, IN
50

3 ❑ 12-1
Drewry's Ltd.
South Bend, IN
50

4 ❑ 12-1
Drewry's Ltd.
South Bend, IN
8

5 ❑ 12-1
Drewry's Ltd.
South Bend, IN
6

6 ❑ 12-2
Drewry's Ltd.
South Bend, IN
8

7 ❑ 12-1
Drewry's Ltd.
South Bend, IN
125

8 ❑ 12-1
Drewry's Ltd.
South Bend, IN
90

9 ❑ 11-2
Silver Springs
Tacoma, WA
275

10 ❑ 12-2
Duquesne
Pittsburgh, PA
40

11 ❑ 12-1
Duquesne
Pittsburgh, PA
40

12 ❑ 12-1
Duquesne
Pittsburgh, PA
30

13 ❑ 12-1
Duquesne
Pittsburgh, PA
35

14 ❑ 12-1
Duquesne
Pittsburgh, PA
8

15 ❑ 12-1
Atlantic
Chicago, IL
30

16 ❑ 12-1
Atlantic
Chicago, IL
30

17 ❑ 12-1
Best
Chicago, IL
35

18 ❑ 12-1
Atlantic
Spokane, WA
30

19 ❑ 11-1
Atlantic
Spokane, WA
35

20 ❑ 12-1
Atlantic
Spokane, WA
35

21 ❑ 12-1
Durst
Spokane, WA
40

22 ❑ 11-1
Silver Springs
Tacoma, WA
35

23 ❑ 12-1-OT
Grace Bros LTD.
Los Angeles, CA
90

Side of 23

25 ❑ 12-2-OT
Grace Bros.
Santa Rosa, CA
85

26 ❑ 12-2-OT
Grace Bros.
Santa Rosa, CA
90

27 ❑ 12-2-OT
Grace Bros.
Santa Rosa, CA
90

Front of 29, 30

29 ❑ 12-2-OT
Grace Bros.
Santa Rosa, CA
90

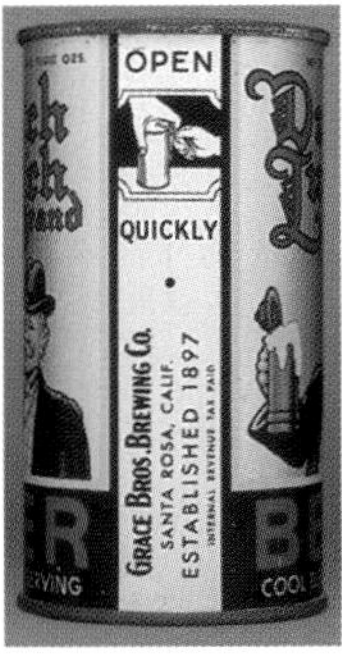

30 ❑ 12-2-OT
Grace Bros.
Santa Rosa, CA
90

31 ❑ 12-2-OT
Grace Bros.
Santa Rosa, CA
DULL GRAY
90

32 ❑ 12-2
Grace Bros.
Santa Rosa, CA
55

33 ❑ 12-2
Grace Bros.
Santa Rosa, CA
55

34 ❑ 12-2
Arizona
Phoenix, AZ
225

35 ❑ 12-2
Arizona
Phoenix, AZ
175

36 ❑ 12-1
Atlas
Chicago, IL
60

37 ❑ 12-1
Atlas
Chicago, IL
60

38 ❑ 12-2
Atlas
Chicago, IL
100

39 ❑ 12-2
Atlas
Chicago, IL
300

40 ❑ 12-2
Drewry's Ltd.
South Bend, IN
225

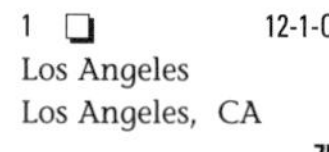

1 ❑ 12-1-OT
Los Angeles
Los Angeles, CA
750

2 ❑ 12-1-OT
Los Angeles
Los Angeles, CA
750

3 ❑ 12-1-OT
Los Angeles
Los Angeles, CA
110

4 ❑ 12-1-OT
Los Angeles
Los Angeles, CA
85

5 ❑ 12-1-OT
Los Angeles
Los Angeles, CA
90

6 ❑ 12-1-T
Los Angeles
Los Angeles, CA
40

7 ❑ 12-1
Los Angeles
Los Angeles, CA
40

8 ❑ 12-1
Los Angeles
Los Angeles, CA
40

9 ❑ 12-2
Los Angeles
Los Angeles, CA
30

10 ❑ 12-2
Los Angeles
Los Angeles, CA
30

11 ❑ 12-1
Los Angeles
Los Angeles, CA
850

12 ❑ 12-2
Los Angeles
Los Angeles, CA
150

13 ❑ 12-2
Pabst
Los Angeles, CA
25

14 ❑ 12-2
Pabst
Los Angeles, CA
20

15 ❑ 12-2
Pabst
Los Angeles, CA
20

16 ❑ 12-2
Pabst
Los Angeles, CA
20

17 ❑ 12-2
Pabst
Los Angeles, CA
8

18 ❑ 12-2
Pabst
Los Angeles, CA
10

19 ❑ 11-2
Pabst
Los Angeles, CA
15

20 ❑ 12-2
Pabst
Los Angeles, CA
12

21 ❑ 12-2
Pabst
Los Angeles, CA
6

22 ❑ 12-2
Pabst
Los Angeles, CA
100

23 ❑ 12-2
Pabst
Los Angeles, CA
110

24 ❑ 11-2
Pabst
Los Angeles, CA
110

25 ❑ 12-2
Pabst
Los Angeles, CA
110

26 ❑ 12-2
Pabst
Peoria Heights, IL
30

27 ❑ 12-1
Best
Chicago, IL
500

28 ❑ 12-1
E & B
Detroit, MI
85

29 ❑ 12-1
E & B
Detroit, MI
65

30 ❑ 12-1
E & B
Detroit, MI
65

31 ❑ 12-2
E & B
Detroit, MI
55

32 ❑ 12-2
Maier
Los Angeles, CA
15

33 ❑ 12-1-T
Edelbrau
New York, NY
1000+

34 ❑ 12-1-T
Edelbrau
New York, NY
1000+

35 ❑ 12-1-T
Edelbrau
New York, NY
1000+

36 ❑ 12-1-T
Edelbrew
New York, NY
400

37 ❑ 12-1-T
Edelbrew
New York, NY
200

38 ❑ 12-1-T
Atlas
Chicago, IL
45

39 ❑ 12-1-T
Schoenhofen Edelweiss
Chicago, IL
40

40 ❑ 12-1
Schoenhofen Edelweiss
Chicago, IL
40

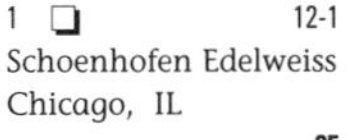

1 ❑ 12-1
Schoenhofen Edelweiss
Chicago, IL
25

2 ❑ 12-1
Schoenhofen Edelweiss
Chicago, IL
50

3 ❑ 12-1
Schoenhofen Edelweiss
Chicago, IL
35

4 ❑ 12-2
Schoenhofen Edelweiss
Chicago, IL
20

5 ❑ 12-2
Schoenhofen Edelweiss
Chicago, IL
15

6 ❑ 12-2
Schoenhofen Edelweiss
Chicago, IL
8

7 ❑ 12-2
Schoenhofen Edelweiss
Chicago, IL
6

8 ❑ 12-2
Schoenhofen Edelweiss
Chicago, IL
85

9 ❑ 12-1
Schoenhofen Edelweiss
Chicago, IL
425

10 ❑ 12-1
Schoenhofen Edelweiss
Chicago, IL
500

11 ❑ 12-2
Drewry's Ltd.
South Bend, IN
8

12 ❑ 12-2
Drewry's Ltd.
South Bend, IN
6

13 ❑ 12-1
Metropolis
Trenton, NJ
80

14 ❑ 12-1-OT
Globe
Baltimore, MD
1000+

15 ❑ 12-1-OT
Globe
Baltimore, MD
1000+

16 ❑ 12-1
Fox
Oconto, WI
40

17 ❑ 12-2
Atlantic
Chicago, IL
175

18 ❑ 12-1
Maier
Los Angeles, CA
125

19 ❑ 12-1
Pacific
Oakland, CA
175

20 ❑ 12-2
Grace Bros.
Santa Rosa, CA
125

21 ❑ 12-2
Maier
Los Angeles, CA
35

22 ❑ 12-2
Grace Bros.
Santa Rosa, CA
55

23 ❑ 12-2
Grace Bros.
Santa Rosa, CA
40

24 ❑ 12-2
Grace Bros.
Santa Rosa, CA
40

25 ❑ 12-2
Grace Bros.
Santa Rosa, CA
200

26 ❑ 12-1
Arizona
Phoenix, AZ
75

27 ❑ 12-2
Maier
Los Angeles, CA
25

28 ❑ 12-1
Grace Bros.
Santa Rosa, CA
55

29 ❑ 12-1
Spearman
Pensacola, FL
45

30 ❑ 12-1
Best
Chicago, IL
100

31 ❑ 12-1-T
Best
Chicago, IL
30

32 ❑ 12-1
Best
Chicago, IL
30

33 ❑ 12-1
Best
Chicago, IL
20

34 ❑ 12-1
Best
Chicago, IL
25

35 ❑ 12-1
Metropolis
Trenton, NJ
35

36 ❑ 12-1
Metropolis
Trenton, NJ
25

37 ❑ 12-1
Metropolis
Trenton, NJ
20

38 ❑ 12-1
Champale
Trenton, NJ
20

39 ❑ 12-1
Embassy Club
Norfolk, VA
40

40 ❑ 12-1
Embassy Club
Norfolk, VA
25

1 ❑ 12-2
Yankee
New York, NY
800

Front of 3, 4

3 ❑ 12-1-OT
Westminster
Chicago, IL
1000+

4 ❑ 12-1-OT
Westminster
Chicago, IL
1000+

5 ❑ 12-1-OT
Manhattan
Chicago, IL
1000+

6 ❑ 12-1-OT
Westminster
Chicago, IL
Text Like 3
1000+

7 ❑ 12-1-OT
Westminster
Chicago, IL
Text Like 4
1000+

8 ❑ 12-1-OT
Manhattan
Chicago, IL
1000+

9 ❑ 12-1
Standard
Cleveland, OH
65

10 ❑ 12-1
Standard
Cleveland, OH
40

11 ❑ 12-1
Standard
Cleveland, OH
40

12 ❑ 12-1
Standard
Cleveland, OH
40

13 ❑ 12-1
Standard
Cleveland, OH
40

14 ❑ 12-2
Essex
Chicago, IL
175

15 ❑ 12-1-OT
Esslinger's
Philadelphia, PA
400

16 ❑ 12-1-T
Esslinger's
Philadelphia, PA
300

17 ❑ 12-1-OT
Esslinger's
Philadelphia, PA
300

18 ❑ 12-1-W
Esslinger's
Philadelphia, PA
600

19 ❑ 12-1-W
Esslinger's
Philadelphia, PA
125

20 ❑ 12-1-T
Esslinger's
Philadelphia, PA
100

21 ❑ 12-1
Esslinger's
Philadelphia, PA
100

22 ❑ 12-1
Esslinger's
Philadelphia, PA
Brown Date
80

23 ❑ 12-1
Esslinger's
Philadelphia, PA
Gold Date
80

24 ❑ 12-1
Esslinger's
Philadelphia, PA
80

25 ❑ 12-1
Esslinger's
Philadelphia, PA
80

26 ❑ 12-1
Esslinger's
Philadelphia, PA
100

27 ❑ 12-1
Esslinger's
Philadelphia, PA
100

28 ❑ 12-1
Esslinger's
Philadelphia, PA
Set 1
75

29 ❑ 12-1
Esslinger's
Philadelphia, PA
Set 2
75

30 ❑ 12-1
Esslinger's
Philadelphia, PA
Set 3
75

31 ❑ 12-1
Esslinger's
Philadelphia, PA
Set 4
75

32 ❑ 12-1
Esslinger's
Philadelphia, PA
Set 5
75

33 ❑ 12-1
Esslinger's
Philadelphia, PA
Set 6
75

34 ❑ 12-1
Esslinger's
Philadelphia, PA
Set 7
75

35 ❑ 12-1
Esslinger's
Philadelphia, PA
Set 8
75

36 ❑ 12-1
Esslinger's
Philadelphia, PA
Set 9
75

37 ❑ 12-1-OT
Esslinger's
Philadelphia, PA
1000+

38 ❑ 12-1
Esslinger's
Philadelphia, PA
Capt. Webb
225

39 ❑ 12-1
Esslinger's
Philadelphia, PA
Human Blood
225

40 ❑ 12-1
Esslinger's
Philadelphia, PA
Jacob Daubert
225

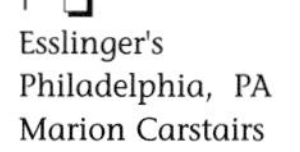
1 ❑ 12-1
Esslinger's
Philadelphia, PA
Marion Carstairs 225

2 ❑ 12-1
Esslinger's
Philadelphia, PA
British New Guin. 225

3 ❑ 12-1
Esslinger's
Philadelphia, PA
Columbia Univ. 225

4 ❑ 12-1
Esslinger's
Philadelphia, PA
8 Member's 225

5 ❑ 12-1
Esslinger's
Philadelphia, PA
1st King All 225

6 ❑ 12-1
Esslinger's
Philadelphia, PA
1st Popular 225

7 ❑ 12-1
Esslinger's
Philadelphia, PA
Jane Austen 225

8 ❑ 12-1
Esslinger's
Philadelphia, PA
Sound Moves 225

9 ❑ 12-1
Esslinger's
Philadelphia, PA
Ted Kluszewski 225

10 ❑ 12-1
Esslinger's
Philadelphia, PA
World's Largest 225

11 ❑ 12-2
Eulberg
Waukesha, WI
175

12 ❑ 12-2-OT
Humboldt
Eureka, CA
1000+

13 ❑ 12-2
Pacific
Oakland, CA
500

14 ❑ 12-2
Maier
Los Angeles, CA
60

15 ❑ 12-2
Pacific
Oakland, CA
175

16 ❑ 12-1-OT
Grace Bros.
Santa Rosa, CA
Enamel 1000+

17 ❑ 12-1-OT
Grace Bros.
Santa Rosa, CA
Metallic 1000+

18 ❑ 12-2
Atlantic
Chicago, IL
200

19 ❑ 12-2
Excell
Chicago, IL
175

20 ❑ 12-2
Atlantic
Spokane, WA
125

21 ❑ 12-1
Wm. Gretz
Philadelphia, PA
700

22 ❑ 12-2
Fischbach
St. Charles, MO
15

23 ❑ 12-2
Fischbach
St. Charles, MO
3

24 ❑ 12-2
Pacific
Oakland, CA
800

25 ❑ 12-1-OT
Falls City
Louisville, KY
Extra Pale 75

26 ❑ 12-1-OW
Falls City
Louisville, KY
35

27 ❑ 12-1-OT
Falls City
Louisville, KY
35

28 ❑ 12-1-O
Falls City
Louisville, KY
35

29 ❑ 12-1-O
Falls City
Louisville, KY
35

30 ❑ 12-1
Falls City
Louisville, KY
10

31 ❑ 12-2
Falls City
Louisville, KY
10

32 ❑ 12-1
Falstaff
San Jose, CA
5

33 ❑ 11-2
Falstaff
San Jose, CA
5

34 ❑ 11-2
Falstaff
San Jose, CA
5

35 ❑ 12-2
Falstaff
San Jose, CA
5

36 ❑ 11-2
Falstaff
San Jose, CA
7

37 ❑ 12-1
Falstaff
Fort Wayne, IN
5

38 ❑ 12-2
Falstaff
Fort Wayne, IN
4

39 ❑ 12-2
Falstaff
Fort Wayne, IN
4

40 ❑ 12-1
Falstaff
New Orleans, LA
40

1 ❑ 12-1
Falstaff
New Orleans, LA
8

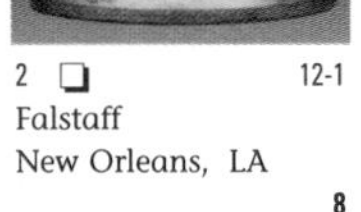

2 ❑ 12-1
Falstaff
New Orleans, LA
8

3 ❑ 10-2
Falstaff
New Orleans, LA
18

4 ❑ 12-2
Falstaff
New Orleans, LA
8

5 ❑ 12-2
Falstaff
New Orleans, LA
8

6 ❑ 12-1-T
Falstaff
St. Louis, MO
20

7 ❑ 12-1
Falstaff
St. Louis, MO
20

8 ❑ 12-1
Falstaff
St. Louis, MO
5

9 ❑ 12-2
Falstaff
St. Louis, MO
4

10 ❑ 12-2
Falstaff
St. Louis, MO
5

11 ❑ 12-1
Falstaff
Omaha, NE
20

12 ❑ 12-1
Falstaff
Omaha, NE
5

13 ❑ 11-2
Falstaff
Omaha, NE
5

14 ❑ 12-2
Falstaff
Omaha, NE
4

15 ❑ 12-2
Falstaff
Omaha, NE
4

16 ❑ 12-2
Falstaff
Cranston, RI
8

17 ❑ 12-1
Falstaff
El Paso, TX
10

18 ❑ 11-2
Falstaff
El Paso, TX
12

19 ❑ 12-2
Falstaff
El Paso, TX
8

20 ❑ 12-1
Falstaff
Galveston, TX
10

21 ❑ 12-2
Falstaff
Galveston, TX
7

22 ❑ 12-2
Falstaff
Galveston, TX
6

23 ❑ 12-2
Falstaff
Galveston, TX
20

24 ❑ 12-2
Fauerbach
Madison, WI
40

25 ❑ 12-1
Fauerbach
Madison, WI
30

Front of 27, 28

27 ❑ 12-1
Anheuser-Busch
St. Louis, MO
1000+

28 ❑ 12-1
Anheuser-Busch
St. Louis, MO
1000+

29 ❑ 12-2
Frank Fehr
Louisville, KY
45

30 ❑ 12-2
Frank Fehr
Louisville, KY
45

31 ❑ 12-2
Frank Fehr
Louisville, KY
80

32 ❑ 12-2
Frank Fehr
Louisville, KY
Blue
15

33 ❑ 12-2
Frank Fehr
Louisville, KY
Black
15

34 ❑ 12-2
Frank Fehr
Louisville, KY
12

35 ❑ 12-2
Frank Fehr
Louisville, KY
18

36 ❑ 12-1-OT
Chr. Feigenspan
Newark, NJ
Long Opener
125

37 ❑ 12-1-OT
Chr. Feigenspan
Newark, NJ
Long Opener
100

38 ❑ 12-1-T
Chr. Feigenspan
Newark, NJ
40

39 ❑ 12-1-T
Chr. Feigenspan
Newark, NJ
75

40 ❑ 12-1-OT
Chr. Feigenspan
Newark, NJ
150

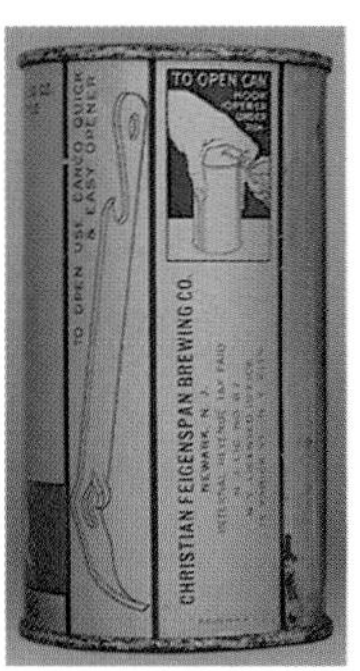

Side of 62-40

2 ❑ 12-1-0T
Chr. Feigenspan
Newark, NJ
150

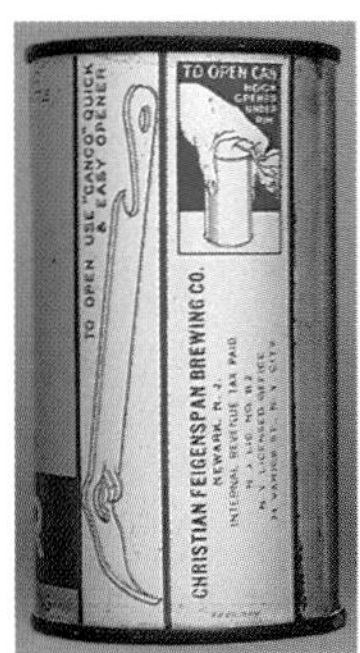

Side of 2

4 ❑ 12-1-T
Chr. Feigenspan
Newark, NJ
60

5 ❑ 12-1-T
Chr. Feigenspan
Newark, NJ
75

6 ❑ 12-1-0T
Chr. Feigenspan
Newark, NJ
1000+

7 ❑ 12-1-T
Chr. Feigenspan
Newark, NJ
750

8 ❑ 12-2
Fesenmeier
Huntington, WV
25

9 ❑ 12-2
Atlas
Chicago, IL
75

10 ❑ 12-2
Drewry's Ltd.
South Bend, IN
60

11 ❑ 12-2
Golden Brew
Lawrence, MA
20

12 ❑ 12-2
Eastern Corp.
Hammonton, NJ
4

13 ❑ 12-2
Eastern Corp.
Hammonton, NJ
50

14 ❑ 12-1-0T
Monarch
Los Angeles, CA
1000+

15 ❑ 11-1-0T
Monarch
Los Angeles, CA
700

16 ❑ 12-1-0T
Monarch
Los Angeles, CA
500

17 ❑ 12-1-T
Monarch
Los Angeles, CA
350

18 ❑ 12-1
Southern
Los Angeles, CA
150

19 ❑ 12-1-0T
Columbia
Tacoma, WA
1000+

20 ❑ 12-1
Atlantic
Chicago, IL
30

21 ❑ 12-1
Atlantic
Chicago, IL
25

22 ❑ 12-1
Queen City
Cumberland, MD
25

23 ❑ 12-2
Queen City
Cumberland, MD
85

24 ❑ 12-2
Queen City
Cumberland, MD
85

25 ❑ 12-2
Fischer
Cumberland, MD
85

26 ❑ 12-1
Cumberland
Cumberland, MD
22

27 ❑ 12-1
Queen City
Cumberland, MD
22

28 ❑ 12-2
Queen City
Cumberland, MD
125

29 ❑ 12-2
Queen City
Cumberland, MD
85

30 ❑ 12-2
Fischer
Cumberland, MD
85

31 ❑ 12-2
Lucky Lager
Azusa, CA
5

32 ❑ 12-2
General Corp.
Azusa, CA
5

33 ❑ 12-2
Lucky Lager
San Francisco, CA
5

34 ❑ 12-2
Lucky
San Francisco, CA
5

35 ❑ 12-2
General Corp.
San Francisco, CA
Metallic
5

36 ❑ 12-2
General Corp.
San Francisco, CA
Enamel
5

37 ❑ 12-2
Fisher
Salt Lake City, UT
135

38 ❑ 12-2
Fisher
Salt Lake City, UT
CNMT 3.2%
30

39 ❑ 11-2
Fisher
Salt Lake City, UT
5

40 ❑ 12-2
Fisher
Salt Lake City, UT
5

1 ❑ 11-2
Fisher
Salt Lake City, UT
7

2 ❑ 12-2
Fisher
Salt Lake City, UT
5

3 ❑ 12-2
Fisher
Salt Lake City, UT
5

4 ❑ 12-2
Lucky Lager
Salt Lake City, UT
5

5 ❑ 12-2
Lucky Lager
Salt Lake City, UT
5

6 ❑ 12-2
General Corp.
Salt Lake City, UT
5

7 ❑ 12-2
Fitger
Duluth, MN
15

8 ❑ 12-2
Fitger
Duluth, MN
15

9 ❑ 12-2
Fitger
Duluth, MN
8

10 ❑ 12-2
Fitger
Duluth, MN
8

11 ❑ 12-2
Fitzgerald Bros.
Willimansett, MA
20

12 ❑ 12-2
Fitzgerald Bros.
Willimansett, MA
18

13 ❑ 12-2
Fitzgerald Bros.
Willimansett, MA
18

14 ❑ 12-2
Fitzgerald Bros.
Willimansett, MA
18

15 ❑ 12-1
Fitzgerald Bros.
Troy, NY
Blue Shield 30

16 ❑ 12-1
Fitzgerald Bros.
Troy, NY
Green Shield 30

17 ❑ 12-2
Fitzgerald Bros.
Troy, NY
30

18 ❑ 12-1
Fitzgerald Bros.
Troy, NY
20

19 ❑ 12-2
Fitzgerald Bros.
Troy, NY
25

20 ❑ 12-1
Fitzgerald Bros.
Troy, NY
650

21 ❑ 10-2
Five Star
New York, NY
200

22 ❑ 12-2
Fort Pitt
Baltimore, MD
50

Front of 24, 25

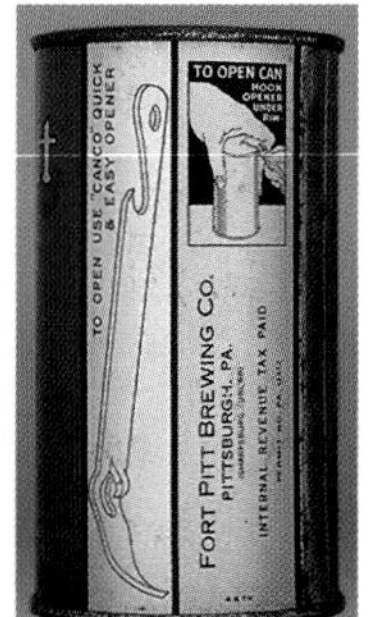

24 ❑ 12-1-OT
Fort Pitt
Pittsburgh, PA
1000+

25 ❑ 12-1-OT
Fort Pitt
Pittsburgh, PA
1000+

Front of 27, 28

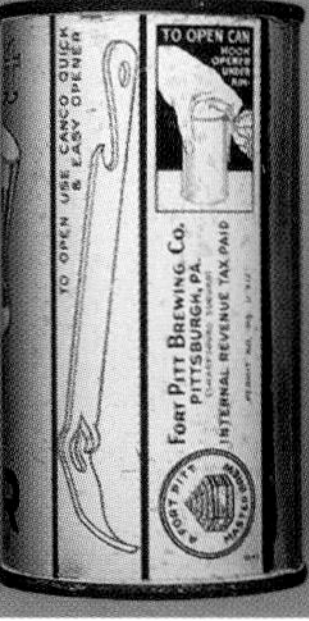

27 ❑ 12-1-OT
Fort Pitt
Pittsburgh, PA
500

28 ❑ 12-1-OT
Fort Pitt
Pittsburgh, PA
400

29 ❑ 12-1
Fort Schuyler
Utica, NY
40

30 ❑ 12-1
Fort Schuy. Div. of W. End
Utica, NY
40

31 ❑ 12-2-T
Globe
San Francisco, CA
800

32 ❑ 12-2-OT
Grace Bros.
Santa Rosa, CA
750

33 ❑ 12-2
Pacific
Oakland, CA
225

34 ❑ 12-2
Atlas
Chicago, IL
375

35 ❑ 12-1
Oconto
Oconto, WI
12

36 ❑ 12-1
Fox Head
Waukesha, WI
8

37 ❑ 12-1-OT
Peter Fox
Chicago, IL
1000+

38 ❑ 12-1-OT
Peter Fox
Chicago, IL
650

39 ❑ 12-1-OT
Peter Fox
Chicago, IL
650

40 ❑ 12-1-W
Peter Fox
Chicago, IL
750

1 ❑ 12-1-0T
Peter Fox
Chicago, IL
Metallic 90

2 ❑ 12-1-0T
Peter Fox
Chicago, IL
Enamel 90

3 ❑ 12-1-0T
Peter Fox
Chicago, IL
Metallic 90

4 ❑ 12-1-0T
Peter Fox
Chicago, IL
Enamel 90

5 ❑ 12-1-T
Peter Fox
Chicago, IL
30

6 ❑ 12-1
Peter Fox
Chicago, IL
25

7 ❑ 12-1
Peter Fox
Chicago, IL
25

8 ❑ 12-1
Peter Fox
Chicago, IL
15

9 ❑ 12-1
Peter Fox
Chicago, IL
200

10 ❑ 12-2
Peter Fox
Chicago, IL
275

11 ❑ 12-1-0T
Fox Deluxe
Grand Rapids, MI
Metallic 110

12 ❑ 12-1-0T
Fox Deluxe
Grand Rapids, MI
Enamel 100

13 ❑ 12-1-W
Fox Deluxe
Grand Rapids, MI
550

14 ❑ 12-1-T
Fox Deluxe
Grand Rapids, MI
45

15 ❑ 12-1
Fox Deluxe
Grand Rapids, MI
50

16 ❑ 12-2
Fox Head
La Crosse, WI
15

17 ❑ 12-2
Fox Head
La Crosse, WI
4

18 ❑ 12-2
G. Heileman
La Crosse, WI
12

19 ❑ 12-2
Fox Head
Waukesha, WI
30

20 ❑ 12-2
Fox Head
Waukesha, WI
20

21 ❑ 12-2
Fox Head
Waukesha, WI
20

22 ❑ 12-2
Fox Head
Waukesha, WI
Black Writing 25

23 ❑ 12-2
Fox Head
Waukesha, WI
Brown Writing 25

24 ❑ 12-2
Peter Fox
Waukesha, WI
20

25 ❑ 12-2
Peter Fox
Waukesha, WI
Red Writing 35

26 ❑ 12-2
Peter Fox
Waukesha, WI
Black Writing 25

27 ❑ 12-2
Peter Fox
Waukesha, WI
Brown Writing 25

28 ❑ 12-2
Peter Fox
Waukesha, WI
225

29 ❑ 12-2
Geo. Wiedemann
Newport, KY
3

30 ❑ 12-2
Fox Head
La Crosse, WI
18

31 ❑ 12-2
Fox Head
La Crosse, WI
18

32 ❑ 12-2
Fox Head
La Crosse, WI
5

33 ❑ 12-2
Fox Head
La Crosse, WI
7

34 ❑ 12-2
Fox Head
La Crosse, WI
12

35 ❑ 12-2
G. Heileman
La Crosse, WI
8

36 ❑ 12-2
G. Heileman
La Crosse, WI
50

37 ❑ 12-2
Fox Head
Sheboygan, WI
25

38 ❑ 12-2
Fox Head
Sheboygan, WI
15

39 ❑ 12-2
Fox Head Div. of Heile.
Sheboygan, WI
8

40 ❑ 12-2
Fox Head Div. of Heile.
Sheboygan, WI
25

1 ❑ 12-2
Fox Head Div. of Heile.
Sheboygan, WI
5

2 ❑ 12-2
Fox Head Div. of Heile.
Sheboygan, WI
12

3 ❑ 12-2
Kingsbury
Sheboygan, WI
45

Front of 5, 6

5 ❑ 12-1
Fox Head
Waukesha, WI
100

6 ❑ 12-1
Fox Head
Waukesha, WI
80

7 ❑ 12-1-T
Fox Head
Waukesha, WI
18

8 ❑ 12-1
Fox Head
Waukesha, WI
18

9 ❑ 12-1
Fox Head
Waukesha, WI
18

10 ❑ 12-1
Fox Head
Waukesha, WI
18

11 ❑ 12-2
Fox Head
Waukesha, WI
25

12 ❑ 12-2
Fox Head
Waukesha, WI
18

13 ❑ 12-2
Fox Head
Waukesha, WI
40

14 ❑ 12-2
Fox Head
Waukesha, WI
18

15 ❑ 12-2
Fox Head
Waukesha, WI
40

16 ❑ 12-2
Fox Head
Waukesha, WI
40

17 ❑ 12-2
Fox Head
Waukesha, WI
40

18 ❑ 12-2
Fox Head Div. of Heile.
Waukesha, WI
65

19 ❑ 12-2
Fox Head Div. of Heile.
Waukesha, WI
80

20 ❑ 12-2
Frankenmuth
Tampa, FL
90

21 ❑ 12-2
Frankenmuth
Tampa, FL
70

22 ❑ 12-2
Frankenmuth
Frankenmuth, MI
75

23 ❑ 12-2
Frankenmuth
Frankenmuth, MI
75

24 ❑ 12-2
Frankenmuth
Frankenmuth, MI
175

25 ❑ 12-1
Frankenmuth
Frankenmuth, MI
65

26 ❑ 12-1
International
Frankenmuth, MI
65

27 ❑ 12-2
Frankenmuth
Frankenmuth, MI
55

28 ❑ 12-2
Frankenmuth
Frankenmuth, MI
55

29 ❑ 12-2
Frankenmuth
Frankenmuth, MI
50

30 ❑ 12-2
Frankenmuth
Frankenmuth, MI
50

31 ❑ 12-1
Frankenmuth
Frankenmuth, MI
60

32 ❑ 12-1
International
Frankenmuth, MI
60

33 ❑ 12-2
Frankenmuth
Frankenmuth, MI
300

34 ❑ 12-1
International
Frankenmuth, MI
225

35 ❑ 12-2
Frankenmuth
Buffalo, NY
75

36 ❑ 12-2
International
Buffalo, NY
75

37 ❑ 12-1
International
Buffalo, NY
65

38 ❑ 12-1
International
Buffalo, NY
65

39 ❑ 12-1
International
Buffalo, NY
65

40 ❑ 12-2
Frankenmuth
Buffalo, NY
40

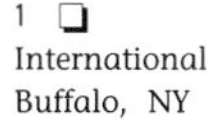
1 ❑ 12-2
International
Buffalo, NY
45

2 ❑ 12-2
International
Buffalo, NY
60

3 ❑ 12-2
International
Buffalo, NY
175

4 ❑ 12-1
International
Findlay, OH
60

5 ❑ 12-1
International
Findlay, OH
60

6 ❑ 12-1-OT
South Side
Chicago, IL
Tall Can 1000+

7 ❑ 12-2
Drewry's Ltd.
South Bend, IN
15

8 ❑ 12-2
Drewry's Ltd.
South Bend, IN
12

9 ❑ 12-1-OT
General Corp.
San Francisco, CA
500

10 ❑ 12-1-T
General Corp.
San Francisco, CA
150

11 ❑ 12-1-T
General Corp.
San Francisco, CA
200

12 ❑ 12-2
Fuhrmann & Schmidt
Shamokin, PA
50

13 ❑ 12-2
Fuhrmann & Schmidt
Shamokin, PA
40

14 ❑ 12-2
Fuhrmann & Schmidt
Shamokin, PA
Enamel 18

15 ❑ 12-2
Fuhrmann & Schmidt
Shamokin, PA
Metallic 15

16 ❑ 12-2
Fuhrmann & Schmidt
Shamokin, PA
12

17 ❑ 12-2
Fuhrmann & Schmidt
Shamokin, PA
1000+

18 ❑ 12-1
August Wagner
Columbus, OH
125

19 ❑ 12-1
August Wagner
Columbus, OH
25

20 ❑ 12-1
Garten Brau
Potosi, WI
5

21 ❑ 12-2
Grace Bros LTD.
Los Angeles, CA
1000+

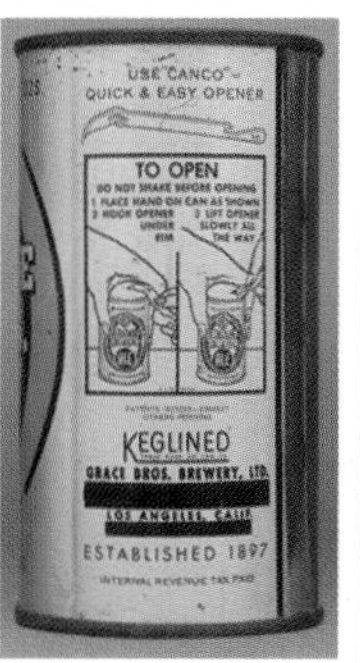

Side of 21

23 ❑ 12-1-OT
Grace Bros LTD.
Los Angeles, CA
300

24 ❑ 12-1-T
Grace Bros LTD.
Los Angeles, CA
450

25 ❑ 12-2
Maier
Los Angeles, CA
20

26 ❑ 12-2
Maier
Los Angeles, CA
60

27 ❑ 12-1-OT
Grace Bros.
Santa Rosa, CA
1000+

28 ❑ 12-1-OT
Grace Bros.
Santa Rosa, CA
1000+

29 ❑ 12-1-OT
Grace Bros.
Santa Rosa, CA
300

Front of 31, 32

31 ❑ 12-1-OT
Grace Bros.
Santa Rosa, CA
300

32 ❑ 12-1-OT
Grace Bros.
Santa Rosa, CA
300

33 ❑ 12-1-OT
Grace Bros.
Santa Rosa, CA
300

34 ❑ 12-1-T
Grace Bros.
Santa Rosa, CA
450

35 ❑ 12-1-T
Grace Bros.
Santa Rosa, CA
65

36 ❑ 12-1
Grace Bros.
Santa Rosa, CA
65

37 ❑ 12-2
Grace Bros.
Santa Rosa, CA
75

38 ❑ 12-2
Grace Bros.
Santa Rosa, CA
60

39 ❑ 12-2
Grace Bros.
Santa Rosa, CA
60

40 ❑ 12-2
Grace Bros.
Santa Rosa, CA
25

1 ❑ 12-2
Grace Bros.
Santa Rosa, CA
20

2 ❑ 12-2
Grace Bros.
Santa Rosa, CA
20

3 ❑ 12-2
Grace Bros.
Santa Rosa, CA
20

4 ❑ 12-2
Grace Bros.
Santa Rosa, CA
20

5 ❑ 12-2
Grace Bros.
Santa Rosa, CA
20

6 ❑ 12-2
Grace Bros.
Santa Rosa, CA
65

7 ❑ 12-2
Grace Bros.
Santa Rosa, CA
70

8 ❑ 12-2
Grace Bros.
Santa Rosa, CA
Orange Goat
65

9 ❑ 12-2
Grace Bros.
Santa Rosa, CA
Yellow Goat
65

10 ❑ 12-2
Grace Bros.
Santa Rosa, CA
65

11 ❑ 12-2
Grace Bros.
Santa Rosa, CA
65

12 ❑ 12-1
North Bay
Santa Rosa, CA
75

13 ❑ 12-2
Colonial
Hammonton, NJ
65

14 ❑ 12-2
Fuhrmann & Schmidt
Shamokin, PA
18

15 ❑ 12-1-OT
Genesee
Rochester, NY
1000+

16 ❑ 12-1-OT
Genesee
Rochester, NY
1000+

17 ❑ 12-1-OT
Genesee
Rochester, NY
75

18 ❑ 12-1-O
Genesee
Rochester, NY
Yellow
75

19 ❑ 12-1-O
Genesee
Rochester, NY
Orange
75

20 ❑ 12-2
Genesee
Rochester, NY
50

21 ❑ 12-2
Genesee
Rochester, NY
125

22 ❑ 12-2
Genesee
Rochester, NY
25

23 ❑ 12-1
Genesee
Rochester, NY
65

24 ❑ 12-2
Genesee
Rochester, NY
15

25 ❑ 12-1-OT
Genesee
Rochester, NY
1000+

26 ❑ 12-1-OT
Genesee
Rochester, NY
900

27 ❑ 12-1-OT
Genesee
Rochester, NY
900

28 ❑ 12-1-OT
Genesee
Rochester, NY
125

29 ❑ 12-1-W
Genesee
Rochester, NY
500

30 ❑ 12-1-T
Genesee
Rochester, NY
110

31 ❑ 12-1-T
Genesee
Rochester, NY
30

32 ❑ 12-2
Genesee
Rochester, NY
25

33 ❑ 12-2
Genesee
Rochester, NY
45

34 ❑ 12-2
Genesee
Rochester, NY
30

35 ❑ 12-2
Genesee
Rochester, NY
12

36 ❑ 12-2
Genesee
Rochester, NY
22

37 ❑ 12-2
Genesee
Rochester, NY
75

38 ❑ 12-2
Genesee
Rochester, NY
15

39 ❑ 12-2
Genesee
Rochester, NY
5

40 ❑ 12-2
Genesee
Rochester, NY
5

1 ❑ 12-1
Wm. Gretz
Philadelphia, PA
1000+

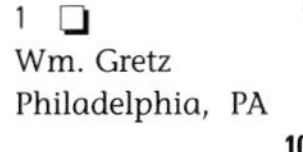

2 ❑ 12-2
Atlas
Chicago, IL
185

3 ❑ 12-2
Drewry's Ltd.
South Bend, IN
160

4 ❑ 12-1
A. Gettelman
Milwaukee, WI
20

5 ❑ 12-2
A. Gettelman
Milwaukee, WI
8

6 ❑ 12-2
A. Gettelman
Milwaukee, WI
8

7 ❑ 12-2
A. Gettelman
Milwaukee, WI
BBQ and TV 75

8 ❑ 12-2
A. Gettelman
Milwaukee, WI
BBQ and TV 75

9 ❑ 12-2
A. Gettelman
Milwaukee, WI
BBQ and TV 75

10 ❑ 12-2
A. Gettelman
Milwaukee, WI
BBQ and TV 75

11 ❑ 12-2
A. Gettelman
Milwaukee, WI
Fishing 75

12 ❑ 12-2
A. Gettelman
Milwaukee, WI
Fishing 75

13 ❑ 12-2
A. Gettelman
Milwaukee, WI
Fishing 75

14 ❑ 12-2
A. Gettelman
Milwaukee, WI
Fishing 75

15 ❑ 12-2
A. Gettelman
Milwaukee, WI
Picnic 75

16 ❑ 12-2
A. Gettelman
Milwaukee, WI
Picnic 75

17 ❑ 12-2
A. Gettelman
Milwaukee, WI
Picnic 75

18 ❑ 12-2
A. Gettelman
Milwaukee, WI
Picnic 75

19 ❑ 12-2
A. Gettelman
Milwaukee, WI
Roll Out the Barrel 75

20 ❑ 12-2
A. Gettelman
Milwaukee, WI
Roll Out the Barrel 75

21 ❑ 12-2
A. Gettelman
Milwaukee, WI
Roll Out the Barrel 75

22 ❑ 12-2
A. Gettelman
Milwaukee, WI
Roll Out the Barrel 75

23 ❑ 12-1
A. Gettelman
Milwaukee, WI
15

24 ❑ 12-2
Gettelman, Div. of Miller
Milwaukee, WI
8

25 ❑ 12-2
Gettelman, Div. of Miller
Milwaukee, WI
8

26 ❑ 12-2
Hofbrau
Allentown, PA
75

27 ❑ 12-2
Fuhrmann & Schmidt
Shamokin, PA
15

28 ❑ 12-2
Lion
Wilkes-Barre, PA
Met. Red 18

29 ❑ 12-2
Lion
Wilkes-Barre, PA
Enam. Red 18

30 ❑ 12-1-OT
Buffalo
Sacramento, CA
1000+

31 ❑ 12-1-T
Buffalo
Sacramento, CA
1000+

32 ❑ 12-2
Gilt Edge
Trenton, NJ
40

33 ❑ 12-2
Gilt Edge
Trenton, NJ
35

34 ❑ 12-2
Hornell
Hornell, NY
50

35 ❑ 12-2
Hornell
Hornell, NY
45

36 ❑ 12-2
Century
Norfolk, VA
45

37 ❑ 12-2
Century
Norfolk, VA
40

38 ❑ 12-2
Gipps
Chicago, IL
50

39 ❑ 12-2
Gipps
Chicago, IL
50

40 ❑ 12-2
Gipps
Chicago, IL
20

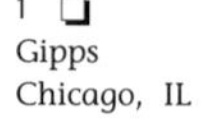
1 ❏ 12-2
Gipps
Chicago, IL
20

2 ❏ 12-2
Gipps
Peoria, IL
50

3 ❏ 12-2
Maier
Los Angeles, CA
30

4 ❏ 12-2
Pacific
Oakland, CA
1000+

5 ❏ 12-2
Grace Bros.
Santa Rosa, CA
35

6 ❏ 12-1-T
General Corp.
San Francisco, CA
900

7 ❏ 12-2
Gluek
Minneapolis, MN
50

8 ❏ 12-1
Gluek
Minneapolis, MN
15

9 ❏ 12-2
Gluek
Minneapolis, MN
12

10 ❏ 12-2
Gluek
Minneapolis, MN
65

11 ❏ 12-2
Gluek
Minneapolis, MN
25

12 ❏ 12-1
Gluek
Minneapolis, MN
20

13 ❏ 12-2
Gluek
Minneapolis, MN
15

14 ❏ 12-2
Gluek
Minneapolis, MN
15

15 ❏ 12-2
Gluek, Div. Heile.
La Crosse, WI
5

16 ❏ 12-2
Gluek, Div. Heile.
La Crosse, WI
4

17 ❏ 12-2
Gluek, Div. Heile.
La Crosse, WI
4

18 ❏ 12-1
Goebel
Oakland, CA
25

19 ❏ 12-1
Goebel
Oakland, CA
25

20 ❏ 12-1
Goebel
Oakland, CA
22

21 ❏ 12-1
Goebel
Oakland, CA
22

22 ❏ 12-1
Goebel
Oakland, CA
22

Front of 24, 25

24 ❏ 12-2
Goebel
Oakland, CA
20

25 ❏ 12-2
Goebel
Oakland, CA
20

26 ❏ 12-1
Goebel
Oakland, CA
Enamel
18

27 ❏ 12-1
Goebel
Oakland, CA
Metallic
18

28 ❏ 12-1
Goebel
Oakland, CA
300

29 ❏ 12-2
Goebel
Oakland, CA
400

30 ❏ 12-1-OT
Goebel
Detroit, MI
250

31 ❏ 12-1-OT
Goebel
Detroit, MI
250

32 ❏ 12-1-OT
Goebel
Detroit, MI
35

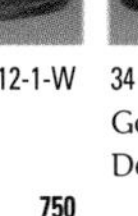
33 ❏ 12-1-W
Goebel
Detroit, MI
750

34 ❏ 12-1-W
Goebel
Detroit, MI
50

35 ❏ 12-1-T
Goebel
Detroit, MI
25

36 ❏ 12-1
Goebel
Detroit, MI
20

37 ❏ 12-1
Goebel
Detroit, MI
Nationally Famous
25

38 ❏ 12-1
Goebel
Detroit, MI
Naturally Famous
35

39 ❏ 12-1
Goebel
Detroit, MI
25

40 ❏ 12-1
Goebel
Detroit, MI
25

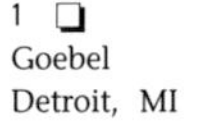

1 ❑ 11-1
Goebel
Detroit, MI
125

2 ❑ 12-1
Goebel
Detroit, MI
Red Outline 20

3 ❑ 12-1
Goebel
Detroit, MI
Gold Outline 20

4 ❑ 12-1
Goebel
Detroit, MI
20

5 ❑ 12-1
Goebel
Detroit, MI
Enamel 20

6 ❑ 12-1
Goebel
Detroit, MI
Metallic 20

7 ❑ 12-1
Goebel
Detroit, MI
20

8 ❑ 12-2
Goebel
Detroit, MI
20

9 ❑ 12-2
Goebel
Detroit, MI
18

10 ❑ 12-1
Goebel
Detroit, MI
18

11 ❑ 12-1
Goebel
Detroit, MI
20

12 ❑ 12-1
Goebel
Detroit, MI
300

13 ❑ 12-1
Goebel
Detroit, MI
400

14 ❑ 12-1
M. K. Goetz
St. Joseph, MO
9

15 ❑ 12-1
M. K. Goetz
St. Joseph, MO
7

16 ❑ 12-1
Goetz Div. of Pearl
St. Joseph, MO
7

17 ❑ 12-1
Pearl
St. Joseph, MO
7

18 ❑ 12-1
Country Tavern
St. Joseph, MO
7

19 ❑ 12-1
M. K. Goetz
St. Joseph, MO
60

20 ❑ 12-1
M. K. Goetz
St. Joseph, MO
10

21 ❑ 12-1
Goetz Div. of Pearl
St. Joseph, MO
10

22 ❑ 12-1
Pearl
St. Joseph, MO
7

23 ❑ 12-2
Cleveland-Sandusky
Cumberland, MD
40

24 ❑ 12-2
Cleveland-Sandusky
Cleveland, OH
125

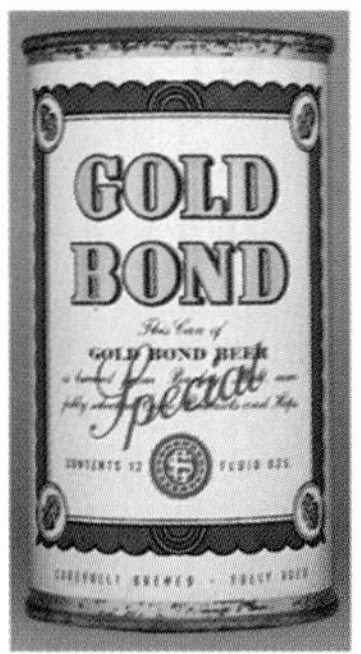

25 ❑ 12-1
Cleveland-Sandusky
Cleveland, OH
200

26 ❑ 12-2
Cleveland-Sandusky
Cleveland, OH
75

27 ❑ 12-2
Cleveland-Sandusky
Cleveland, OH
85

28 ❑ 12-2
Cleveland-Sandusky
Cleveland, OH
35

29 ❑ 12-2
Cleveland-Sandusky
Cleveland, OH
250

30 ❑ 12-2
Gold Brau
Chicago, IL
20

31 ❑ 12-2
Gold Brau
Chicago, IL
20

32 ❑ 12-2
Atlas
Chicago, IL
300

33 ❑ 12-2
9-0-5
Chicago, IL
50

34 ❑ 12-2
Drewry's Ltd.
South Bend, IN
45

35 ❑ 12-2
Lederer
Chicago, IL
75

36 ❑ 12-2
Queen City
Cumberland, MD
60

37 ❑ 12-2
Cumberland
Cumberland, MD
90

38 ❑ 12-2
Tennessee
Memphis, TN
85

39 ❑ 12-2
Atlantic
Chicago, IL
75

40 ❑ 12-2
Gold Label
Pueblo, CO
45

1 ❑ 12-2
Walter
Pueblo, CO
18

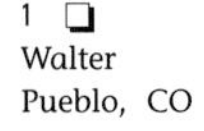

2 ❑ 12-2
Walter
Pueblo, CO
50

3 ❑ 12-2
Walter
Pueblo, CO
10

4 ❑ 12-2
Walter
Pueblo, CO
10

5 ❑ 12-1
Walter
Pueblo, CO
Enamel 10

6 ❑ 12-1
Walter
Pueblo, CO
Metallic 10

7 ❑ 12-1
Walter
Pueblo, CO
12

8 ❑ 12-2-T
General Corp.
San Francisco, CA
350

9 ❑ 12-1-OT
Grace Bros.
Santa Rosa, CA
350

10 ❑ 12-1-OT
Grace Bros.
Santa Rosa, CA
300

11 ❑ 12-1-OT
Grace Bros.
Santa Rosa, CA
300

12 ❑ 12-2-W
Grace Bros.
Santa Rosa, CA
325

13 ❑ 12-2-T
Grace Bros.
Santa Rosa, CA
325

14 ❑ 12-1-OT
Indianapolis
Indianapolis, IN
Orange 1000+

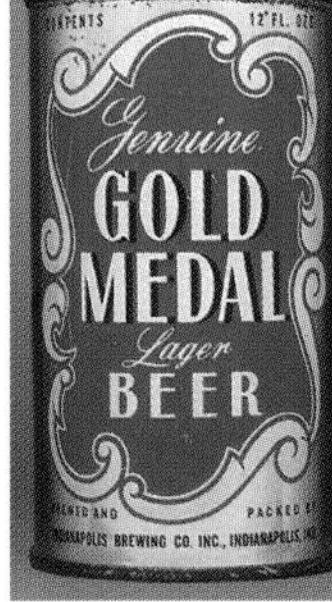

15 ❑ 12-1-OT
Indianapolis
Indianapolis, IN
Dk. Red 1000+

16 ❑ 12-1
Lebanon Valley
Lebanon, PA
500

17 ❑ 12-2
Florida
Miami, FL
900

18 ❑ 12-2-OT
South Side
Chicago, IL
Tall Can 1000+

19 ❑ 12-1
Harvard
Lowell, MA
250

20 ❑ 12-1
Harvard
Lowell, MA
225

21 ❑ 12-1
Harvard
Lowell, MA
175

22 ❑ 12-1
Harvard
Lowell, MA
150

23 ❑ 12-1
Harvard
Lowell, MA
1000+

24 ❑ 12-2
Maier
Los Angeles, CA
7

25 ❑ 12-2
Maier
Los Angeles, CA
7

26 ❑ 12-2
Grace Bros.
Santa Rosa, CA
30

27 ❑ 12-2
Grace Bros.
Santa Rosa, CA
30

28 ❑ 12-2
Grace Bros.
Santa Rosa, CA
30

29 ❑ 12-2
Grace Bros.
Santa Rosa, CA
30

30 ❑ 12-2
Diamond Spring
Lawrence, MA
65

31 ❑ 12-2
Diamond Spring
Lawrence, MA
45

32 ❑ 12-1-OT
Grace Bros LTD.
Los Angeles, CA
1000+

33 ❑ 12-2
Southern
Los Angeles, CA
25

34 ❑ 12-2
Maier
Los Angeles, CA
25

35 ❑ 12-2
Maier
Los Angeles, CA
12

36 ❑ 12-2
Grace Bros.
Santa Rosa, CA
8

37 ❑ 12-2
Maier
Los Angeles, CA
125

38 ❑ 12-2
Maier
Los Angeles, CA
125

39 ❑ 11-2
Regal Pale
San Francisco, CA
110

40 ❑ 12-2-OT
Golden West
Oakland, CA
350

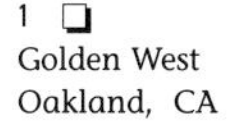

1 ❑ 12-1-T
Golden West
Oakland, CA
125

2 ❑ 12-1-T
Golden West
Oakland, CA
125

3 ❑ 12-1-T
Golden West
Oakland, CA
125

4 ❑ 12-2-T
Golden West
Oakland, CA
110

5 ❑ 12-2-T
Golden West
Oakland, CA
600

6 ❑ 12-2-OT
Golden West
Oakland, CA
300

7 ❑ 12-1-T
Golden West
Oakland, CA
Trademark
175

8 ❑ 12-1-T
Golden West
Oakland, CA
No Trademark
175

9 ❑ 12-2-T
Golden West
Oakland, CA
175

10 ❑ 12-2-T
Golden West
Oakland, CA
65

11 ❑ 12-2-T
Golden West
Oakland, CA
150

12 ❑ 12-2
Pacific
Oakland, CA
125

13 ❑ 12-2
Jos. Huber
Monroe, WI
40

14 ❑ 12-2
Maier
Los Angeles, CA
10

15 ❑ 12-2
Maier
Los Angeles, CA
10

16 ❑ 12-2
Maier
Los Angeles, CA
10

17 ❑ 12-2
Mitchell
El Paso, TX
35

18 ❑ 12-2
Maier
Los Angeles, CA
75

19 ❑ 12-2
Pacific
Oakland, CA
800

20 ❑ 12-2
Grace Bros.
Santa Rosa, CA
50

21 ❑ 12-2
Grace Bros.
Santa Rosa, CA
50

22 ❑ 12-2
Grace Bros.
Santa Rosa, CA
50

23 ❑ 12-2
Southern
Los Angeles, CA
50

24 ❑ 12-2
Maier
Los Angeles, CA
40

25 ❑ 11-2
Becker Products
Ogden, UT
35

26 ❑ 12-2
Becker Products
Ogden, UT
35

27 ❑ 12-2
Becker
Ogden, UT
35

28 ❑ 12-1-T
Edelbrau
New York, NY
1000+

29 ❑ 12-1-T
Edelbrau
New York, NY
1000+

30 ❑ 12-1-W
Edelbrau
New York, NY
1000+

31 ❑ 12-1-T
Edelbrau
New York, NY
1000+

32 ❑ 12-2
Schoenhofen Edelweiss
Chicago, IL
40

33 ❑ 12-2
Drewry's Ltd.
South Bend, IN
40

34 ❑ 12-2
Maier
Los Angeles, CA
65

35 ❑ 12-2
Maier
Los Angeles, CA
65

36 ❑ 12-2
Tivoli
Denver, CO
12

37 ❑ 12-2
Pacific
Oakland, CA
450

38 ❑ 12-1
Minneapolis
Minneapolis, MN
15

39 ❑ 12-1
Minneapolis
Minneapolis, MN
15

40 ❑ 12-2
Minneapolis
Minneapolis, MN
5

1 ❑ 12-2
Minneapolis
Minneapolis, MN
5

2 ❑ 12-2
Minneapolis
Minneapolis, MN
5

3 ❑ 12-2
Minneapolis
Minneapolis, MN
5

4 ❑ 12-2
Grand Lager
St. Charles, MO
50

5 ❑ 12-1-OT
Gulf
Houston, TX
DULL GRAY 250

6 ❑ 12-1-OT
Gulf
Houston, TX
175

7 ❑ 12-1-OT
Gulf
Houston, TX
175

8 ❑ 12-1-W
Gulf
Houston, TX
750

9 ❑ 12-1-W
Gulf
Houston, TX
150

10 ❑ 12-1-T
Gulf
Houston, TX
50

11 ❑ 12-1
Gulf
Houston, TX
600

12 ❑ 12-1
Gulf
Houston, TX
30

13 ❑ 12-2
Gulf
Houston, TX
30

14 ❑ 12-1
Gulf
Houston, TX
35

15 ❑ 12-1
Gulf
Houston, TX
35

16 ❑ 12-1
Gulf
Houston, TX
35

17 ❑ 12-1
Gulf
Houston, TX
40

18 ❑ 12-2
Gilt Edge
Trenton, NJ
150

19 ❑ 12-2
Gilt Edge
Trenton, NJ
25

20 ❑ 12-1-T
Sicks' Great Falls
Great Falls, MT
N.O. 4% 100

21 ❑ 12-1
Great Falls
Great Falls, MT
20

22 ❑ 12-1
Great Falls
Great Falls, MT
N.I.E.O. 4% 20

23 ❑ 12-2
Great Falls
Great Falls, MT
N.I.E.O. 4% 10

24 ❑ 12-2
Great Falls
Great Falls, MT
N.I.E.O. 4% 10

25 ❑ 12-2
Great Falls
Great Falls, MT
10

26 ❑ 12-2
Great Falls
Great Falls, MT
8

27 ❑ 12-1
Schoenhofen Edelweiss
Chicago, IL
65

28 ❑ 12-1
Schoenhofen Edelweiss
Chicago, IL
65

29 ❑ 12-2
Schoenhofen Edelweiss
Chicago, IL
20

30 ❑ 12-2
Schoenhofen Edelweiss
Chicago, IL
18

31 ❑ 12-2
Drewry's Ltd.
South Bend, IN
20

32 ❑ 12-2
Drewry's Ltd.
South Bend, IN
18

33 ❑ 12-2
Jacob Ruppert
New York, NY
40

34 ❑ 12-2
Esslinger's
Philadelphia, PA
40

35 ❑ 12-2
Esslinger's
Philadelphia, PA
40

36 ❑ 12-1
Wm. Gretz
Philadelphia, PA
125

37 ❑ 12-1
Wm. Gretz
Philadelphia, PA
150

38 ❑ 12-1
Wm. Gretz
Philadelphia, PA
150

39 ❑ 12-1
Wm. Gretz
Philadelphia, PA
150

Front of 75-1 to 75-12

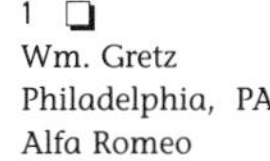
1 ❑ 12-1
Wm. Gretz
Philadelphia, PA
Alfa Romeo 250

2 ❑ 12-1
Wm. Gretz
Philadelphia, PA
Austin Healey 250

3 ❑ 12-1
Wm. Gretz
Philadelphia, PA
Corvette 250

4 ❑ 12-1
Wm. Gretz
Philadelphia, PA
Fiat 600 250

5 ❑ 12-1
Wm. Gretz
Philadelphia, PA
Fiat 1100 250

6 ❑ 12-1
Wm. Gretz
Philadelphia, PA
Ford Thunderbird 250

7 ❑ 12-1
Wm. Gretz
Philadelphia, PA
Isetta BMW 250

8 ❑ 12-1
Wm. Gretz
Philadelphia, PA
Mercedes Benz 250

9 ❑ 12-1
Wm. Gretz
Philadelphia, PA
MGA Convertible 250

10 ❑ 12-1
Wm. Gretz
Philadelphia, PA
MGA Hardtop 250

11 ❑ 12-1
Wm. Gretz
Philadelphia, PA
Porsche 1500 250

12 ❑ 12-1
Wm. Gretz
Philadelphia, PA
Saab 93 250

Front of 14-25

14 ❑ 12-1
Wm. Gretz
Philadelphia, PA
Alfa Romeo 175

15 ❑ 12-1
Wm. Gretz
Philadelphia, PA
Austin Healey 175

16 ❑ 12-1
Wm. Gretz
Philadelphia, PA
Corvette 175

17 ❑ 12-1
Wm. Gretz
Philadelphia, PA
Fiat 600 175

18 ❑ 12-1
Wm. Gretz
Philadelphia, PA
Fiat 1100 175

19 ❑ 12-1
Wm. Gretz
Philadelphia, PA
Ford Thunderbird 175

20 ❑ 12-1
Wm. Gretz
Philadelphia, PA
Isetta BMW 175

21 ❑ 12-1
Wm. Gretz
Philadelphia, PA
Mercedes Benz 175

22 ❑ 12-1
Wm. Gretz
Philadelphia, PA
MGA Convertible 175

23 ❑ 12-1
Wm. Gretz
Philadelphia, PA
MGA Hardtop 175

24 ❑ 12-1
Wm. Gretz
Philadelphia, PA
Porsche 1500 175

25 ❑ 12-1
Wm. Gretz
Philadelphia, PA
Saab 93 175

26 ❑ 12-1
Wm. Gretz
Philadelphia, PA
A Bird... 200

27 ❑ 12-1
Wm. Gretz
Philadelphia, PA
A Hot Time 200

28 ❑ 12-1
Wm. Gretz
Philadelphia, PA
After The Ball 200

29 ❑ 12-1
Wm. Gretz
Philadelphia, PA
Casey Jones 200

30 ❑ 12-1
Wm. Gretz
Philadelphia, PA
Everybody Works 200

31 ❑ 12-1
Wm. Gretz
Philadelphia, PA
Hello Ma Baby 200

32 ❑ 12-1
Wm. Gretz
Philadelphia, PA
I Want A Girl 200

33 ❑ 12-1
Wm. Gretz
Philadelphia, PA
I Wonder Who's 200

34 ❑ 12-1
Wm. Gretz
Philadelphia, PA
Ida Sweet 200

35 ❑ 12-1
Wm. Gretz
Philadelphia, PA
In The Good Old 200

36 ❑ 12-1
Wm. Gretz
Philadelphia, PA
Kentucky Babe 250

37 ❑ 12-1
Wm. Gretz
Philadelphia, PA
Let Me Call 200

38 ❑ 12-1
Wm. Gretz
Philadelphia, PA
Meet Me Tonight 200

39 ❑ 12-1
Wm. Gretz
Philadelphia, PA
My Gal Sal 200

40 ❑ 12-1
Wm. Gretz
Philadelphia, PA
My Wild Irish 200

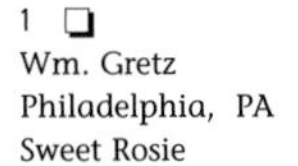

1 ❑ 12-1
Wm. Gretz
Philadelphia, PA
Sweet Rosie **200**

2 ❑ 12-1
Wm. Gretz
Philadelphia, PA
Take Me Out **200**

3 ❑ 12-1
Wm. Gretz
Philadelphia, PA
Take Me Out **200**

4 ❑ 12-1
Wm. Gretz
Philadelphia, PA
The Sidewalks **200**

5 ❑ 12-1
Wm. Gretz
Philadelphia, PA
Wait Til The Sun **200**

6 ❑ 12-1
Wm. Gretz
Philadelphia, PA
When You Were **200**

7 ❑ 12-1
Wm. Gretz
Philadelphia, PA
45

8 ❑ 12-2
Wm. Gretz
Philadelphia, PA
45

9 ❑ 12-1
Wm. Gretz
Philadelphia, PA
800

10 ❑ 12-1
Wm. Gretz
Philadelphia, PA
500

11 ❑ 12-1-T
Griesedieck Bros.
St. Louis, MO
50

12 ❑ 12-1-T
Griesedieck Bros.
St. Louis, MO
18

13 ❑ 12-1
Griesedieck Bros.
St. Louis, MO
15

14 ❑ 12-1
Griesedieck Bros.
St. Louis, MO
12

15 ❑ 12-1
Griesedieck Bros.
St. Louis, MO
Blue **75**

16 ❑ 12-1
Griesedieck Bros.
St. Louis, MO
Dk. Blue **75**

17 ❑ 12-1
Griesedieck Bros.
St. Louis, MO
Copper **75**

18 ❑ 12-1
Griesedieck Bros.
St. Louis, MO
Green **75**

19 ❑ 12-1
Griesedieck Bros.
St. Louis, MO
Lt. Pink **75**

20 ❑ 12-1
Griesedieck Bros.
St. Louis, MO
Pink **75**

21 ❑ 12-1
Griesedieck Bros.
St. Louis, MO
Dk. Pink **75**

22 ❑ 12-1
Griesedieck Bros.
St. Louis, MO
20

23 ❑ 12-1
Griesedieck Bros.
St. Louis, MO
Blue **90**

24 ❑ 12-1
Griesedieck Bros.
St. Louis, MO
Copper **90**

25 ❑ 12-1
Griesedieck Bros.
St. Louis, MO
Brown **90**

26 ❑ 12-1
Griesedieck Bros.
St. Louis, MO
Emerald Green **90**

27 ❑ 12-1
Griesedieck Bros.
St. Louis, MO
Med. Olive Grn. **90**

28 ❑ 12-1
Griesedieck Bros.
St. Louis, MO
Dk. Olive Grn. **90**

29 ❑ 12-1
Griesedieck Bros.
St. Louis, MO
Orange **90**

30 ❑ 12-1
Griesedieck Bros.
St. Louis, MO
Purple **90**

31 ❑ 12-1
Griesedieck Bros.
St. Louis, MO
Dk. Pink **90**

Front of 33, 34

33 ❑ 12-1
Griesedieck Bros.
St. Louis, MO
10

34 ❑ 12-1
Griesedieck Bros.
St. Louis, MO
10

35 ❑ 12-1
Griesedieck Bros.
St. Louis, MO
Lt. Blue **55**

36 ❑ 12-1
Griesedieck Bros.
St. Louis, MO
Med. Blue **55**

37 ❑ 12-1
Griesedieck Bros.
St. Louis, MO
Dk. Blue **55**

38 ❑ 12-1
Griesedieck Bros.
St. Louis, MO
Lt. Copper **55**

39 ❑ 12-1
Griesedieck Bros.
St. Louis, MO
Dk. Copper **55**

40 ❑ 12-1
Griesedieck Bros.
St. Louis, MO
Med. Brown **55**

1 ❑ 12-1
Griesedieck Bros.
St. Louis, MO
Dk. Brown 55

2 ❑ 12-1
Griesedieck Bros.
St. Louis, MO
Teal 55

3 ❑ 12-1
Griesedieck Bros.
St. Louis, MO
Med. Green 55

4 ❑ 12-1
Griesedieck Bros.
St. Louis, MO
Dk. Green 55

5 ❑ 12-1
Griesedieck Bros.
St. Louis, MO
Lt. Olive Grn. 55

6 ❑ 12-1
Griesedieck Bros.
St. Louis, MO
Dk. Olive Grn. 55

7 ❑ 12-1
Griesedieck Bros.
St. Louis, MO
Orange 55

8 ❑ 12-1
Griesedieck Bros.
St. Louis, MO
Lt. Purple 55

9 ❑ 12-1
Griesedieck Bros.
St. Louis, MO
Med. Purple 55

10 ❑ 12-1
Griesedieck Bros.
St. Louis, MO
Dk. Purple 55

11 ❑ 12-1
Griesedieck Bros.
St. Louis, MO
Red 55

12 ❑ 12-2
Goebel
Oakland, CA
#55047 175

13 ❑ 12-2
Goebel
Oakland, CA
#55048 175

14 ❑ 12-2
Goebel
Oakland, CA
#55049 175

15 ❑ 12-2
Goebel
Oakland, CA
#55050 175

16 ❑ 12-2
Goebel
Oakland, CA
#55051 175

17 ❑ 12-2
Goebel
Oakland, CA
#55052 175

18 ❑ 12-2
Goebel
Oakland, CA
#55055 175

19 ❑ 12-2
Goebel
Oakland, CA
#55056 175

20 ❑ 12-2
Goebel
Oakland, CA
#55057 175

21 ❑ 12-2
Goebel
Oakland, CA
#55059 175

22 ❑ 12-2
Goebel
Oakland, CA
#55060 175

23 ❑ 12-2
Goebel
Oakland, CA
#44042 175

24 ❑ 12-2
Goebel
Oakland, CA
#44046 175

25 ❑ 12-2
Goebel
Oakland, CA
#44047 175

26 ❑ 12-2
Goebel
Oakland, CA
#44050 175

27 ❑ 12-2
Goebel
Oakland, CA
#44053 175

28 ❑ 12-2
Goebel
Oakland, CA
#44054 175

29 ❑ 12-2
Goebel
Oakland, CA
#44055 175

30 ❑ 12-2
Goebel
Oakland, CA
#44059 175

31 ❑ 12-2
Goebel
Oakland, CA
#44060 175

32 ❑ 12-2
Goebel
Detroit, MI
#55042 175

33 ❑ 12-2
Goebel
Detroit, MI
#55045 175

34 ❑ 12-2
Goebel
Detroit, MI
#55047 175

35 ❑ 12-2
Goebel
Detroit, MI
#55049 175

36 ❑ 12-2
Goebel
Detroit, MI
#55050 175

37 ❑ 12-2
Goebel
Detroit, MI
#55055 175

38 ❑ 12-2
Goebel
Detroit, MI
#55057 175

39 ❑ 12-2
Goebel
Detroit, MI
#55061 175

40 ❑ 12-2
Goebel
Detroit, MI
#44042 175

1 ❑ 12-2
Goebel
Detroit, MI
#44044 175

2 ❑ 12-2
Goebel
Detroit, MI
#44046 175

3 ❑ 12-2
Goebel
Detroit, MI
#44047 175

4 ❑ 12-2
Goebel
Detroit, MI
#44048 175

5 ❑ 12-2
Goebel
Detroit, MI
#44049 175

6 ❑ 12-2
Goebel
Detroit, MI
#44050 175

7 ❑ 12-2
Goebel
Detroit, MI
#44052 175

8 ❑ 12-2
Goebel
Detroit, MI
#44053 175

9 ❑ 12-2
Goebel
Detroit, MI
#44055 175

10 ❑ 12-2
Goebel
Detroit, MI
#44056 175

11 ❑ 12-2
Goebel
Detroit, MI
#44057 175

12 ❑ 12-2
Goebel
Detroit, MI
#44060 175

13 ❑ 12-1-OT
Gunther
Baltimore, MD
1000+

14 ❑ 12-1-T
Gunther
Baltimore, MD
600

15 ❑ 12-1-T
Gunther
Baltimore, MD
500

16 ❑ 12-1
Gunther
Baltimore, MD
400

17 ❑ 12-2
Gunther
Baltimore, MD
60

18 ❑ 12-1-OT
Gunther
Baltimore, MD
1000+

19 ❑ 12-1-T
Gunther
Baltimore, MD
125

20 ❑ 12-1-T
Gunther
Baltimore, MD
60

21 ❑ 12-1-T
Gunther
Baltimore, MD
35

22 ❑ 12-1-T
Gunther
Baltimore, MD
25

23 ❑ 12-1-T
Gunther
Baltimore, MD
25

24 ❑ 12-1
Gunther
Baltimore, MD
25

25 ❑ 12-1
Gunther
Baltimore, MD
20

26 ❑ 12-1
Gunther
Baltimore, MD
20

27 ❑ 12-1
Gunther
Baltimore, MD
"Baltimore" 12

28 ❑ 12-1
Gunther
Baltimore, MD
"Of Baltimore" 12

29 ❑ 12-1-T
Gunther
Baltimore, MD
1000+

30 ❑ 12-1
Gunther
Baltimore, MD
1000+

31 ❑ 12-2
Gunther
Baltimore, MD
175

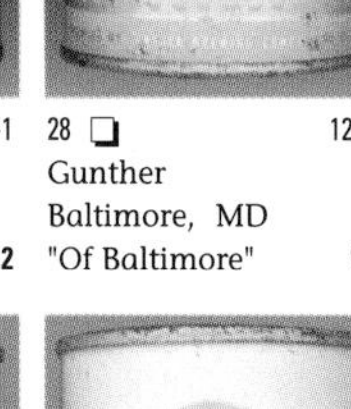
32 ❑ 12-2
Haberle Congress
Rochester, NY
100

33 ❑ 12-2
Haberle Congress
Syracuse, NY
100

34 ❑ 12-1
Hacker
Lawrence, MA
1000+

35 ❑ 12-1
Hacker
Lawrence, MA
750

36 ❑ 12-2
Haffenreffer
Boston, MA
30

37 ❑ 12-1
Haffenreffer
Boston, MA
65

38 ❑ 12-1
Pilsener
Cleveland, OH
275

39 ❑ 12-2
Hals
Baltimore, MD
110

40 ❑ 11-2
Theo. Hamm
Los Angeles, CA
15

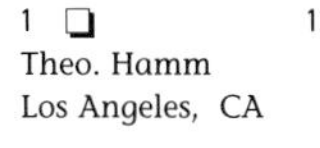
1 ❑ 11-2
Theo. Hamm
Los Angeles, CA
15

2 ❑ 11-2
Theo. Hamm
Los Angeles, CA
12

3 ❑ 11-2
Theo. Hamm
Los Angeles, CA
12

4 ❑ 12-2
Theo. Hamm
San Francisco, CA
12

5 ❑ 11-2
Theo. Hamm
San Francisco, CA
8

6 ❑ 12-2
Theo. Hamm
San Francisco, CA
8

7 ❑ 11-2
Theo. Hamm
San Francisco, CA
5

8 ❑ 11-2
Theo. Hamm
San Francisco, CA
5

9 ❑ 11-2
Theo. Hamm
San Francisco, CA
3

10 ❑ 12-2
Theo. Hamm
Baltimore, MD
15

11 ❑ 12-2
Theo. Hamm
Baltimore, MD
12

12 ❑ 12-2
Theo. Hamm
Baltimore, MD
12

13 ❑ 12-1-OT
Theo. Hamm
St. Paul, MN
100

14 ❑ 12-1-OT
Theo. Hamm
St. Paul, MN
100

15 ❑ 12-1-OT
Theo. Hamm
St. Paul, MN
DULL GRAY
90

16 ❑ 12-1-OT
Theo. Hamm
St. Paul, MN
75

17 ❑ 12-1-W
Theo. Hamm
St. Paul, MN
600

18 ❑ 12-2-T
Theo. Hamm
St. Paul, MN
20

19 ❑ 12-2
Theo. Hamm
St. Paul, MN
20

20 ❑ 12-2
Theo. Hamm
St. Paul, MN
10

21 ❑ 12-2
Theo. Hamm
St. Paul, MN
5

22 ❑ 12-2
Theo. Hamm
St. Paul, MN
5

23 ❑ 12-2
Theo. Hamm
St. Paul, MN
3

24 ❑ 12-2
Theo. Hamm
St. Paul, MN
75

25 ❑ 12-2
Theo. Hamm
St. Paul, MN
3

26 ❑ 12-2
Theo. Hamm
St. Paul, MN
3

27 ❑ 12-2
Theo. Hamm
St. Paul, MN
3

28 ❑ 12-2
Theo. Hamm
St. Paul, MN
3

29 ❑ 12-2
Theo. Hamm
St. Paul, MN
3

30 ❑ 12-2
Theo. Hamm
Houston, TX
20

31 ❑ 12-2
Theo. Hamm
Houston, TX
20

32 ❑ 12-1-T
Hampden
Willimansett, MA
700

33 ❑ 12-1-T
Hampden
Willimansett, MA
40

34 ❑ 12-1
Hampden
Willimansett, MA
40

35 ❑ 12-1
Hampden
Willimansett, MA
20

36 ❑ 12-1
Hampden-Harvard
Willimansett, MA
15

37 ❑ 12-1
Hampden
Willimansett, MA
200

38 ❑ 12-1
Hampden
Willimansett, MA
250

39 ❑ 12-1
Hampden
Willimansett, MA
70

40 ❑ 12-1
Hampden
Willimansett, MA
20

1 ❑ 12-1
Hampden-Harvard
Willimansett, MA
15

2 ❑ 12-1
Hampden-Harvard
Willimansett, MA
10

3 ❑ 12-1
Hampden-Harvard
Willimansett, MA
8

4 ❑ 12-1
James Hanley
Providence, RI
135

5 ❑ 12-2
James Hanley
Providence, RI
75

6 ❑ 12-1
James Hanley
Providence, RI
135

7 ❑ 12-2
James Hanley
Providence, RI
90

8 ❑ 12-2
James Hanley
Providence, RI
80

9 ❑ 12-2
James Hanley
Providence, RI
80

10 ❑ 12-1
James Hanley
Cranston, RI
125

11 ❑ 12-1
James Hanley
Cranston, RI
10

12 ❑ 12-1
Maier
Los Angeles, CA
40

13 ❑ 12-2
North Bay
Santa Rosa, CA
325

14 ❑ 12-2
Grace Bros.
Santa Rosa, CA
300

15 ❑ 12-2
Grace Bros.
Santa Rosa, CA
135

16 ❑ 12-1-OT
Best
Chicago, IL
200

17 ❑ 12-1-OT
Best
Chicago, IL
130

18 ❑ 12-1-OW
Best
Chicago, IL
200

19 ❑ 12-1-OT
Best
Chicago, IL
140

20 ❑ 12-1-OT
Best
Chicago, IL
DULL GRAY
150

21 ❑ 12-1
Best
Chicago, IL
50

22 ❑ 12-1
Best
Chicago, IL
30

23 ❑ 12-2
Hapsburg
Chicago, IL
20

24 ❑ 12-1-OT
Harvard
Lowell, MA
DULL GRAY
75

25 ❑ 12-1-OT
Harvard
Lowell, MA
75

26 ❑ 12-1-T
Harvard
Lowell, MA
200

27 ❑ 12-1-T
Harvard
Lowell, MA
40

Front of 29, 30

29 ❑ 12-1
Harvard
Lowell, MA
60

30 ❑ 12-1
Harvard
Lowell, MA
40

31 ❑ 12-1
Harvard
Lowell, MA
45

32 ❑ 12-1-OT
Harvard
Lowell, MA
100

33 ❑ 12-1-OW
Harvard
Lowell, MA
125

34 ❑ 12-1-T
Harvard
Lowell, MA
40

Front of 36, 37

36 ❑ 12-1
Harvard
Lowell, MA
60

37 ❑ 12-1
Harvard
Lowell, MA
40

38 ❑ 12-1
Harvard
Lowell, MA
45

39 ❑ 12-1
Hampden-Harvard
Willimansett, MA
18

40 ❑ 12-1
Hampden-Harvard
Willimansett, MA
150

1 ❑ 12-2
Maier
Los Angeles, CA
35

2 ❑ 12-1
Hedrick
Albany, NY
20

3 ❑ 12-2
Heidel Brau
La Crosse, WI
15

4 ❑ 12-2
Heidel Brau
La Crosse, WI
5

5 ❑ 12-2
G. Heileman
La Crosse, WI
15

6 ❑ 12-2
G. Heileman
La Crosse, WI
5

7 ❑ 12-2
Heidel Brau
Sioux City, IA
35

8 ❑ 12-2
Sioux City
Sioux City, IA
35

9 ❑ 12-2
Heidel Brau
Sheboygan, WI
40

10 ❑ 12-2
Columbia
Tacoma, WA
35

11 ❑ 12-2
Heidelberg
Tacoma, WA
20

12 ❑ 12-2
Heidelberg
Tacoma, WA
20

13 ❑ 12-2
Heidelberg
Tacoma, WA
20

14 ❑ 12-2
Heidelberg
Tacoma, WA
25

15 ❑ 12-2
Heidelberg
Tacoma, WA
25

16 ❑ 12-2
Heidelberg
Tacoma, WA
18

17 ❑ 11-1
Carling
Tacoma, WA
18

18 ❑ 11-1
Carling
Tacoma, WA
15

19 ❑ 12-1
Carling
Tacoma, WA
15

20 ❑ 12-2
G. Heileman
La Crosse, WI
20

21 ❑ 12-2
G. Heileman
La Crosse, WI
20

22 ❑ 12-2
G. Heileman
La Crosse, WI
20

23 ❑ 12-2
G. Heileman
La Crosse, WI
20

24 ❑ 12-2
G. Heileman
La Crosse, WI
15

25 ❑ 12-2
G. Heileman
La Crosse, WI
15

26 ❑ 12-2
G. Heileman
La Crosse, WI
10

27 ❑ 12-2
G. Heileman
La Crosse, WI
5

28 ❑ 12-2
G. Heileman
La Crosse, WI
55

29 ❑ 12-2
Heim-Brau
Monroe, WI
35

30 ❑ 12-1
Joseph Hensler
Newark, NJ
65

31 ❑ 12-1
Joseph Hensler
Newark, NJ
65

32 ❑ 12-1
Joseph Hensler
Newark, NJ
175

33 ❑ 12-2
Tivoli
Denver, CO
150

34 ❑ 12-2
Schoenhofen Edelweiss
Chicago, IL
125

35 ❑ 12-2
Schoenhofen Edelweiss
Chicago, IL
60

36 ❑ 12-2
Drewry's Ltd.
South Bend, IN
40

37 ❑ 12-2
Chr. Heurich
Washington, DC
55

38 ❑ 12-2
Jos. Huber
Monroe, WI
25

39 ❑ 12-2
Jos. Huber
Monroe, WI
Metallic
5

40 ❑ 12-2
Jos. Huber
Monroe, WI
Enamel
5

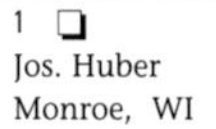

1 ❑ 12-2
Jos. Huber
Monroe, WI
5

2 ❑ 12-2
Tivoli
Denver, CO
35

3 ❑ 12-2
Grace Bros.
Santa Rosa, CA
125

4 ❑ 12-2
Grace Bros.
Santa Rosa, CA
125

5 ❑ 12-2
Grace Bros.
Santa Rosa, CA
125

6 ❑ 12-1-T
Sicks' Missoula
Missoula, MT
CNO 4%
100

7 ❑ 12-1-T
Missoula
Missoula, MT
CNO 4%
40

Front of 9, 10

9 ❑ 12-1
Missoula
Missoula, MT
CNO 4%
40

10 ❑ 12-1
Missoula
Missoula, MT
CNO 4%
40

11 ❑ 12-2
Missoula
Missoula, MT
10

12 ❑ 12-2
Missoula
Missoula, MT
8

13 ❑ 12-2
Missoula
Missoula, MT
8

14 ❑ 12-1-OT
Peter Fox
Chicago, IL
150

15 ❑ 12-1-OT
Best
Chicago, IL
125

16 ❑ 12-1-O
Best
Chicago, IL
50

17 ❑ 12-1
Best
Chicago, IL
50

18 ❑ 12-1
Best
Chicago, IL
50

19 ❑ 12-1
Empire
Chicago, IL
50

20 ❑ 12-1
United States
Chicago, IL
50

21 ❑ 12-1
Walter
Pueblo, CO
175

22 ❑ 12-2
L. F. Neuweiler's Sons
Allentown, PA
75

23 ❑ 12-2
Maier
Los Angeles, CA
5

24 ❑ 12-2
Maier
Los Angeles, CA
5

25 ❑ 12-2
Maier
Los Angeles, CA
5

26 ❑ 12-2
Hofbrau
Allentown, PA
20

27 ❑ 12-2
Allied
Chicago, IL
35

28 ❑ 12-2
Berghoff
Fort Wayne, IN
35

29 ❑ 12-2
Berghoff
Fort Wayne, IN
35

30 ❑ 12-2
Hoffman
Pueblo, CO
Gold Trim
75

31 ❑ 12-2
Hoffman
Pueblo, CO
Silver Trim
60

32 ❑ 12-2
Hoffman
Pueblo, CO
25

33 ❑ 12-2
Walter
Pueblo, CO
5

34 ❑ 12-2
Southern
Los Angeles, CA
300

35 ❑ 12-2
Hofbrau
Allentown, PA
75

36 ❑ 12-2
Potosi
Potosi, WI
30

37 ❑ 12-2
Holiday
Potosi, WI
40

38 ❑ 12-2
Holiday
Potosi, WI
60

39 ❑ 12-2
Holiday
Potosi, WI
10

40 ❑ 12-2
Holiday
Potosi, WI
15

1 ❑ 12-2
Diamond Spring
Lawrence, MA
50

2 ❑ 12-2
Diamond Spring
Lawrence, MA
40

3 ❑ 12-2
Diamond Spring
Lawrence, MA
18

4 ❑ 12-2
Maier
Los Angeles, CA
50

5 ❑ 12-2
Maier
Los Angeles, CA
40

6 ❑ 12-1
Eastern Brewing
Hammonton, NJ
75

7 ❑ 12-2
Eastern Brewing
Hammonton, NJ
75

8 ❑ 12-1
Eastern Beverage
Hammonton, NJ
35

9 ❑ 12-1
Eastern Brewing
Hammonton, NJ
35

10 ❑ 12-2
Eastern Brewing
Hammonton, NJ
5

11 ❑ 12-2
Pacific
Oakland, CA
1000+

12 ❑ 12-2
Atlas
Chicago, IL
100

13 ❑ 12-2
Atlas
Chicago, IL
50

14 ❑ 12-2
Atlas
Chicago, IL
90

15 ❑ 12-2
Drewry's Ltd.
South Bend, IN
45

16 ❑ 12-2
Drewry's Ltd.
South Bend, IN
40

17 ❑ 12-2
Drewry's Ltd.
South Bend, IN
10

18 ❑ 12-2
Drewry's Ltd.
South Bend, IN
70

19 ❑ 12-1-OT
Star
Vancouver, WA
900

20 ❑ 12-1-OT
Star
Vancouver, WA
DNCMT (4%) PC 700

21 ❑ 12-1-OT
Star
Vancouver, WA
DNCMT 4% 800

22 ❑ 12-1-OT
Interstate
Vancouver, WA
DNCMT 4% 800

23 ❑ 12-1-OT
Interstate
Vancouver, WA
DNCMT 4% 900

24 ❑ 12-1-OT
Globe
Baltimore, MD
Hopfheiser 1000+

25 ❑ 12-2
Pacific
Oakland, CA
550

26 ❑ 12-1
Horlacher
Allentown, PA
8

27 ❑ 12-2
Horlacher
Allentown, PA
8

28 ❑ 12-1-OT
Horluck
Seattle, WA
DNCMT 4% 1000+

29 ❑ 12-1-OT
Horluck
Seattle, WA
DNCMT 4% 1000+

30 ❑ 12-1-OT
Horluck
Seattle, WA
DNCMT 4% 1000+

31 ❑ 12-1-OT
Jacob Hornung
Philadelphia, PA
1000+

32 ❑ 12-1-O
Jacob Hornung
Philadelphia, PA
1000+

33 ❑ 12-1-OT
Jacob Hornung
Philadelphia, PA
1000+

34 ❑ 12-1-OT
Jacob Hornung
Philadelphia, PA
300

35 ❑ 12-1-OT
Jacob Hornung
Philadelphia, PA
275

36 ❑ 12-1-OT
Jacob Hornung
Philadelphia, PA
Tall "Hornung" 225

37 ❑ 12-1-OT
Jacob Hornung
Philadelphia, PA
225

38 ❑ 12-1-OW
Jacob Hornung
Philadelphia, PA
225

39 ❑ 12-1-O
Jacob Hornung
Philadelphia, PA
225

40 ❑ 12-1
Jacob Hornung
Philadelphia, PA
375

1 ❑ 12-1-OT
Jacob Hornung
Philadelphia, PA
1000+

2 ❑ 12-1-OT
Jacob Hornung
Philadelphia, PA
1000+

3 ❑ 12-1
Metropolis
Trenton, NJ
50

4 ❑ 12-1-T
Metropolis
New York, NY
60

5 ❑ 12-1-T
Metropolis
New York, NY
50

6 ❑ 12-1
Metropolis
New York, NY
50

7 ❑ 12-1-T
Old Dutch
New York, NY
65

8 ❑ 12-2
Horlacher
Allentown, PA
50

9 ❑ 12-2
Jos. Huber
Monroe, WI
Maroon 5

10 ❑ 12-2
Jos. Huber
Monroe, WI
Red 5

11 ❑ 12-2
Hudepohl
Cincinnati, OH
30

12 ❑ 12-2
Hudepohl
Cincinnati, OH
18

13 ❑ 12-2
Hudepohl
Cincinnati, OH
20

14 ❑ 12-2
Hudepohl
Cincinnati, OH
20

15 ❑ 12-2
Hudepohl
Cincinnati, OH
15

16 ❑ 12-2
Hudepohl
Cincinnati, OH
15

17 ❑ 12-2-OT
Hull
New Haven, CT
325

18 ❑ 12-2-OT
Hull
New Haven, CT
325

19 ❑ 12-1
Hull
New Haven, CT
50

20 ❑ 12-1
Hull
New Haven, CT
250

21 ❑ 12-1
Hull
New Haven, CT
275

22 ❑ 12-2-OT
Hull
New Haven, CT
1000+

23 ❑ 12-2-OT
Hull
New Haven, CT
600

24 ❑ 12-1
Hull
New Haven, CT
50

25 ❑ 12-2
Hull
New Haven, CT
15

26 ❑ 12-2
Hull
New Haven, CT
8

27 ❑ 12-2
Hull
New Haven, CT
900

28 ❑ 12-2
Hull
New Haven, CT
45

29 ❑ 12-2-OT
Humboldt
Eureka, CA
1000+

30 ❑ 12-1-T
Hyde Park Assn.
St. Louis, MO
125

31 ❑ 12-1-T
Hyde Park Assn.
St. Louis, MO
50

32 ❑ 12-1
Hyde Park Assn.
St. Louis, MO
50

33 ❑ 12-1
Hyde Park Assn.
St. Louis, MO
40

34 ❑ 12-1
Hyde Park Assn.
St. Louis, MO
40

35 ❑ 12-2
Metz
Pueblo, CO
50

36 ❑ 12-2
Walter
Pueblo, CO
15

37 ❑ 12-1-T
Indianapolis
Indianapolis, IN
350

Front of 39, 40

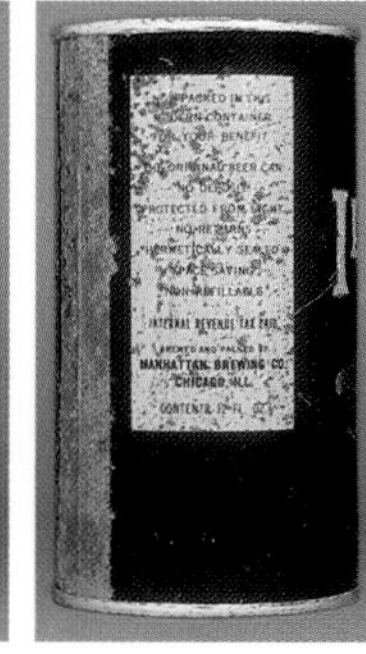

39 ❑ 12-1-OT
Manhattan
Chicago, IL
1000+

40 ❑ 12-1-OT
No Mandatory
1000+

1 ❑ 12-1-T
Grace Bros LTD.
Los Angeles, CA
60

2 ❑ 12-2-T
Grace Bros LTD.
Los Angeles, CA
60

3 ❑ 12-1-W
Grace Bros LTD.
Los Angeles, CA
80

4 ❑ 12-1-T
Southern
Los Angeles, CA
60

5 ❑ 12-1
Southern
Los Angeles, CA
60

6 ❑ 12-1
Southern
Los Angeles, CA
60

7 ❑ 12-1
Maier
Los Angeles, CA
500

8 ❑ 12-2
Maier
Los Angeles, CA
700

9 ❑ 12-1-OT
Weibel
New Haven, CT
1000+

10 ❑ 12-2
Imperial
Baltimore, MD
700

11 ❑ 12-2
Imperial
Baltimore, MD
125

12 ❑ 12-1-W
Jacob Hornung
Philadelphia, PA
1000+

13 ❑ 12-1
Cerveceria India
Mayaguez, PR
600

14 ❑ 12-1
Cerveceria India
Mayaguez, PR
600

15 ❑ 10-1
Cerveceria India
Mayaguez, PR
100

16 ❑ 12-1
Cerveceria India
Mayaguez, PR
150

17 ❑ 12-2
International
Tampa, FL
40

18 ❑ 12-2
International
Tampa, FL
25

19 ❑ 12-2
International
Covington, KY
40

20 ❑ 12-2
International
Covington, KY
25

21 ❑ 12-2
International
Covington, KY
55

22 ❑ 12-2
International
Buffalo, NY
45

23 ❑ 12-2
International
Buffalo, NY
35

24 ❑ 12-2
International
Buffalo, NY
150

25 ❑ 12-2
International
Buffalo, NY
40

26 ❑ 12-2
International
Buffalo, NY
20

27 ❑ 12-2
International
Buffalo, NY
55

28 ❑ 12-2
International
Findlay, OH
Enamel
35

29 ❑ 12-2
International
Findlay, OH
Metallic
55

30 ❑ 12-2
International
Findlay, OH
20

31 ❑ 12-2
International
Findlay, OH
20

32 ❑ 12-2-OT
Pittsburgh
Pittsburgh, PA
1000+

33 ❑ 12-1
Pittsburgh
Pittsburgh, PA
175

34 ❑ 12-1
Pittsburgh
Pittsburgh, PA
200

35 ❑ 12-1
Pittsburgh
Pittsburgh, PA
150

36 ❑ 12-1
Pittsburgh
Pittsburgh, PA
125

37 ❑ 12-1
Pittsburgh
Pittsburgh, PA
Almost a century...
60

38 ❑ 12-1
Pittsburgh
Pittsburgh, PA
Over a century...
60

39 ❑ 12-1
Pittsburgh
Pittsburgh, PA
35

40 ❑ 12-1
Iroquois Bev.
Buffalo, NY
65

1 ❑ 12-1
International
Buffalo, NY
55

2 ❑ 12-1
International
Buffalo, NY
50

3 ❑ 12-2
Div. of Iroquois Indus.
Buffalo, NY
1

4 ❑ 12-1
Jaguar
Rochester, NY
75

5 ❑ 12-1
Standard Rochester
Rochester, NY
40

6 ❑ 12-1
Jax
Jacksonville, FL
1000+

7 ❑ 12-1
Jax
Jacksonville, FL
500

8 ❑ 12-1-T
Jackson
New Orleans, LA
225

9 ❑ 12-1-T
Jackson
New Orleans, LA
225

10 ❑ 12-1
Jackson
New Orleans, LA
225

11 ❑ 12-1
Jackson
New Orleans, LA
Gold Trim
25

12 ❑ 12-1
Jackson
New Orleans, LA
Silver Trim
25

13 ❑ 12-1
Jackson
New Orleans, LA
63 Years
18

14 ❑ 12-1
Jackson
New Orleans, LA
65 Years
18

15 ❑ 12-1
Jackson
New Orleans, LA
15

16 ❑ 12-1
Jackson
New Orleans, LA
10

17 ❑ 12-2
Jackson
New Orleans, LA
10

18 ❑ 12-2
Jackson
New Orleans, LA
10

19 ❑ 10-2
Jackson
New Orleans, LA
15

20 ❑ 12-2
Jackson
New Orleans, LA
5

21 ❑ Front Like 20 12-2
Jackson
New Orleans, LA
200

22 ❑ 12-2
Jackson
New Orleans, LA
15

Side of 22

24 ❑ 12-2
Jackson
New Orleans, LA
8

25 ❑ 12-2
Jackson
New Orleans, LA
50

26 ❑ 12-2
Grace Bros.
Santa Rosa, CA
150

27 ❑ 12-2
Grace Bros.
Santa Rosa, CA
150

28 ❑ 12-2
Cold Spring
Cold Spring, MN
500

29 ❑ 12-1
Cold Spring
Cold Spring, MN
375

30 ❑ 12-2
Cold Spring
Cold Spring, MN
375

31 ❑ 12-2
Jester
Chicago, IL
400

32 ❑ 12-2
Jester
Chicago, IL
400

33 ❑ 12-2
Westminster
Chicago, IL
10

34 ❑ 12-2
Canadian Ace
Chicago, IL
15

35 ❑ 12-2
United States
Chicago, IL
8

36 ❑ 12-2
United States
Chicago, IL
8

37 ❑ 12-2
United States
Chicago, IL
8

38 ❑ 12-2
Atlas
Chicago, IL
200

39 ❑ 12-2
Chas. D. Kaier
Mahanoy City, PA
10

40 ❑ 12-2
Horlacher
Allentown, PA
35

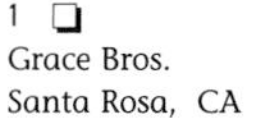

1 ❑ 12-2
Grace Bros.
Santa Rosa, CA
55

2 ❑ 12-2
Grace Bros.
Santa Rosa, CA
50

3 ❑ 12-2
Grace Bros.
Santa Rosa, CA
55

4 ❑ 12-2
Ph. Schneider
Trinidad, CO
90

5 ❑ 12-1
Duluth
Duluth, MN
20

6 ❑ 12-1
Duluth, Div. G. Heile.
La Crosse, WI
4

7 ❑ 12-2
Best
Chicago, IL
7

8 ❑ 12-2
Schoenhofen Edelweiss
Chicago, IL
7

9 ❑ 12-2
Drewry's Ltd.
South Bend, IN
7

10 ❑ 12-1
Spearman
Pensacola, FL
75

11 ❑ 12-1
Atlantic
Chicago, IL
45

12 ❑ 12-1
Atlantic
Chicago, IL
45

13 ❑ 12-1
Atlantic
Chicago, IL
25

14 ❑ 12-1
Best
Chicago, IL
22

15 ❑ 12-1
Empire
Chicago, IL
22

16 ❑ 12-1
United States
Chicago, IL
22

17 ❑ 12-1
Storz
Omaha, NE
12

18 ❑ 12-2
Best
Chicago, IL
50

19 ❑ 12-1
Keeley
Chicago, IL
250

20 ❑ 12-2
Best
Chicago, IL
30

21 ❑ 12-2
Cumberland
Cumberland, MD
45

22 ❑ 12-2
Cumberland
Cumberland, MD
30

23 ❑ 12-2
Maier
Los Angeles, CA
25

24 ❑ 12-2
Maier
Los Angeles, CA
12

25 ❑ 12-2
Maier
Los Angeles, CA
12

26 ❑ 12-2
Goebel
Oakland, CA
35

27 ❑ 12-2
Jacob Ruppert
New York, NY
30

28 ❑ 12-2
Jacob Ruppert
New York, NY
30

29 ❑ 12-2
Esslinger
Philadelphia, PA
Metallic
30

30 ❑ 12-2
Esslinger
Philadelphia, PA
Enamel
30

31 ❑ 12-1-OT
G. Krueger
Newark, NJ
1000+

32 ❑ 12-2
Frank Fehr
Louisville, KY
25

33 ❑ 12-2
Frank Fehr
Louisville, KY
25

34 ❑ 12-2
Frank Fehr
Louisville, KY
25

35 ❑ 12-2
Oertel
Louisville, KY
25

36 ❑ 12-2
Oertel
Louisville, KY
STOUT
90

37 ❑ 12-2
Maier
Los Angeles, CA
75

38 ❑ 12-2
Kings
Chicago, IL
Metallic
20

39 ❑ 12-2
Kings
Chicago, IL
Enamel
20

40 ❑ 12-2
Queen City
Cumberland, MD
85

1 ❑ 12-2-OT
Kings
New York, NY
600

2 ❑ 12-2-OT
Kings
New York, NY
750

3 ❑ 12-1-T
Grace Bros.
Santa Rosa, CA
1000+

4 ❑ 12-2
Grace Bros.
Santa Rosa, CA
125

5 ❑ 12-2
Grace Bros.
Santa Rosa, CA
100

6 ❑ 12-2
Grace Bros.
Santa Rosa, CA
100

7 ❑ 12-2
Kingsbury
Sioux City, IA
35

8 ❑ 12-2
G. Heileman
Newport, KY
3

9 ❑ 12-2
Kingsbury
Sheboygan, WI
8

10 ❑ 12-2
Kingsbury
Sheboygan, WI
8

11 ❑ 12-2
Kingsbury
Sheboygan, WI
3

12 ❑ 12-2
Kingsbury
Sheboygan, WI
3

13 ❑ 12-2
Kingsbury
Sheboygan, WI
35

14 ❑ 12-2
Kingsbury
Sheboygan, WI
35

15 ❑ 12-2
Kingsbury
Sheboygan, WI
10

16 ❑ 12-1
Kingsbury
Sheboygan, WI
63 Calories
3

17 ❑ 12-1
Kingsbury
Sheboygan, WI
70 Calories
3

18 ❑ 12-1
Kingsbury
Sheboygan, WI
73 Calories
3

19 ❑ 12-1
Kingsbury
Sheboygan, WI
3

20 ❑ 12-1
Schoenhofen Edelweiss
Chicago, IL
150

21 ❑ 12-1
Drewry's Ltd.
South Bend, IN
150

22 ❑ 12-1
Fred Koch
Dunkirk, NY
8

23 ❑ 12-2
Erie
Erie, PA
55

24 ❑ 12-1
Erie
Erie, PA
60

25 ❑ 12-2
Erie
Erie, PA
16

26 ❑ 12-2
Erie
Erie, PA
25

27 ❑ 12-2
Erie
Erie, PA
5

28 ❑ 12-1
Prima Bismarck
Chicago, IL
15

29 ❑ 12-1
Bismarck
Chicago, IL
10

30 ❑ 12-1
Canadian Ace
Chicago, IL
10

31 ❑ 12-1
Koenig Brau
Chicago, IL
7

32 ❑ 12-1
Maier
Los Angeles, CA
30

33 ❑ 12-1
Grace Bros.
Santa Rosa, CA
35

34 ❑ 12-1
Metz
Pueblo, CO
50

35 ❑ 12-1
Kol
Tampa, FL
55

36 ❑ 12-1
Atlantic
Chicago, IL
18

37 ❑ 12-1
Atlas
Chicago, IL
20

38 ❑ 12-1
Atlas
Chicago, IL
20

39 ❑ 12-1
Best
Chicago, IL
18

40 ❑ 12-1
Kol
Chicago, IL
22

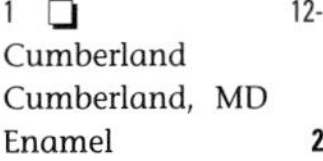

1 ❑ 12-1
Cumberland
Cumberland, MD
Enamel 20

2 ❑ 12-1
Cumberland
Cumberland, MD
Metallic 20

3 ❑ 12-2
Cumberland
Cumberland, MD
50

4 ❑ 12-1
Metz
Omaha, NE
30

5 ❑ 12-1
My
Omaha, NE
30

6 ❑ 12-1
Metropolis
Trenton, NJ
30

7 ❑ 12-1
Geo. F. Stein
Buffalo, NY
55

8 ❑ 12-2
Kol
Findlay, OH
50

9 ❑ 12-1
Gulf
Houston, TX
50

10 ❑ 11-1
Atlantic
Spokane, WA
35

11 ❑ 12-1
Atlantic
Spokane, WA
335

12 ❑ 11-1
Silver Springs
Tacoma, WA
4

13 ❑ 12-1
Wisconsin
Burlington, WI
50

14 ❑ 12-1
Wisconsin
Waukesha, WI
35

15 ❑ 12-1
Atlas
Chicago, IL
65

16 ❑ 12-1
Schoenhofen Edelweiss
Chicago, IL
65

17 ❑ 12-1
Schoenhofen Edelweiss
Chicago, IL
65

18 ❑ 12-1-OT
Grace Bros LTD.
Los Angeles, CA
400

19 ❑ 12-1-OT
Grace Bros.
Santa Rosa, CA
400

20 ❑ 12-1
Geo. Muehlebach
Kansas City, MO
150

21 ❑ 12-2
G. Krueger
Wilmington, DE
60

22 ❑ 12-1-O
G. Krueger
Newark, NJ
Pre Tax 500

23 ❑ 12-1-OT
G. Krueger
Newark, NJ
IRTP on back 500

24 ❑ 12-1-OT
G. Krueger
Newark, NJ
IRTP bf 500

25 ❑ 12-1-OT
G. Krueger
Newark, NJ
500

26 ❑ 12-1-OT
G. Krueger
Newark, NJ
250

27 ❑ 12-1-OT
G. Krueger
Newark, NJ
75

28 ❑ 12-1-OT
G. Krueger
Newark, NJ
70

29 ❑ 12-1-W
G. Krueger
Newark, NJ
300

30 ❑ 12-1-T
G. Krueger
Newark, NJ
75

31 ❑ 12-1-T
G. Krueger
Newark, NJ
35

32 ❑ 12-1-T
G. Krueger
Newark, NJ
35

33 ❑ 12-1
G. Krueger
Newark, NJ
35

34 ❑ 12-2
G. Krueger
Newark, NJ
40

35 ❑ 12-2
G. Krueger
Newark, NJ
40

36 ❑ 12-1
G. Krueger
Newark, NJ
35

37 ❑ 12-2
G. Krueger
Newark, NJ
40

38 ❑ 12-2
G. Krueger
Newark, NJ
50

39 ❑ 12-1
G. Krueger
Newark, NJ
45

40 ❑ 12-1
G. Krueger
Newark, NJ
35

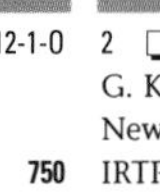

1 ❑ 12-1-0
G. Krueger
Newark, NJ
Pre Tax 750

2 ❑ 12-1-OT
G. Krueger
Newark, NJ
IRTP on back 750

3 ❑ 12-1-OT
G. Krueger
Newark, NJ
IRTP bf 750

4 ❑ 12-1-OT
G. Krueger
Newark, NJ
750

5 ❑ 12-1-OT
G. Krueger
Newark, NJ
375

6 ❑ 12-1-OT
G. Krueger
Newark, NJ
200

7 ❑ 12-1-OW
G. Krueger
Newark, NJ
200

8 ❑ 12-1-OT
G. Krueger
Newark, NJ
200

9 ❑ 12-1-W
G. Krueger
Newark, NJ
750

10 ❑ 12-1-W
G. Krueger
Newark, NJ
300

11 ❑ 12-1-T
G. Krueger
Newark, NJ
125

12 ❑ 12-1-T
G. Krueger
Newark, NJ
35

13 ❑ 12-1-W
G. Krueger
Newark, NJ
35

14 ❑ 12-1
G. Krueger
Newark, NJ
35

15 ❑ 12-2
G. Krueger
Newark, NJ
Black 40

16 ❑ 12-2
G. Krueger
Newark, NJ
Brown 40

17 ❑ 12-1
G. Krueger
Newark, NJ
30

18 ❑ 12-2
G. Krueger
Newark, NJ
30

19 ❑ 12-2
G. Krueger
Newark, NJ
25

20 ❑ 12-2
G. Krueger
Newark, NJ
30

21 ❑ 12-2
G. Krueger
Newark, NJ
30

22 ❑ 12-2
G. Krueger
Newark, NJ
30

23 ❑ 12-1
G. Krueger
Newark, NJ
12

24 ❑ 12-1
G. Krueger
Newark, NJ
12

25 ❑ 12-1-OT
G. Krueger
Newark, NJ
1000+

26 ❑ 12-1-OT
G. Krueger
Newark, NJ
1000+

27 ❑ 12-1-0
G. Krueger
Newark, NJ
1000+

28 ❑ 12-2
G. Krueger
Newark, NJ
650

29 ❑ 12-1
G. Krueger
Cranston, RI
45

30 ❑ 12-1
G. Krueger
Cranston, RI
25

31 ❑ 12-2
G. Krueger
Cranston, RI
20

32 ❑ 12-1
G. Krueger
Cranston, RI
12

33 ❑ 12-2
G. Krueger
Cranston, RI
15

34 ❑ 12-1
G. Krueger
Cranston, RI
25

35 ❑ 12-2
G. Krueger
Cranston, RI
6

36 ❑ 12-1-T
San Francisco
San Francisco, CA
1000+

37 ❑ 12-2
Grace Bros.
Santa Rosa, CA
500

38 ❑ 12-1
Canadian Ace
Chicago, IL
15

39 ❑ 12-1
Pilsen
Chicago, IL
12

40 ❑ 12-1
Pilsen
Chicago, IL
12

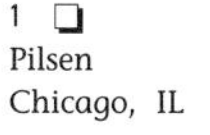
1 ❑ 12-1
Pilsen
Chicago, IL
12

2 ❑ 12-2
Grace Bros.
Santa Rosa, CA
65

3 ❑ 12-2
Grace Bros.
Santa Rosa, CA
65

4 ❑ 12-2
Grace Bros.
Santa Rosa, CA
65

5 ❑ 12-2
Eagle
Catasauqua, PA
60

6 ❑ 12-1
Lebanon Valley
Lebanon, PA
85

7 ❑ 12-1
Lebanon Valley
Lebanon, PA
60

8 ❑ 12-1-OT
Regal Amber
San Francisco, CA
1000+

9 ❑ 12-1-OT
San Francisco
San Francisco, CA
1000+

10 ❑ 12-2
Jacob Leinenkugel
Chippewa Falls, WI
20

11 ❑ 12-2
Jacob Leinenkugel
Chippewa Falls, WI
10

12 ❑ 12-2
Jacob Leinenkugel
Chippewa Falls, WI
10

13 ❑ 12-2
Jacob Leinenkugel
Chippewa Falls, WI
6

14 ❑ 12-1
Leisy
Chicago, IL
70

15 ❑ 12-1
Leisy
Chicago, IL
80

16 ❑ 12-2
Leisy
Chicago, IL
60

17 ❑ 12-2
Drewry's Ltd.
South Bend, IN
60

18 ❑ 12-1
Leisy
Cleveland, OH
70

19 ❑ 12-1
Leisy
Cleveland, OH
70

20 ❑ 12-1
Leisy
Cleveland, OH
80

21 ❑ 12-2
Leisy
Cleveland, OH
45

22 ❑ 12-1
Leisy
Cleveland, OH
45

23 ❑ 12-2
Leisy
Cleveland, OH
45

24 ❑ 12-2
Leisy
Cleveland, OH
45

25 ❑ 12-2
Leisy
Cleveland, OH
100

26 ❑ 12-1-T
Liebmann
New York, NY
1000+

27 ❑ 12-1-OT
Liebmann
New York, NY
1000+

28 ❑ 12-1-T
Rainier
San Francisco, CA
600

29 ❑ 12-2
Colonial
Hammonton, NJ
60

30 ❑ 12-2
Colonial
Hammonton, NJ
300

31 ❑ 12-1-T
Greater New York
New York, NY
300

32 ❑ 12-1-T
Greater New York
New York, NY
400

33 ❑ 12-1-OT
Lion
New York, NY
300

34 ❑ 12-1-T
Lion
New York, NY
175

35 ❑ 12-1-OT
Lion
New York, NY
650

36 ❑ 12-1-OT
Lion
New York, NY
600

37 ❑ 12-1-T
Lion
New York, NY
200

38 ❑ 12-1-T
Lion
New York, NY
1000+

39 ❑ 12-1-T
Pilser
New York, NY
1000+

40 ❑ 12-1-T
General Corp.
San Francisco, CA
1000+

1 ❑ 12-1-T
Southern
Los Angeles, CA
1000+

2 ❑ 12-2
Horlacher
Allentown, PA
75

3 ❑ 12-2
Horlacher
Allentown, PA
75

4 ❑ 12-2
ABC
Los Angeles, CA
60

5 ❑ 12-2
Maier
Los Angeles, CA
40

6 ❑ 12-2-T
Globe
San Francisco, CA
1000+

7 ❑ 12-2
Lone Star
Oklahoma City, OK
20

8 ❑ 12-2
Lone Star
Oklahoma City, OK
10

9 ❑ 12-2-T
Lone Star
San Antonio, TX
50

10 ❑ 12-2
Lone Star
San Antonio, TX
Thin Letters 40

11 ❑ 12-2
Lone Star
San Antonio, TX
Thick Letters 40

12 ❑ 12-2
Lone Star
San Antonio, TX
22

13 ❑ 12-2
Lone Star
San Antonio, TX
15

14 ❑ 12-2
Lone Star
San Antonio, TX
15

15 ❑ 12-2
Lone Star
San Antonio, TX
6

16 ❑ 12-1-0T
Manhattan
Chicago, IL
1000+

17 ❑ 12-1-0T
Lubeck
Toledo, OH
1000+

18 ❑ 12-1
Lubeck
Chicago, IL
Metallic 25

19 ❑ 12-1
Lubeck
Chicago, IL
Enamel 25

20 ❑ 12-1
Lubeck
Chicago, IL
30

21 ❑ 12-1
Lubeck
Chicago, IL
25

22 ❑ 12-1
Canadian Ace
Chicago, IL
30

23 ❑ 12-1
Canadian Ace
Chicago, IL
25

24 ❑ 12-1-T
Lucky Lager
Azusa, CA
15

25 ❑ 12-1
Lucky Lager
Azusa, CA
12

26 ❑ 12-1
Lucky Lager
Azusa, CA
10

27 ❑ 12-1
Lucky Lager
Azusa, CA
10

28 ❑ 11-2
Lucky Lager
Azusa, CA
5

29 ❑ 11-2
Lucky Lager
Azusa, CA
5

30 ❑ 12-2
Lucky Lager
Azusa, CA
5

31 ❑ 12-2
Lucky Lager
Azusa, CA
6

32 ❑ 12-2
Lucky Lager
Azusa, CA
6

33 ❑ 12-2
Lucky Lager
Azusa, CA
6

34 ❑ 12-2
Lucky Lager
Azusa, CA
6

35 ❑ 12-2
Lucky Lager
Azusa, CA
6

36 ❑ 12-1-0T
General Corp.
San Francisco, CA
1000+

37 ❑ 12-1-0T
General Corp.
San Francisco, CA
IRTP bf. 1000+

38 ❑ 12-1-0T
General Corp.
San Francisco, CA
600

39 ❑ 12-1-0T
General Corp.
San Francisco, CA
600

40 ❑ 12-1-T
General Corp.
San Francisco, CA
1000+

1 ❑ 12-1-T
General Corp.
San Francisco, CA
1000+

2 ❑ 12-1-T
General Corp.
San Francisco, CA
1000+

3 ❑ 11-2
Lucky Lager
San Francisco, CA
Enamel **75**

4 ❑ 11-2
Lucky Lager
San Francisco, CA
Enamel **75**

5 ❑ 11-2
Lucky Lager
San Francisco, CA
Metallic **75**

6 ❑ 12-2
Lucky Lager
San Francisco, CA
Metallic **100**

7 ❑ 12-1-0T
General Corp.
San Francisco, CA
50

8 ❑ 12-1-0T
General Corp.
San Francisco, CA
40

9 ❑ 12-1-0T
General Corp.
San Francisco, CA
30

10 ❑ 12-1-0W
General Corp.
San Francisco, CA
30

11 ❑ 12-1-T
General Corp.
San Francisco, CA
20

12 ❑ 12-1-T
General Corp.
San Francisco, CA
30

13 ❑ 12-1-W
General Corp.
San Francisco, CA
90

14 ❑ 12-1-T
General Corp.
San Francisco, CA
20

15 ❑ 12-1-T
Lucky Lager
San Francisco, CA
15

16 ❑ 12-1
Lucky Lager
San Francisco, CA
12

17 ❑ 12-2
Lucky Lager
San Francisco, CA
10

18 ❑ 12-2
Lucky Lager
San Francisco, CA
10

19 ❑ 12-2
Lucky Lager
San Francisco, CA
5

20 ❑ 11-2
Lucky Lager
San Francisco, CA
5

21 ❑ 11-2
Lucky Lager
San Francisco, CA
5

22 ❑ 12-2
Lucky Lager
San Francisco, CA
5

23 ❑ 12-2
Lucky Lager
San Francisco, CA
5

24 ❑ 12-2
Lucky Lager
San Francisco, CA
5

25 ❑ 12-2
Lucky Lager
San Francisco, CA
5

26 ❑ 12-2
Lucky Lager
San Francisco, CA
5

27 ❑ 12-2
Lucky Lager
San Francisco, CA
5

28 ❑ 12-2
Lucky Lager
San Francisco, CA
5

29 ❑ 12-2
Lucky Lager
Salt Lake City, UT
5

30 ❑ 12-2
Lucky Lager
Salt Lake City, UT
6

31 ❑ 12-2
Lucky Lager
Salt Lake City, UT
6

32 ❑ 12-2
Lucky Lager
Salt Lake City, UT
6

33 ❑ 12-2
Lucky Lager
Salt Lake City, UT
6

34 ❑ 12-2
Lucky Lager
Salt Lake City, UT
6

35 ❑ 12-1-T
Interstate
Vancouver, WA
20

36 ❑ 12-1-T
Lucky Lager
Vancouver, WA
50

37 ❑ 12-1
Lucky Lager
Vancouver, WA
10

38 ❑ 12-1
Lucky Lager
Vancouver, WA
10

39 ❑ 12-2
Lucky Lager
Vancouver, WA
5

40 ❑ 11-2
Lucky Lager
Vancouver, WA
5

1 ❑ 11-2
Lucky Lager
Vancouver, WA
5

2 ❑ 12-2
Lucky Lager
Vancouver, WA
5

3 ❑ 12-2
Lucky Lager
Vancouver, WA
7

4 ❑ 12-2
Lucky Lager
Vancouver, WA
7

5 ❑ 12-2
Lucky Lager
Vancouver, WA
7

6 ❑ 12-2
Grace Bros.
Santa Rosa, CA
135

7 ❑ 12-2
Grace Bros.
Santa Rosa, CA
135

8 ❑ 12-2
Grace Bros.
Santa Rosa, CA
150

Front of 10,11

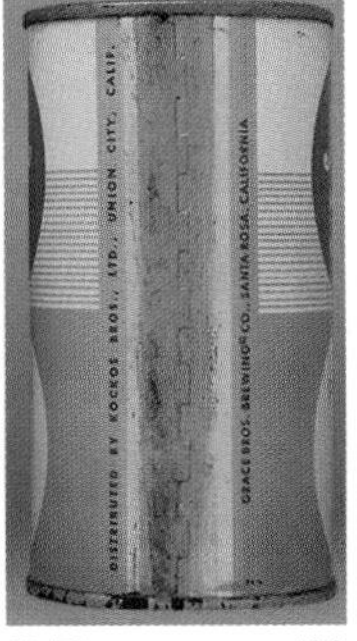
10 ❑ 12-2
Grace Bros.
Santa Rosa, CA
75

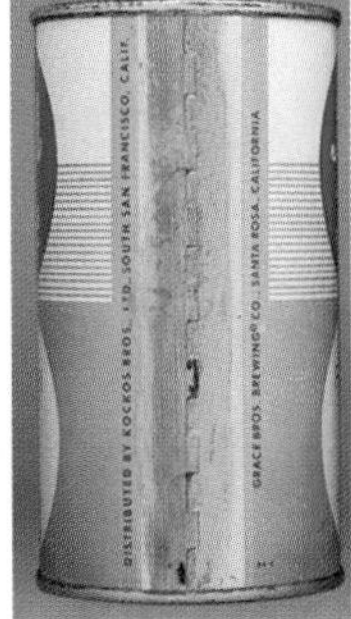
11 ❑ 12-2
Grace Bros.
Santa Rosa, CA
75

12 ❑ 12-2
Maier
Los Angeles, CA
500

13 ❑ 12-1
Maier
Los Angeles, CA
90

14 ❑ 12-1
Maier
Los Angeles, CA
60

15 ❑ 12-2
Maier
Los Angeles, CA
125

16 ❑ 12-2
Maier
Los Angeles, CA
5

17 ❑ 12-2
Maier
Los Angeles, CA
5

18 ❑ 12-2
Geo. Muehlebach
Kansas City, MO
200

19 ❑ 12-2
Best
Chicago, IL
400

20 ❑ 12-2
Malt Marrow
Chicago, IL
200

21 ❑ 12-1
Cremo
New Britain, CT
600

22 ❑ 12-1-OT
Manhattan
Chicago, IL
1000+

23 ❑ 12-1-OT
Manhattan
Chicago, IL
110

24 ❑ 12-1-OT
Manhattan
Chicago, IL
1000+

25 ❑ 12-1
Lebanon Valley
Lebanon, PA
500

26 ❑ 12-2
Old Reading
Reading, PA
12

27 ❑ 12-2
Reading
Reading, PA
5

28 ❑ 12-1
Maier
Los Angeles, CA
18

29 ❑ 12-1
Maier
Los Angeles, CA
18

30 ❑ 12-2
Maier
Los Angeles, CA
18

31 ❑ 12-1
Maier
Los Angeles, CA
18

32 ❑ 12-2
Maier
Los Angeles, CA
18

33 ❑ 12-2
Cumberland
Cumberland, MD
85

34 ❑ 12-1
Marlin
Orlando, FL
1000+

35 ❑ 12-1
Marlin
Orlando, FL
1000+

36 ❑ 12-2
Fuhrmann & Schmidt
Shamokin, PA
65

37 ❑ 12-2
Burgermeister
San Francisco, CA
15

38 ❑ 12-1
West End
Utica, NY
40

39 ❑ 12-2
Best
Chicago, IL
250

40 ❑ 12-2
United States
Chicago, IL
175

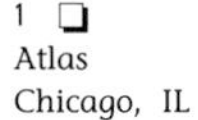

1 ❑ 12-2
Atlas
Chicago, IL
50

2 ❑ 12-2
Drewry's Ltd.
South Bend, IN
45

3 ❑ 12-1
Maier
Los Angeles, CA
125

4 ❑ 12-1-OT
Peter Hand
Chicago, IL
Metallic Gold 65

5 ❑ 12-1-OT
Peter Hand
Chicago, IL
Dull Gold 65

6 ❑ 12-1-W
Peter Hand
Chicago, IL
800

7 ❑ 12-1-T
Peter Hand
Chicago, IL
35

8 ❑ 12-1-T
Peter Hand
Chicago, IL
25

9 ❑ 12-1
Peter Hand
Chicago, IL
25

10 ❑ 12-1
Peter Hand
Chicago, IL
30

11 ❑ 12-1
Peter Hand
Chicago, IL
No Date 90

12 ❑ 12-1
Peter Hand
Chicago, IL
No Date 90

13 ❑ 12-1
Peter Hand
Chicago, IL
No Date 90

14 ❑ 12-1
Peter Hand
Chicago, IL
No Date 90

15 ❑ 12-1
Peter Hand
Chicago, IL
No Date 90

16 ❑ 12-1
Peter Hand
Chicago, IL
No Date 90

17 ❑ 12-1
Peter Hand
Chicago, IL
1952 90

18 ❑ 12-1
Peter Hand
Chicago, IL
1952 90

19 ❑ 12-1
Peter Hand
Chicago, IL
1952 90

20 ❑ 12-1
Peter Hand
Chicago, IL
1952 90

21 ❑ 12-1
Peter Hand
Chicago, IL
1952 90

22 ❑ 12-1
Peter Hand
Chicago, IL
1952 90

23 ❑ 12-1
Peter Hand
Chicago, IL
1952 90

24 ❑ 12-1
Peter Hand
Chicago, IL
1952 90

25 ❑ 12-1
Peter Hand
Chicago, IL
Astronomy 90

26 ❑ 12-1
Peter Hand
Chicago, IL
Baseball 90

27 ❑ 12-1
Peter Hand
Chicago, IL
Bowling 90

28 ❑ 12-1
Peter Hand
Chicago, IL
Gardening 90

29 ❑ 12-1
Peter Hand
Chicago, IL
Glasses and Bottles 90

30 ❑ 12-1
Peter Hand
Chicago, IL
Golf 90

31 ❑ 12-1
Peter Hand
Chicago, IL
Hats 90

32 ❑ 12-1
Peter Hand
Chicago, IL
Horse Racing 90

33 ❑ 12-1
Peter Hand
Chicago, IL
Music Instruments 90

34 ❑ 12-1
Peter Hand
Chicago, IL
Pipes 90

35 ❑ 12-1
Peter Hand
Chicago, IL
Playing Card Suits 90

36 ❑ 12-1
Peter Hand
Chicago, IL
Ranch 90

37 ❑ 12-1
Peter Hand
Chicago, IL
Sailboats 90

38 ❑ 12-1
Peter Hand
Chicago, IL
Sports 90

39 ❑ 12-1
Peter Hand
Chicago, IL
Square Dancing 90

40 ❑ 12-1
Peter Hand
Chicago, IL
Duck Hunting 100

1 ❑ 12-1
Peter Hand
Chicago, IL
Eskimos 100

2 ❑ 12-1
Peter Hand
Chicago, IL
Houses 100

3 ❑ 12-1
Peter Hand
Chicago, IL
Hockey 100

4 ❑ 12-1
Peter Hand
Chicago, IL
Ice Boating 100

5 ❑ 12-1
Peter Hand
Chicago, IL
Ice Fishing 100

6 ❑ 12-1
Peter Hand
Chicago, IL
Ice Skating 100

7 ❑ 12-1
Peter Hand
Chicago, IL
Inn 100

8 ❑ 12-1
Peter Hand
Chicago, IL
Penguins 100

9 ❑ 12-1
Peter Hand
Chicago, IL
Pine Boughs 100

10 ❑ 12-1
Peter Hand
Chicago, IL
Rabbit Hunting 100

11 ❑ 12-1
Peter Hand
Chicago, IL
Reindeer 100

12 ❑ 12-1
Peter Hand
Chicago, IL
Skiing 100

13 ❑ 12-1
Peter Hand
Chicago, IL
Sleighing 100

14 ❑ 12-1
Peter Hand
Chicago, IL
Snowflakes 100

15 ❑ 12-1
Peter Hand
Chicago, IL
Snowflakes 100

16 ❑ 12-1
Peter Hand
Chicago, IL
Snowmen 100

17 ❑ 12-1
Peter Hand
Chicago, IL
Tobogganing 100

18 ❑ 12-1
Peter Hand
Chicago, IL
Trees 100

19 ❑ 12-1
Peter Hand
Chicago, IL
Yule Log 100

20 ❑ 12-1
Peter Hand
Chicago, IL
China 125

21 ❑ 12-1
Peter Hand
Chicago, IL
France 125

22 ❑ 12-1
Peter Hand
Chicago, IL
Germany 125

23 ❑ 12-1
Peter Hand
Chicago, IL
Greece 125

24 ❑ 12-1
Peter Hand
Chicago, IL
Gypsies 125

25 ❑ 12-1
Peter Hand
Chicago, IL
Hawaii 125

26 ❑ 12-1
Peter Hand
Chicago, IL
Holland 125

27 ❑ 12-1
Peter Hand
Chicago, IL
Ireland 125

28 ❑ 12-1
Peter Hand
Chicago, IL
Italy 125

29 ❑ 12-1
Peter Hand
Chicago, IL
Poland 125

30 ❑ 12-1
Peter Hand
Chicago, IL
Russia 125

31 ❑ 12-1
Peter Hand
Chicago, IL
Scandinavia 125

32 ❑ 12-1
Peter Hand
Chicago, IL
Scotland 125

33 ❑ 12-1
Peter Hand
Chicago, IL
South America 125

34 ❑ 12-1
Peter Hand
Chicago, IL
South Seas 125

35 ❑ 12-1
Peter Hand
Chicago, IL
Spain 125

36 ❑ 12-1
Peter Hand
Chicago, IL
Switzerland 125

37 ❑ 12-1
Peter Hand
Chicago, IL
Thailand 125

38 ❑ 12-1
Peter Hand
Chicago, IL
Turkey 125

39 ❑ 12-1
Peter Hand
Chicago, IL
United States 125

40 ❑ 12-1
Peter Hand
Chicago, IL
China 90

1 ❏ 12-1
Peter Hand
Chicago, IL
France 90

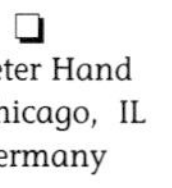

2 ❏ 12-1
Peter Hand
Chicago, IL
Germany 90

3 ❏ 12-1
Peter Hand
Chicago, IL
Greece 90

4 ❏ 12-1
Peter Hand
Chicago, IL
Gypsies 90

5 ❏ 12-1
Peter Hand
Chicago, IL
Hawaii 90

6 ❏ 12-1
Peter Hand
Chicago, IL
Holland 90

7 ❏ 12-1
Peter Hand
Chicago, IL
Ireland 90

8 ❏ 12-1
Peter Hand
Chicago, IL
Italy 90

9 ❏ 12-1
Peter Hand
Chicago, IL
Poland 90

10 ❏ 12-1
Peter Hand
Chicago, IL
Russia 90

11 ❏ 12-1
Peter Hand
Chicago, IL
Scandinavia 90

12 ❏ 12-1
Peter Hand
Chicago, IL
Scotland 90

13 ❏ 12-1
Peter Hand
Chicago, IL
South America 90

14 ❏ 12-1
Peter Hand
Chicago, IL
South Seas 90

15 ❏ 12-1
Peter Hand
Chicago, IL
Spain 90

16 ❏ 12-1
Peter Hand
Chicago, IL
Switzerland 90

17 ❏ 12-1
Peter Hand
Chicago, IL
Thailand 90

18 ❏ 12-1
Peter Hand
Chicago, IL
Turkey 90

19 ❏ 12-1
Peter Hand
Chicago, IL
United States 90

20 ❏ 12-1
Peter Hand
Chicago, IL
France 90

21 ❏ 12-1
Peter Hand
Chicago, IL
Gypsies 90

22 ❏ 12-1
Peter Hand
Chicago, IL
Scotland 90

23 ❏ 12-1
Peter Hand
Chicago, IL
Thailand 90

24 ❏ 12-1
Peter Hand
Chicago, IL
Turkey 90

25 ❏ 12-2
Peter Hand
Chicago, IL
60

26 ❏ 12-2
Peter Hand
Chicago, IL
60

27 ❏ 12-2
Peter Hand
Chicago, IL
60

28 ❏ 12-2
Peter Hand
Chicago, IL
60

29 ❏ 12-2
Peter Hand
Chicago, IL
60

30 ❏ 12-2
Peter Hand
Chicago, IL
60

31 ❏ 12-2
Peter Hand
Chicago, IL
60

32 ❏ 12-2
Peter Hand
Chicago, IL
60

33 ❏ 12-2
Peter Hand
Chicago, IL
60

34 ❏ 12-2
Peter Hand
Chicago, IL
60

35 ❏ 12-2
Peter Hand
Chicago, IL
50

36 ❏ 12-2
Peter Hand
Chicago, IL
50

37 ❏ 12-2
Peter Hand
Chicago, IL
50

38 ❏ 12-2
Peter Hand
Chicago, IL
50

39 ❏ 12-2
Peter Hand
Chicago, IL
50

40 ❏ 12-2
Peter Hand
Chicago, IL
50

1 ❑ 12-2
Peter Hand
Chicago, IL
50

2 ❑ 12-2
Peter Hand
Chicago, IL
45

3 ❑ 12-2
Peter Hand
Chicago, IL
45

4 ❑ 12-2
Peter Hand
Chicago, IL
45

5 ❑ 12-2
Peter Hand
Chicago, IL
45

6 ❑ 12-2
Peter Hand
Chicago, IL
45

7 ❑ 12-2
Peter Hand
Chicago, IL
45

8 ❑ 12-2
Peter Hand
Chicago, IL
45

9 ❑ 12-2
Peter Hand
Chicago, IL
45

10 ❑ 12-2
Peter Hand
Chicago, IL
45

11 ❑ 12-2
Peter Hand
Chicago, IL
45

12 ❑ 12-2
Peter Hand
Chicago, IL
45

13 ❑ 12-2
Peter Hand
Chicago, IL
45

14 ❑ 12-2
Peter Hand
Chicago, IL
45

Front of 16-25

16 ❑ 12-1
Peter Hand
Chicago, IL
Barbecue 110

17 ❑ 12-1
Peter Hand
Chicago, IL
Beachball 110

18 ❑ 12-1
Peter Hand
Chicago, IL
Boat Model 110

19 ❑ 12-1
Peter Hand
Chicago, IL
Bowling 110

20 ❑ 12-1
Peter Hand
Chicago, IL
Fishing 110

21 ❑ 12-1
Peter Hand
Chicago, IL
Golfing 110

22 ❑ 12-1
Peter Hand
Chicago, IL
Moonlight 110

23 ❑ 12-1
Peter Hand
Chicago, IL
Skiing 110

24 ❑ 12-1
Peter Hand
Chicago, IL
Snowman 110

25 ❑ 12-1
Peter Hand
Chicago, IL
Square Dancing 110

Front of 27-36

27 ❑ 12-1
Peter Hand
Chicago, IL
Barbecue 110

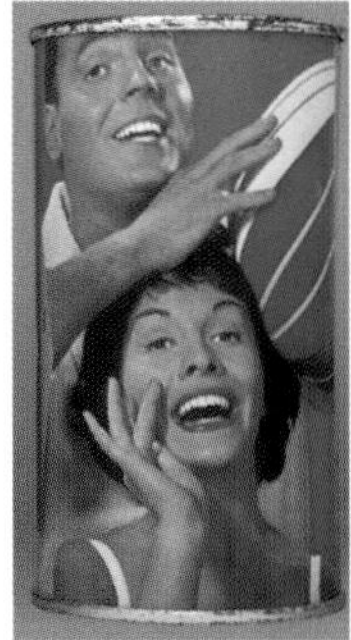
28 ❑ 12-1
Peter Hand
Chicago, IL
Beachball 110

29 ❑ 12-1
Peter Hand
Chicago, IL
Boat Model 110

30 ❑ 12-1
Peter Hand
Chicago, IL
Bowling 110

31 ❑ 12-1
Peter Hand
Chicago, IL
Fishing 110

32 ❑ 12-1
Peter Hand
Chicago, IL
Golfing 110

33 ❑ 12-1
Peter Hand
Chicago, IL
Moonlight 110

34 ❑ 12-1
Peter Hand
Chicago, IL
Skiing 110

35 ❑ 12-1
Peter Hand
Chicago, IL
Snowman 110

36 ❑ 12-1
Peter Hand
Chicago, IL
Square Dancing 110

37 ❑ 12-3
Peter Hand
Chicago, IL
6

38 ❑ 12-3
Peter Hand
Chicago, IL
6

39 ❑ 12-3
Peter Hand
Chicago, IL
6

40 ❑ 12-3
Peter Hand
Chicago, IL
6

1 ❑ 12-1
Peter Hand
Chicago, IL
500

2 ❑ 12-2
Peter Hand
Chicago, IL
65

3 ❑ 12-1
Peter Hand
Chicago, IL
15

4 ❑ 12-1
Peter Hand
Chicago, IL
15

5 ❑ 12-1
Peter Hand
Chicago, IL
6

6 ❑ 12-3
Peter Hand
Chicago, IL
6

7 ❑ 12-3
Meister Brau
Chicago, IL
4

8 ❑ 12-3
Meister Brau
Chicago, IL
12

9 ❑ 12-3
Meister Brau
Chicago, IL
6

10 ❑ 12-3
Meister Brau
Chicago, IL
6

11 ❑ 12-2
Pacific
Oakland, CA
350

12 ❑ 12-2
Metropolis
Trenton, NJ
3

13 ❑ 12-2
Metz
Pueblo, CO
25

14 ❑ 12-2
Walter
Pueblo, CO
15

15 ❑ 12-1
Metz
Omaha, NE
65

16 ❑ 12-1
Metz
Omaha, NE
18

17 ❑ 12-1
Metz
Omaha, NE
15

18 ❑ 12-1
Metz
Omaha, NE
15

19 ❑ 12-2
Metz
Omaha, NE
25

20 ❑ 12-2
Metz
Omaha, NE
Enamel 15

21 ❑ 12-2
Metz
Omaha, NE
Metallic 15

22 ❑ 12-1
Metz
Omaha, NE
250

23 ❑ 12-1-T
Anheuser-Busch
St. Louis, MO
Paper Label 750

24 ❑ 12-2
Mountain
Denver, CO
40

25 ❑ 12-2
Tivoli
Denver, CO
35

26 ❑ 12-2
Tivoli
Denver, CO
25

27 ❑ 12-1-OT
Miller
Milwaukee, WI
75

28 ❑ 12-1-OW
Miller
Milwaukee, WI
100

29 ❑ 12-1-OT
Miller
Milwaukee, WI
75

30 ❑ 12-1-OW
Miller
Milwaukee, WI
100

31 ❑ 12-1-OT
Miller
Milwaukee, WI
75

32 ❑ 12-1-T
Miller
Milwaukee, WI
18

Front of 34, 35, 36

34 ❑ 12-1
Miller
Milwaukee, WI
12

35 ❑ 12-1
Miller
Milwaukee, WI
12

36 ❑ 12-1
Miller
Milwaukee, WI
12

37 ❑ 12-1
Miller
Milwaukee, WI
6

38 ❑ 12-1
Miller
Milwaukee, WI
5

39 ❑ 10-2
Miller
Milwaukee, WI
15

40 ❑ 12-1
Miller
Milwaukee, WI
4

1 ❑ 12-1
Miller
Milwaukee, WI
4

2 ❑ 12-2
Miller
Milwaukee, WI
3

3 ❑ 12-2
Waukee
Hammonton, NJ
5

4 ❑ 12-2
Waukee
Hammonton, NJ
5

5 ❑ 12-2
Gettelman
Milwaukee, WI
700

6 ❑ 12-2
A. Gettelman
Milwaukee, WI
25

7 ❑ 12-2
A. Gettelman
Milwaukee, WI
50

8 ❑ 12-2
A. Gettelman
Milwaukee, WI
4

9 ❑ 12-2
Gettelman, Div. of Miller
Milwaukee, WI
4

10 ❑ 12-2
West Bend Lithia
West Bend, WI
300

11 ❑ 12-1-T
Mitchell
El Paso, TX
100

12 ❑ 12-1
Mitchell
El Paso, TX
100

13 ❑ 12-2
Mitchell
El Paso, TX
100

14 ❑ 12-2
Mitchell
El Paso, TX
"Harry" In Loop
75

15 ❑ 12-2
Mitchell
El Paso, TX
75

16 ❑ 12-2
Mitchell
El Paso, TX
90

17 ❑ 12-2
Monarch
Chicago, IL
30

18 ❑ 12-2
Monarch
Chicago, IL
20

19 ❑ 12-2
Monarch
Chicago, IL
20

20 ❑ 12-1-T
Grace Bros LTD.
Los Angeles, CA
1000+

21 ❑ 12-1-OT
Grace Bros.
Santa Rosa, CA
1000+

22 ❑ 12-1-OT
Grace Bros.
Santa Rosa, CA
1000+

Side of 22

24 ❑ 12-2
Monticello
Norfolk, VA
70

25 ❑ 12-2
Monticello
Norfolk, VA
35

26 ❑ 12-2
Monticello
Norfolk, VA
50

27 ❑ 12-2
Monticello
Norfolk, VA
35

28 ❑ 12-2
Wisconsin
Waukesha, WI
375

29 ❑ 12-1
Geo. Muehlebach
Kansas City, MO
20

30 ❑ 12-1
Geo. Muehlebach
Kansas City, MO
Lg. Date
20

31 ❑ 12-1
Geo. Muehlebach
Kansas City, MO
Enamel Gold
20

32 ❑ 12-1
Geo. Muehlebach
Kansas City, MO
Metalic Gold
20

33 ❑ 12-1
Geo. Muehlebach
Kansas City, MO
20

34 ❑ 12-1
Burkhardt
Akron, OH
225

35 ❑ 12-2
Burkhardt
Akron, OH
150

36 ❑ 12-2
Burger
Akron, OH
100

37 ❑ 12-1-OT
Wehle
West Haven, CT
700

38 ❑ 12-1-OT
Wehle
West Haven, CT
White Background
500

39 ❑ 12-1-OT
Wehle
West Haven, CT
Silver Background
400

40 ❑ 12-1-T
Wehle
West Haven, CT
250

1 ❑ 12-1-OT
Wehle
West Haven, CT
700

2 ❑ 12-2
Chr. Feigenspan
Newark, NJ
3

3 ❑ 12-2
Chr. Feigenspan
Newark, NJ
Metallic Gold
3

4 ❑ 12-2
Chr. Feigenspan
Newark, NJ
Dull Gold
3

5 ❑ 12-2
Pittsburgh
Pittsburgh, PA
10

6 ❑ 12-2
Metz
Pueblo, CO
60

7 ❑ 12-2
Walter
Pueblo, CO
50

8 ❑ 12-2
Metz
Omaha, NE
40

9 ❑ 12-2
My
Omaha, NE
40

10 ❑ 12-1-OT
Narragansett
Cranston, RI
150

11 ❑ 12-1-OT
Narragansett
Cranston, RI
150

12 ❑ 12-1-T
Narragansett
Cranston, RI
110

13 ❑ 12-1-T
Narragansett
Cranston, RI
110

14 ❑ 12-1-OT
Narragansett
Cranston, RI
225

15 ❑ 12-1-T
Narragansett
Cranston, RI
175

16 ❑ 12-1T
Narragansett
Cranston, RI
150

17 ❑ 12-1
Narragansett
Cranston, RI
150

18 ❑ 12-1-T
Narragansett
Cranston, RI
125

19 ❑ 12-1
Narragansett
Cranston, RI
80

20 ❑ 12-1
Narragansett
Cranston, RI
60

21 ❑ 12-1
Narragansett
Cranston, RI
30

22 ❑ 12-1
Narragansett
Cranston, RI
20

23 ❑ 12-1-OT
Narragansett
Cranston, RI
750

24 ❑ 12-1-T
Narragansett
Cranston, RI
350

25 ❑ 12-1-W
Narragansett
Cranston, RI
800

26 ❑ 12-1-T
Narragansett
Cranston, RI
65

27 ❑ 12-1
Narragansett
Cranston, RI
65

28 ❑ 12-1
Narragansett
Cranston, RI
18

29 ❑ 12-1
Narragansett
Cranston, RI
15

30 ❑ 12-2
Narragansett
Cranston, RI
15

31 ❑ 12-2
Narragansett
Cranston, RI
12

32 ❑ 12-2
National
Miami, FL
30

33 ❑ 12-2
National
Orlando, FL
28

34 ❑ 12-2
National
Orlando, FL
25

35 ❑ 12-2
National
Orlando, FL
20

36 ❑ 12-2
National
Orlando, FL
White Text
15

37 ❑ 12-2
National
Orlando, FL
White Text
15

38 ❑ 12-2
National
Orlando, FL
Silver Text
15

39 ❑ 12-2
National
Orlando, FL
350

40 ❑ 12-2
National
Orlando, FL
175

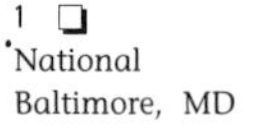

1 ❑ 12-1
National
Baltimore, MD
18

2 ❑ 12-2
National
Baltimore, MD
18

3 ❑ 12-1
National
Baltimore, MD
500

4 ❑ 12-1
National
Baltimore, MD
18

5 ❑ 12-1
National
Baltimore, MD
18

6 ❑ 12-1
National
Baltimore, MD
18

7 ❑ 12-2
National
Baltimore, MD
15

8 ❑ 12-2
National
Baltimore, MD
15

9 ❑ 12-2
National
Baltimore, MD
15

10 ❑ 12-2
National
Baltimore, MD
White Text 10

11 ❑ 12-2
National
Baltimore, MD
Met. Silver 10

12 ❑ 12-2
National
Baltimore, MD
Enam. Silver 10

13 ❑ Back like 12 12-2
National
Baltimore, MD
35

14 ❑ 12-2
National
Baltimore, MD
5

15 ❑ 12-2
National
Baltimore, MD
5

16 ❑ 12-1
National
Baltimore, MD
500

17 ❑ 12-2
National
Baltimore, MD
Gold Text 300

18 ❑ 12-2
National
Baltimore, MD
White Text 300

19 ❑ 12-2
National
Baltimore, MD
White Goat 150

20 ❑ 12-2
National
Baltimore, MD
Silver Goat 150

21 ❑ 12-1
National
Detroit, MI
200

22 ❑ 12-1
National
Detroit, MI
20

23 ❑ 12-1
National
Detroit, MI
20

24 ❑ 12-2
National
Detroit, MI
18

25 ❑ 12-2
National
Detroit, MI
18

26 ❑ 12-2
National
Detroit, MI
12

27 ❑ 12-2
Fischbach
St. Charles, MO
2

28 ❑ 12-1-T
Ambrosia
Chicago, IL
40

29 ❑ 12-1
Ambrosia
Chicago, IL
35

30 ❑ 12-1
Ambrosia
Chicago, IL
18

31 ❑ 12-1-0T
L. F. Neuweiler's Sons
Allentown, PA
Long Opener 275

32 ❑ 12-1-0T
L. F. Neuweiler's Sons
Allentown, PA
DULL GRAY 225

33 ❑ 12-1-0T
L. F. Neuweiler's Sons
Allentown, PA
225

34 ❑ 12-1-0
L. F. Neuweiler's Sons
Allentown, PA
200

35 ❑ 12-2
L. F. Neuweiler's Sons
Allentown, PA
75

36 ❑ 12-1-0T
L. F. Neuweiler's Sons
Allentown, PA
Long Opener 250

37 ❑ 12-1-0T
L. F. Neuweiler's Sons
Allentown, PA
190

38 ❑ 12-1-0
L. F. Neuweiler's Sons
Allentown, PA
175

39 ❑ 12-1
L. F. Neuweiler's Sons
Allentown, PA
175

40 ❑ 12-1
L. F. Neuweiler's Sons
Allentown, PA
15

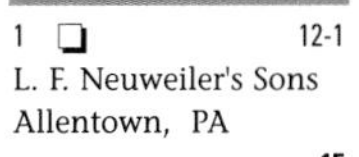

1 ❑ 12-1
L. F. Neuweiler's Sons
Allentown, PA
15

2 ❑ 12-2
L. F. Neuweiler's Sons
Allentown, PA
18

3 ❑ 12-2
L. F. Neuweiler's Sons
Allentown, PA
18

4 ❑ 12-2
L. F. Neuweiler's Sons
Allentown, PA
15

5 ❑ 12-1-0T
L. F. Neuweiler's Sons
Allentown, PA
1000+

6 ❑ 12-1-0T
Commonwealth
Springfield, MA
Yellow Shoreline 1000+

7 ❑ 12-1-0T
Commonwealth
Springfield, MA
White Shoreline 500

8 ❑ 12-1-0T
Commonwealth
Springfield, MA
500

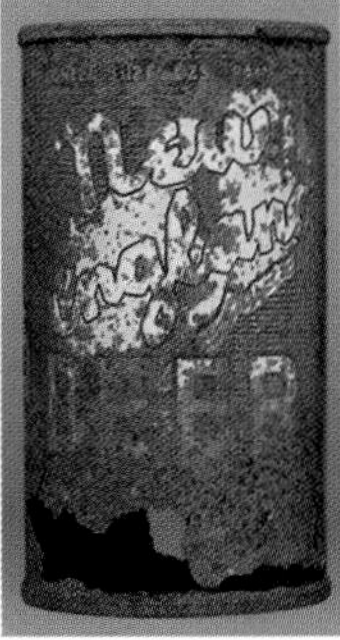

9 ❑ 12-1-0T
Commonwealth
Springfield, MA
1000+

10 ❑ 12-1
Cleveland-Sandusky
Cleveland, OH
300

11 ❑ 12-1-T
Greater New York
New York, NY
1000+

12 ❑ 12-1-T
Greater New York
New York, NY
1000+

13 ❑ 12-1-T
Greater New York
New York, NY
1000+

14 ❑ 12-2
Maier
Los Angeles, CA
1000+

15 ❑ 12-1
Gold Brau
Chicago, IL
65

16 ❑ 12-1
Atlas
Chicago, IL
60

17 ❑ 12-1
9-0-5
Chicago, IL
20

18 ❑ 12-1
9-0-5
Chicago, IL
20

19 ❑ 12-2
9-0-5
Chicago, IL
10

20 ❑ 12-2
9-0-5
Chicago, IL
10

21 ❑ 12-1
9-0-5
Chicago, IL
85

Side of 21

23 ❑ 12-2
Drewry's Ltd.
Chicago, IL
8

24 ❑ 12-2
Drewry's Ltd.
Chicago, IL
8

25 ❑ 12-1
Sterling
Evansville, IN
8

26 ❑ 12-2
9-0-5
South Bend, IN
10

27 ❑ 12-2
Drewry's Ltd.
South Bend, IN
10

28 ❑ 12-2
Drewry's Ltd.
South Bend, IN
8

29 ❑ 12-2
Drewry's Ltd.
South Bend, IN
8

30 ❑ 12-1
Pfeiffer
Detroit, MI
50

31 ❑ 12-1
Jacob Schmidt
St. Paul, MN
20

32 ❑ 12-1
Jacob Schmidt
St. Paul, MN
20

33 ❑ 12-1
Associated, DBA Jac. Sch.
St. Paul, MN
20

34 ❑ 12-2
Northern
Superior, WI
15

35 ❑ 12-2
Northern
Superior, WI
25

36 ❑ 12-2
Northern
Superior, WI
15

37 ❑ 12-2
Regional
Willimansett, MA
40

38 ❑ 12-1-0T
Grace Bros.
Santa Rosa, CA
1000+

39 ❑ 12-1-0T
Grace Bros.
Santa Rosa, CA
1000+

40 ❑ 12-2-T
Globe
San Francisco, CA
1000+

1 ❑ 12-2
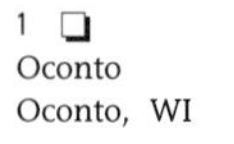
Oconto
Oconto, WI
8

2 ❑ 12-2
Oertel
Louisville, KY
22

3 ❑ 12-2
Oertel
Louisville, KY
22

4 ❑ 12-2
Oertel
Louisville, KY
22

5 ❑ 12-2
Oertel
Louisville, KY
Metallic
15

6 ❑ 12-2
Oertel
Louisville, KY
Enamel
15

7 ❑ 12-2
Oertel
Louisville, KY
35

8 ❑ 12-2
Oertel
Louisville, KY
35

9 ❑ 12-1-OT
Salem Assn.
Salem, OR
1000+

10 ❑ 12-1-OT
Salem Assn.
Salem, OR
DNCMT 4%
1000+

11 ❑ 12-1
Metropolis
Trenton, NJ
20

12 ❑ 12-1
Harvard
Lowell, MA
55

13 ❑ 12-1
Harvard
Lowell, MA
65

14 ❑ 12-1
Harvard
Lowell, MA
175

15 ❑ 12-1
Harvard
Lowell, MA
125

16 ❑ 12-1
Eastern Brewing
Hammonton, NJ
85

17 ❑ 12-1
Eastern Brewing
Hammonton, NJ
85

18 ❑ 12-2
Eastern Brewing
Hammonton, NJ
25

19 ❑ 12-2
Eastern Brewing
Hammonton, NJ
5

20 ❑ 12-2
Eastern Brewing
Hammonton, NJ
5

21 ❑ 12-2
Eastern Brewing
Hammonton, NJ
5

22 ❑ 12-1
Eastern Beverage
Hammonton, NJ
50

23 ❑ 12-1
Eastern Brewing
Hammonton, NJ
45

24 ❑ 12-2
Eastern Brewing
Hammonton, NJ
20

25 ❑ 12-2
Eastern Brewing
Hammonton, NJ
5

26 ❑ 12-2
Eastern Brewing
Hammonton, NJ
3

27 ❑ 12-1
Eastern Brewing
Hammonton, NJ
90

28 ❑ 12-2
Eastern Brewing
Hammonton, NJ
40

29 ❑ 12-2
Eastern Brewing
Hammonton, NJ
7

30 ❑ 12-2
Eastern Brewing
Hammonton, NJ
60

31 ❑ 12-2
Old Bohemian
Trenton, NJ
35

32 ❑ 12-1-OT
Forest City
Cleveland, OH
450

33 ❑ 12-1-OT
Forest City
Cleveland, OH
225

34 ❑ 12-1-OT
Forest City
Cleveland, OH
175

35 ❑ 12-1
Mount Carbon
Pottsville, PA
300

36 ❑ 12-2
Oconto
Oconto, WI
30

37 ❑ 12-1-OT
Centlivre
Fort Wayne, IN
45

38 ❑ 12-1-OT
Centlivre
Fort Wayne, IN
45

39 ❑ 12-1-O
Centlivre
Fort Wayne, IN
45

40 ❑ 12-2
Centlivre
Fort Wayne, IN
Lt. Blue
75

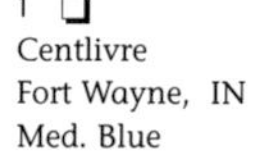

1 ❑ 12-2
Centlivre
Fort Wayne, IN
Med. Blue 75

2 ❑ 12-2
Centlivre
Fort Wayne, IN
Copper 75

3 ❑ 12-2
Centlivre
Fort Wayne, IN
Yellow 75

4 ❑ 12-2
Centlivre
Fort Wayne, IN
Green 75

5 ❑ 12-2
Centlivre
Fort Wayne, IN
Dk. Red 75

6 ❑ 12-2
Centlivre
Fort Wayne, IN
Med. Purple 75

7 ❑ 12-2
Centlivre
Fort Wayne, IN
Dk. Purple 75

8 ❑ 12-2
Centlivre
Fort Wayne, IN
Lazy Aged on Side 75

9 ❑ 12-2
Centlivre
Fort Wayne, IN
Blue 60

10 ❑ 12-2
Centlivre
Fort Wayne, IN
Green 40

11 ❑ 12-2
Centlivre
Fort Wayne, IN
Pink 60

12 ❑ 12-2
Centlivre
Fort Wayne, IN
Orange 60

13 ❑ 12-2
Centlivre
Fort Wayne, IN
Purple 60

14 ❑ 12-2
Centlivre
Fort Wayne, IN
Red 60

15 ❑ 12-1-0T
Centlivre
Fort Wayne, IN
55

16 ❑ 12-1-0T
Centlivre
Fort Wayne, IN
55

17 ❑ 12-1-0
Centlivre
Fort Wayne, IN
55

18 ❑ 12-2
Centlivre
Fort Wayne, IN
12

19 ❑ 12-2
Centlivre
Fort Wayne, IN
200

20 ❑ 12-2
Centlivre
Fort Wayne, IN
30

21 ❑ 12-2
Old Crown
Fort Wayne, IN
12

22 ❑ 12-2
Old Crown
Fort Wayne, IN
5

23 ❑ 12-2
Old Crown
Fort Wayne, IN
20

24 ❑ 12-2
Maier
Los Angeles, CA
20

25 ❑ 12-2
Maier
Los Angeles, CA
20

26 ❑ 12-2
Maier
Los Angeles, CA
20

27 ❑ 12-1-0T
Old Dutch
Brooklyn, NY
Enamel 600

28 ❑ 12-1-0T
Old Dutch
Brooklyn, NY
Metallic 600

29 ❑ 12-1-0T
Old Dutch
Brooklyn, NY
800

30 ❑ 12-1-0T
Old Dutch
Brooklyn, NY
800

31 ❑ 12-1-0T
Old Dutch
Brooklyn, NY
800

32 ❑ 12-1-0T
Old Dutch
Brooklyn, NY
800

33 ❑ 12-1-0T
Old Dutch
Brooklyn, NY
800

34 ❑ 12-1-0T
Old Dutch
Brooklyn, NY
Enamel 275

35 ❑ 12-1-0T
Old Dutch
Brooklyn, NY
Metallic 200

36 ❑ 12-1-0W
Old Dutch
Brooklyn, NY
800

37 ❑ 12-1-T
Old Dutch
Brooklyn, NY
1000+

38 ❑ 12-1-T
Metropolis
New York, NY
50

39 ❑ 12-1
Metropolis
Trenton, NJ
40

40 ❑ 12-2
Hornell
Hornell, NY
40

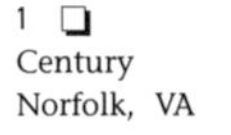

1 ❑ 12-1
Century
Norfolk, VA
45

2 ❑ 12-1
International
Findlay, OH
25

3 ❑ 12-1
International
Findlay, OH
20

4 ❑ 12-1
Krantz
Findlay, OH
18

5 ❑ 12-2
Eagle
Catasauqua, PA
90

6 ❑ 12-2
Eagle
Catasauqua, PA
65

7 ❑ 12-2
Eagle
Catasauqua, PA
45

8 ❑ 12-2
Eagle
Catasauqua, PA
55

9 ❑ 12-1
Fox Head
Waukesha, WI
175

10 ❑ 12-2
Cumberland
Cumberland, MD
60

11 ❑ 12-2
Cumberland
Cumberland, MD
Lg. Contents
35

12 ❑ 12-2
Cumberland
Cumberland, MD
Sm. Contents
35

13 ❑ 12-2
Cumberland
Cumberland, MD
35

14 ❑ 12-2
Cumberland
Cumberland, MD
8

15 ❑ 12-1
Chr. Heurich
Washington, DC
175

16 ❑ 12-1
Chr. Heurich
Washington, DC
Lt. Brown
55

17 ❑ 12-1
Chr. Heurich
Washington, DC
55

18 ❑ 12-1-T
North Bay
Santa Rosa, CA
275

19 ❑ 12-1
North Bay
Santa Rosa, CA
275

20 ❑ 12-2
Grace Bros.
Santa Rosa, CA
35

21 ❑ 12-2
Grace Bros.
Santa Rosa, CA
20

22 ❑ 12-2
Grace Bros.
Santa Rosa, CA
20

23 ❑ 12-2
Grace Bros.
Santa Rosa, CA
20

24 ❑ 12-2
Maier
Los Angeles, CA
20

25 ❑ 12-1
Renner
Fort Wayne, IN
7

26 ❑ 12-2
Queen City
Cumberland, MD
30

27 ❑ 12-2
Queen City
Cumberland, MD
10

28 ❑ 12-2
Queen City
Cumberland, MD
10

29 ❑ 12-2
Queen City
Cumberland, MD
10

30 ❑ 12-2
Queen City
Cumberland, MD
10

31 ❑ 12-2
Queen City
Cumberland, MD
1901 in Label
10

32 ❑ 12-2
Queen City
Cumberland, MD
No Date in Label
10

33 ❑ 12-2
Queen City
Cumberland, MD
5

34 ❑ 12-1
Colonial
Hammonton, NJ
60

35 ❑ 12-2
Colonial
Hammonton, NJ
3

36 ❑ 12-2
Eastern Brewing Corp.
Hammonton, NJ
60

37 ❑ 12-1
Eagle
Catasauqua, PA
40

38 ❑ 12-1
Lebanon Valley
Lebanon, PA
40

39 ❑ 12-1
Lebanon Valley
Lebanon, PA
40

40 ❑ 12-2
Maier
Los Angeles, CA
25

1 ❑ 12-2
Tivoli
Denver, CO
55

2 ❑ 12-2
Ph. Schneider
Trinidad, CO
45

3 ❑ 12-2-T
St. Claire
San Jose, CA
1000+

4 ❑ 12-2
Grace Bros.
Santa Rosa, CA
275

5 ❑ 12-2
Grace Bros.
Santa Rosa, CA
275

6 ❑ 12-1-OT
Manhattan
Chicago, IL
600

7 ❑ 12-1-OT
Manhattan
Chicago, IL
600

8 ❑ 12-1-OT
Food City
Battle Creek, MI
650

9 ❑ 12-2
Maier
Los Angeles, CA
6

10 ❑ 12-2
Maier
Los Angeles, CA
6

11 ❑ 12-1
Cremo
New Britain, CT
350

12 ❑ 12-2
Hull
New Haven, CT
125

13 ❑ 12-2
Lebanon Valley
Lebanon, PA
350

14 ❑ 12-2
Jos. Schlitz
Tampa, FL
25

15 ❑ 12-2
Jos. Schlitz
Tampa, FL
20

16 ❑ 12-2
Jos. Schlitz
Tampa, FL
7

17 ❑ 12-2
Jos. Schlitz
Tampa, FL
7

18 ❑ 12-2
Jos. Schlitz
Tampa, FL
10

19 ❑ 12-2
Jos. Schlitz
Kansas City, MO
30

20 ❑ 12-1-T
Jos. Schlitz
Milwaukee, WI
375

21 ❑ 12-1-T
Jos. Schlitz
Milwaukee, WI
350

22 ❑ 12-1-T
Jos. Schlitz
Milwaukee, WI
350

23 ❑ 12-1-T
Jos. Schlitz
Milwaukee, WI
55

24 ❑ 12-1-T
Jos. Schlitz
Milwaukee, WI
45

25 ❑ 12-2
Jos. Schlitz
Milwaukee, WI
175

26 ❑ 12-2
Jos. Schlitz
Milwaukee, WI
35

27 ❑ 12-2
Jos. Schlitz
Milwaukee, WI
55

28 ❑ 12-2
Jos. Schlitz
Milwaukee, WI
150

29 ❑ 12-2
Jos. Schlitz
Milwaukee, WI
20

30 ❑ 12-2
Jos. Schlitz
Milwaukee, WI
5

31 ❑ 12-2
Jos. Schlitz
Milwaukee, WI
5

32 ❑ 12-2
Jos. Schlitz
Milwaukee, WI
7

33 ❑ 12-2
Jos. Schlitz
Milwaukee, WI
7

34 ❑ 12-2
Jos. Schlitz
Milwaukee, WI
8

35 ❑ 12-2
Jos. Schlitz
Milwaukee, WI
8

36 ❑ 12-2
Los Angeles
Los Angeles, CA
100

37 ❑ 12-2
Pabst
Los Angeles, CA
100

38 ❑ 12-1
Hornell
Trenton, NJ
75

39 ❑ 12-2
Hornell
Hornell, NY
550

40 ❑ 12-1
Hornell
Hornell, NY
75

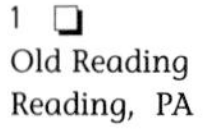

1 ❑ 12-2
Old Reading
Reading, PA
125

2 ❑ 12-2
Old Reading
Reading, PA
110

3 ❑ 12-2
Old Reading
Reading, PA
80

4 ❑ 12-1-OT
ABC
St. Louis, MO
1000+

5 ❑ 12-1-OT
ABC
St. Louis, MO
1000+

6 ❑ 12-2
Lami
St. Louis, MO
85

7 ❑ 12-2
Lami
St. Louis, MO
85

8 ❑ 12-2
Lami
St. Louis, MO
85

9 ❑ 12-1
G. Heileman
La Crosse, WI
20

10 ❑ 12-1
G. Heileman
La Crosse, WI
30

11 ❑ 11-2
G. Heileman
La Crosse, WI
75

12 ❑ 11-2
G. Heileman
La Crosse, WI
75

13 ❑ 12-2
G. Heileman
La Crosse, WI
18

14 ❑ 12-2
G. Heileman
La Crosse, WI
18

15 ❑ 12-2
G. Heileman
La Crosse, WI
18

16 ❑ 12-2
G. Heileman
La Crosse, WI
12

17 ❑ 12-2
G. Heileman
La Crosse, WI
35

18 ❑ 12-2
G. Heileman
La Crosse, WI
Green
18

19 ❑ 12-2
G. Heileman
La Crosse, WI
Blue
18

20 ❑ 12-2
G. Heileman
La Crosse, WI
25

21 ❑ 12-2
G. Heileman
La Crosse, WI
10

22 ❑ 12-2
G. Heileman
La Crosse, WI
3

23 ❑ 12-1
Enterprise
Fall River, MA
90

24 ❑ 12-2
Enterprise
Fall River, MA
50

25 ❑ 12-2
Warsaw
Warsaw, IL
12

26 ❑ 12-1
Cleveland-Sandusky
Cumberland, MD
50

27 ❑ 12-1
Cleveland-Sandusky
Buffalo, NY
65

28 ❑ 12-1
Cleveland-Sandusky
Cleveland, OH
50

29 ❑ 12-1
West Bend Lithia
West Bend, WI
Dark Blue
8

30 ❑ 12-1
West Bend Lithia
West Bend, WI
Light Blue
8

31 ❑ 12-2
Jax Ice
Jacksonville, FL
400

32 ❑ 12-2
Pacific
Oakland, CA
65

33 ❑ 12-2
Grace Bros.
Santa Rosa, CA
35

34 ❑ 12-2
Grace Bros.
Santa Rosa, CA
35

35 ❑ 12-2
Old Vienna
Chicago, IL
Metallic
18

36 ❑ 12-2
Old Vienna
Chicago, IL
Enamel
18

37 ❑ 12-1-OT
Manhattan
Chicago, IL
750

38 ❑ 11-1
Br. Corp. of Ore.
Portland, OR
45

39 ❑ 11-1
Blitz-Weinhard
Portland, OR
55

40 ❑ 12-1
Bohemian
Spokane, WA
45

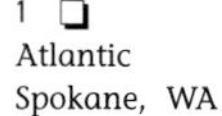
1 ❑ 12-1
Atlantic
Spokane, WA
45

2 ❑ 11-1
Bohemian
Spokane, WA
45

3 ❑ 12-1
Bohemian
Spokane, WA
45

4 ❑ 12-2
Maier
Los Angeles, CA
8

5 ❑ 12-2
Century
Norfolk, VA
135

6 ❑ 12-1
Mountain
Roanoke, VA
135

7 ❑ 12-1
Olympia
Olympia, WA
5

8 ❑ 12-2
Olympia
Olympia, WA
5

9 ❑ 11-2
Olympia
Olympia, WA
4

10 ❑ 12-2
Olympia
Olympia, WA
5

11 ❑ 12-2
A. Gettelman
Milwaukee, WI
30

12 ❑ 12-2
A. Gettelman
Milwaukee, WI
30

13 ❑ 12-2
A. Gettelman
Milwaukee, WI
18

14 ❑ 12-2
Gettelman, Div. of Miller
Milwaukee, WI
35

15 ❑ 12-2
Gettelman, Div. of Miller
Milwaukee, WI
18

16 ❑ 12-2
Orbit
Miami, FL
55

17 ❑ 12-2
Orbit
Tampa, FL
45

18 ❑ 12-1
Henry F. Ortlieb
Philadelphia, PA
35

19 ❑ 12-2
Henry F. Ortlieb
Philadelphia, PA
Enamel
40

20 ❑ 12-2
Henry F. Ortlieb
Philadelphia, PA
Metallic
35

21 ❑ 12-2
Henry F. Ortlieb
Philadelphia, PA
Pink Hops
40

22 ❑ 12-1
Henry F. Ortlieb
Philadelphia, PA
12

23 ❑ 12-1
Henry F. Ortlieb
Philadelphia, PA
15

24 ❑ 12-1-OT
Wehle
West Haven, CT
Long Opener
700

25 ❑ 12-1-OT
Wehle
West Haven, CT
700

26 ❑ 12-2
Pabst
Los Angeles, CA
12

27 ❑ 12-2
Pabst
Los Angeles, CA
12

28 ❑ 12-2
Pabst
Los Angeles, CA
12

29 ❑ 12-2
Los Angeles
Los Angeles, CA
Subsidiary of Pabst
20

30 ❑ 12-2
Pabst
Los Angeles, CA
25

31 ❑ 12-2
Pabst
Los Angeles, CA
15

32 ❑ 12-2
Pabst
Los Angeles, CA
7

33 ❑ 12-2
Pabst
Los Angeles, CA
7

34 ❑ 12-2
Pabst
Los Angeles, CA
10

35 ❑ 12-2
Pabst
Los Angeles, CA
45

36 ❑ 12-2
Pabst
Los Angeles, CA
60

37 ❑ 12-2
Pabst
Los Angeles, CA
12

38 ❑ 12-1-OT
Premier-Pabst
Peoria Heights, IL
250

39 ❑ 12-1-OT
Pabst
Peoria Heights, IL
250

40 ❑ 12-1-T
Pabst
Peoria Heights, IL
450

1 ❑ 12-2
Pabst
Peoria Heights, IL
10

2 ❑ 12-1-OT
Pabst
Peoria Heights, IL
25

3 ❑ 12-1-OT
Premier-Pabst
Peoria Heights, IL
25

4 ❑ 12-1-OT
Pabst
Peoria Heights, IL
25

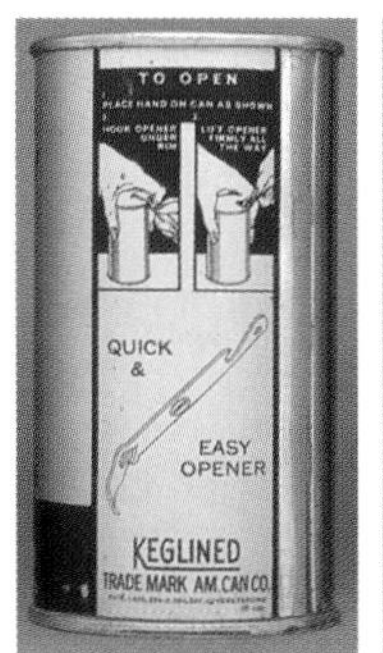

Side of 4

6 ❑ 12-1-T
Pabst
Peoria Heights, IL
96 Years
18

7 ❑ 12-1-T
Pabst
Peoria Heights, IL
97 Years
18

8 ❑ 12-1-W
Pabst
Peoria Heights, IL
550

9 ❑ 12-1-W
Pabst
Peoria Heights, IL
400

10 ❑ 12-1-T
Pabst
Peoria Heights, IL
15

11 ❑ 12-1
Pabst
Peoria Heights, IL
15

12 ❑ 12-2
Pabst
Peoria Heights, IL
12

13 ❑ 12-2
Pabst
Peoria Heights, IL
7

14 ❑ 12-2
Pabst
Peoria Heights, IL
7

15 ❑ 12-2
Pabst
Peoria Heights, IL
7

16 ❑ 12-2
Pabst
Peoria Heights, IL
8

17 ❑ 12-2
Pabst
Peoria Heights, IL
8

18 ❑ 12-2
Pabst
Peoria Heights, IL
3

19 ❑ 12-2
Pabst
Peoria Heights, IL
3

20 ❑ 12-1-OT
Premier-Pabst
Peoria Heights, IL
400

21 ❑ 12-2
Pabst
Peoria Heights, IL
85

22 ❑ 12-2
Pabst
Peoria Heights, IL
45

23 ❑ 12-2
Pabst
Peoria Heights, IL
60

24 ❑ 12-2
Pabst
Peoria Heights, IL
10

25 ❑ 12-1-T
Pabst
Newark, NJ
18

26 ❑ 12-2
Pabst
Newark, NJ
12

27 ❑ 12-2
Pabst
Newark, NJ
8

28 ❑ 12-2
Pabst
Newark, NJ
20

29 ❑ 12-2
Pabst
Newark, NJ
10

30 ❑ 12-2
Pabst
Newark, NJ
3

31 ❑ 12-2
Pabst
Newark, NJ
85

32 ❑ 12-2
Pabst
Newark, NJ
60

33 ❑ 12-2
Pabst
Newark, NJ
9

34 ❑ 12-1-OT
Premier-Pabst
Milwaukee, WI
250

35 ❑ 12-1-OT
Premier-Pabst
Milwaukee, WI
250

36 ❑ 12-1-OT
Premier-Pabst
Milwaukee, WI
500

37 ❑ 12-1-OT
Pabst
Milwaukee, WI
250

38 ❑ 12-1-OW
Pabst
Milwaukee, WI
250

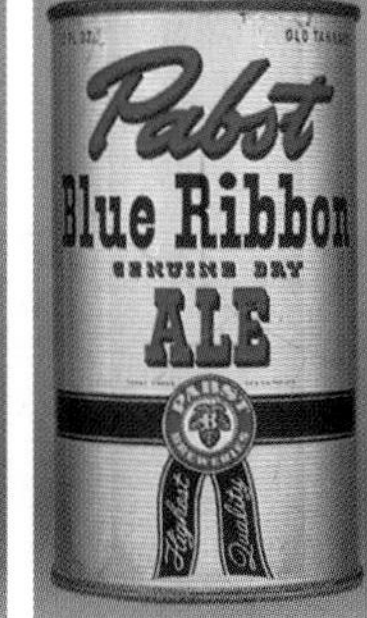

39 ❑ 12-1-T
Pabst
Milwaukee, WI
450

40 ❑ 12-2
Pabst
Milwaukee, WI
70

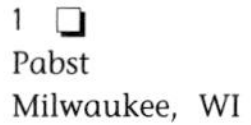

1 ❑ 12-2
Pabst
Milwaukee, WI
70

2 ❑ 12-2
Pabst
Milwaukee, WI
70

3 ❑ 12-2
Pabst
Milwaukee, WI
85

4 ❑ 12-2
Pabst
Milwaukee, WI
10

5 ❑ 12-2
Pabst
Milwaukee, WI
10

6 ❑ 12-2
Pabst
Milwaukee, WI
Enamel
7

7 ❑ 12-1-OT
Premier-Pabst
Milwaukee, WI
110

8 ❑ 12-1-OT
Premier- Pabst
Milwaukee, WI
110

9 ❑ 12-1-OT
Premier-Pabst
Milwaukee, WI
75

10 ❑ 12-1-OT
Premier-Pabst
Milwaukee, WI
110

11 ❑ 12-1-OT
Premier-Pabst
Milwaukee, WI
60

12 ❑ 12-1-OT
Premier-Pabst
Milwaukee, WI
60

13 ❑ 12-1-OT
Premier-Pabst
Milwaukee, WI
25

14 ❑ 12-1-OT
Premier-Pabst
Milwaukee, WI
25

15 ❑ 12-1-OT
Premier- Pabst
Milwaukee, WI
25

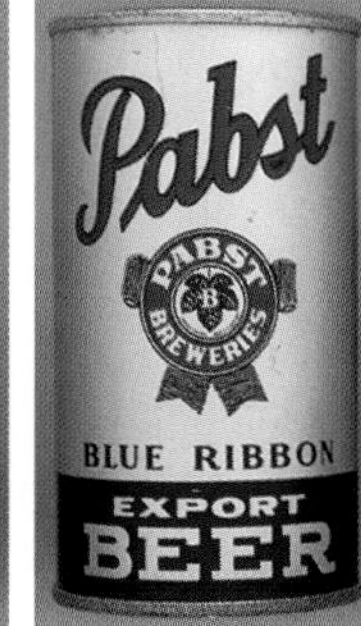

16 ❑ 12-1-OT
Pabst
Milwaukee, WI
25

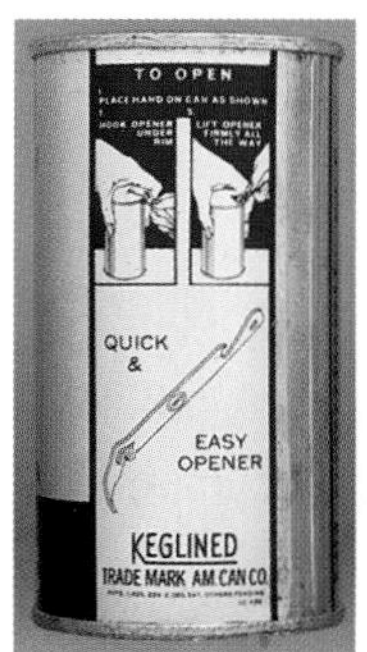

Side of 16

Front of 19, 20, 21

19 ❑ 12-1-T
Pabst
Milwaukee, WI
18

20 ❑ 12-1-T
Pabst
Milwaukee, WI
25

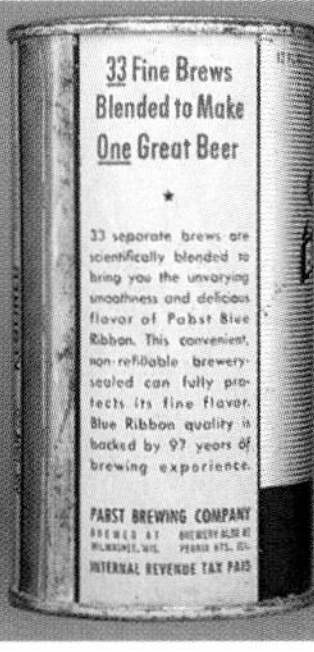

21 ❑ 12-1-T
Pabst
Milwaukee, WI
18

22 ❑ 12-1-W
Pabst
Milwaukee, WI
550

23 ❑ 12-1-W
Pabst
Milwaukee, WI
400

Front of 25, 26

25 ❑ 12-1-W
Pabst
Milwaukee, WI
17

26 ❑ 12-1-W
Pabst
Milwaukee, WI
17

Front of 28, 29

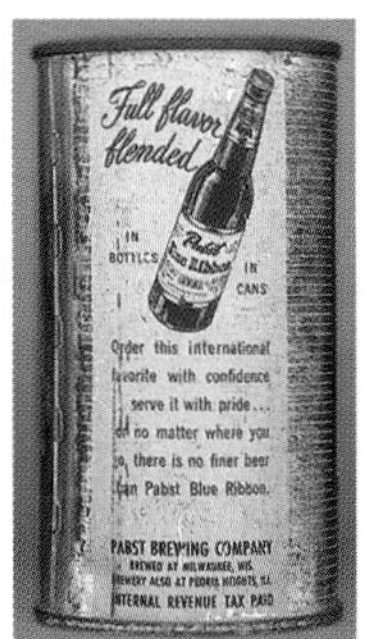

28 ❑ 12-1-T
Pabst
Milwaukee, WI
15

29 ❑ 12-1-T
Pabst
Milwaukee, WI
15

30 ❑ 12-1
Pabst
Milwaukee, WI
15

31 ❑ 12-2
Pabst
Milwaukee, WI
12

32 ❑ 12-2
Pabst
Milwaukee, WI
7

33 ❑ 12-2
Pabst
Milwaukee, WI
7

34 ❑ 12-2
Pabst
Milwaukee, WI
7

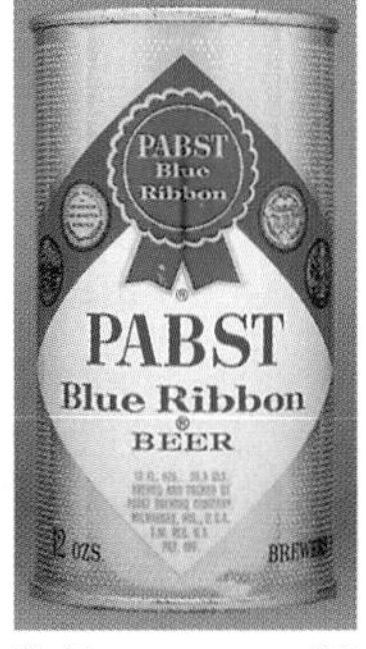

35 ❑ 12-2
Pabst
Milwaukee, WI
Test Can
250

36 ❑ 12-2
Pabst
Milwaukee, WI
8

37 ❑ 10-2
Pabst
Milwaukee, WI
8

38 ❑ 12-2
Pabst
Milwaukee, WI
15

39 ❑ 10-2
Pabst
Milwaukee, WI
8

40 ❑ 12-2
Pabst
Milwaukee, WI
3

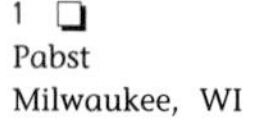

1 ❑ 12-2
Pabst
Milwaukee, WI
3

2 ❑ 12-2
Pabst
Milwaukee, WI
3

3 ❑ 12-1-0T
Premier-Pabst
Milwaukee, WI
350

4 ❑ 12-1-0T
Pabst
Milwaukee, WI
350

5 ❑ 12-1-0T
Pabst
Milwaukee, WI
400

6 ❑ 12-2
Pabst
Milwaukee, WI
85

7 ❑ 12-2
Pabst
Milwaukee, WI
45

8 ❑ 12-2
Pabst
Milwaukee, WI
60

9 ❑ 12-2
Pabst
Milwaukee, WI
7

10 ❑ 12-1-T
Rainier
San Francisco, CA
500

11 ❑ 12-2
Maier
Los Angeles, CA
10

12 ❑ 12-2
Maier
Los Angeles, CA
10

13 ❑ 12-2
Maier
Los Angeles, CA
10

14 ❑ 12-2
Maier
Los Angeles, CA
6

15 ❑ 12-1-T
Monarch
Los Angeles, CA
1000+

16 ❑ 12-2
Fuhrmann & Schmidt
Shamokin, PA
350

17 ❑ 12-2
Hofbrau
Allentown, PA
3

18 ❑ 12-1
Peter Fox
Chicago, IL
200

19 ❑ 12-1-T
Fox Deluxe
Grand Rapids, MI
175

20 ❑ 12-1
Fox Deluxe
Grand Rapids, MI
600

21 ❑ 12-1
Fox Deluxe
Grand Rapids, MI
225

22 ❑ 12-1
Fox Deluxe
Grand Rapids, MI
225

23 ❑ 12-2
Paul Bunyan
Waukesha, WI
100

24 ❑ 12-2
Paul Bunyan
Waukesha, WI
'Country'
100

25 ❑ 12-2
Paul Bunyan Div.
Waukesha, WI
'State' Black Trim
100

26 ❑ 12-2
Paul Bunyan Div.
Waukesha, WI
'State' Green Trim
100

27 ❑ 12-2
Class "A"
Trenton, NJ
125

28 ❑ 12-2
Class "A"
Trenton, NJ
10

29 ❑ 12-2
Horlacher
Allentown, PA
4

30 ❑ 12-2
Wm. Gretz
Philadelphia, PA
15

31 ❑ 12-2
Esslinger's
Philadelphia, PA
15

32 ❑ 12-2
Pearl
St. Joseph, MO
25

33 ❑ 12-2
Pearl
St. Joseph, MO
12

34 ❑ 12-1
San Antonio Assn
San Antonio, TX
35

35 ❑ 12-1
San Antonio Assn
San Antonio, TX
35

36 ❑ 12-1
San Antonio Assn
San Antonio, TX
35

37 ❑ 12-1
San Antonio Assn
San Antonio, TX
35

38 ❑ 12-2
Pearl
San Antonio, TX
30

39 ❑ 12-2
Pearl
San Antonio, TX
18

40 ❑ 12-2
Pearl
San Antonio, TX
25

FLATS

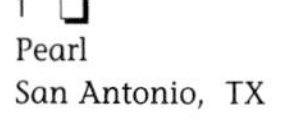
1 ❑ 12-2
Pearl
San Antonio, TX
20

2 ❑ 12-2
Pearl
San Antonio, TX
18

3 ❑ 12-2
Pearl
San Antonio, TX
20

4 ❑ 12-2
Pearl
San Antonio, TX
8

5 ❑ 12-1
Potosi
Potosi, WI
75

6 ❑ 12-1
Horlacher
Allentown, PA
700

7 ❑ 12-2
Colonial
Hammonton, NJ
65

8 ❑ 12-2
People's
Oshkosh, WI
8

9 ❑ 12-1
Horlacher
Allentown, PA
18

10 ❑ 12-1-OT
Peter Doelger
Harrison, NJ
1000+

11 ❑ 12-1-OT
Peter Doelger
Harrison, NJ
400

12 ❑ 12-1-W
Peter Doelger
Harrison, NJ
400

13 ❑ 12-1-T
Peter Doelger
Harrison, NJ
400

14 ❑ 12-1-OT
Peter Doelger
Harrison, NJ
1000+

15 ❑ 12-1-OT
Peter Hand
Chicago, IL
400

Front of 17-22

17 ❑ 12-1
Peter Hand
Chicago, IL
90

18 ❑ 12-1
Peter Hand
Chicago, IL
90

19 ❑ 12-1
Peter Hand
Chicago, IL
90

20 ❑ 12-1
Peter Hand
Chicago, IL
90

21 ❑ 12-1
Peter Hand
Chicago, IL
90

22 ❑ 12-1
Peter Hand
Chicago, IL
90

Front of 24-33

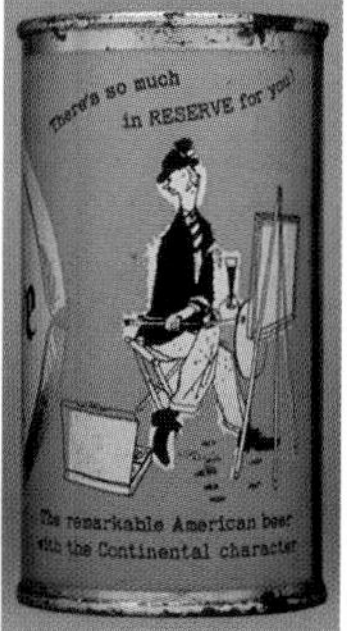

24 ❑ 12-1
Peter Hand
Chicago, IL
80

25 ❑ 12-1
Peter Hand
Chicago, IL
80

26 ❑ 12-1
Peter Hand
Chicago, IL
80

27 ❑ 12-1
Peter Hand
Chicago, IL
80

28 ❑ 12-1
Peter Hand
Chicago, IL
80

29 ❑ 12-1
Peter Hand
Chicago, IL
80

30 ❑ 12-1
Peter Hand
Chicago, IL
80

31 ❑ 12-1
Peter Hand
Chicago, IL
80

32 ❑ 12-1
Peter Hand
Chicago, IL
80

33 ❑ 12-1
Peter Hand
Chicago, IL
80

34 ❑ 12-1
Peter Hand
Chicago, IL
35

35 ❑ 12-1
Peter Hand
Chicago, IL
30

36 ❑ 12-2
Associated
Evansville, IN
10

37 ❑ 12-2
Pfeiffer
South Bend, IN
10

38 ❑ 12-1-W
Pfeiffer
Detroit, MI
225

39 ❑ 12-1-T
Pfeiffer
Detroit, MI
12

40 ❑ 12-1
Pfeiffer
Detroit, MI
8

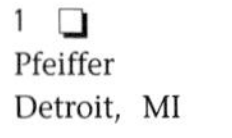

1 ❑ 12-1
Pfeiffer
Detroit, MI
8

2 ❑ 12-1
Pfeiffer
Detroit, MI
8

3 ❑ 12-2
Pfeiffer
Detroit, MI
10

4 ❑ 12-2
Pfeiffer
Detroit, MI
10

5 ❑ 12-2
Pfeiffer
Detroit, MI
10

6 ❑ 12-1
Pfeiffer
Detroit, MI
Metallic
90

7 ❑ 12-1
Pfeiffer
Detroit, MI
Metallic
90

8 ❑ 12-1
Pfeiffer
Detroit, MI
Metallic
80

9 ❑ 12-1
Pfeiffer
Detroit, MI
Metallic
90

10 ❑ 12-1
Pfeiffer
Detroit, MI
Metallic
80

11 ❑ 12-1
Pfeiffer
Detroit, MI
Metallic
90

12 ❑ 12-1
Pfeiffer
Detroit, MI
Metallic
80

13 ❑ 12-1
Pfeiffer
Detroit, MI
Metallic
90

14 ❑ 12-1
Pfeiffer
Detroit, MI
Metallic
90

15 ❑ 12-1
Pfeiffer
Detroit, MI
Dull
90

16 ❑ 12-1
Pfeiffer
Detroit, MI
Dull
90

17 ❑ 12-1
Pfeiffer
Detroit, MI
Dull
80

18 ❑ 12-1
Pfeiffer
Detroit, MI
Dull
90

19 ❑ 12-1
Pfeiffer
Detroit, MI
Dull
80

20 ❑ 12-1
Pfeiffer
Detroit, MI
Dull
90

21 ❑ 12-1
Pfeiffer
Detroit, MI
Dull
80

22 ❑ 12-1
Pfeiffer
Detroit, MI
Dull
90

23 ❑ 12-1
Pfeiffer
Detroit, MI
Dull
90

24 ❑ 12-2
Pfeiffer
Detroit, MI
10

25 ❑ 12-2
Pfeiffer
Detroit, MI
10

26 ❑ 12-2
Pfeiffer
Detroit, MI
10

27 ❑ 12-2
Pfeiffer
Detroit, MI
12

28 ❑ 12-1
Pfeiffer
Detroit, MI
150

29 ❑ 12-2
Pfeiffer
Detroit, MI
15

30 ❑ 12-2
Pfeiffer
St. Paul, MN
12

31 ❑ 12-2
Pfeiffer
St. Paul, MN
12

32 ❑ 12-2
Associated, DBA Jac. Sch.
St. Paul, MN
12

33 ❑ 12-2
G. Heileman
La Crosse, WI
6

34 ❑ 12-2
Phoenix
Tampa, FL
90

35 ❑ 12-2
International
Tampa, FL
125

36 ❑ 12-2
Phoenix
Buffalo, NY
60

37 ❑ 12-2
International
Buffalo, NY
60

38 ❑ 12-1-T
Haffenreffer
Boston, MA
110

39 ❑ 12-1
Haffenreffer
Boston, MA
125

40 ❑ 12-1
Haffenreffer
Boston, MA
300

1 ❑ 12-1
Haffenreffer
Boston, MA
300

2 ❑ 12-2
Haffenreffer
Boston, MA
35

3 ❑ 12-2
Haffenreffer
Boston, MA
35

4 ❑ 12-1
Haffenreffer
Boston, MA
175

5 ❑ 12-1
Haffenreffer
Boston, MA
275

6 ❑ 12-1
Haffenreffer
Boston, MA
275

7 ❑ 12-1
Haffenreffer
Boston, MA
90

8 ❑ 12-1
Haffenreffer
Boston, MA
175

9 ❑ 12-2
Piel Bros.
Willimansett, MA
15

10 ❑ 12-2
Piel Bros.
Willimansett, MA
15

11 ❑ 12-2
Piel Bros.
Willimansett, MA
12

12 ❑ 12-2
Piel Bros.
Willimansett, MA
7

13 ❑ 12-1-T
Piel Bros.
New York, NY
35

14 ❑ 12-1
Piel Bros.
New York, NY
18

15 ❑ 12-1
Piel Bros.
Brooklyn, NY
18

16 ❑ 12-1-T
Piel's, Inc.
Stapleton, NY
40

17 ❑ 12-1
Piel's, Inc.
Staten Island, NY
30

18 ❑ 12-1
Piel Bros.
Brooklyn, NY
12

19 ❑ 12-2
Piel Bros.
Brooklyn, NY
12

20 ❑ 12-1
Piel Bros.
Staten Island, NY
12

21 ❑ 12-2
Piel Bros.
Brooklyn, NY
10

22 ❑ 12-2
Piel Bros.
Brooklyn, NY
10

23 ❑ 12-2
Piel Bros.
Staten Island, NY
10

24 ❑ 12-2
Piel Bros.
Staten Island, NY
10

25 ❑ 12-2
Piel Bros.
Brooklyn, NY
4

26 ❑ 12-2
Piel Bros.
Brooklyn, NY
4

27 ❑ 12-2
Piel Bros.
Brooklyn, NY
4

28 ❑ 12-2
Piel Bros.
Brooklyn, NY
4

29 ❑ 12-2
Piel Bros.
Brooklyn, NY
4

30 ❑ 12-2
Tivoli
Denver, CO
1000+

31 ❑ 12-2
Walter
Pueblo, CO
175

32 ❑ 12-2
Walter
Pueblo, CO
600

33 ❑ 12-2
Walter
Pueblo, CO
80

34 ❑ 12-2
Walter
Pueblo, CO
80

35 ❑ 12-2
Walter
Pueblo, CO
90

36 ❑ 12-2
Walter
Pueblo, CO
30

37 ❑ 12-2-T
Rainier
San Francisco, CA
1000+

38 ❑ 12-2-T
St. Claire
San Jose, CA
1000+

39 ❑ 12-2
Storz
Omaha, NE
5

40 ❑ 12-1-OT
San Francisco
San Francisco, CA
Form. Milw.
175

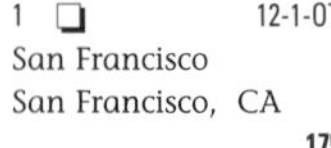
1 ❑ 12-1-OT
San Francisco
San Francisco, CA
175

2 ❑ 12-1
Metropolis
Trenton, NJ
60

3 ❑ 12-1
Metropolis
Trenton, NJ
20

4 ❑ 12-1
Metropolis
Trenton, NJ
20

5 ❑ 12-1-T
Metropolis
New York, NY
65

6 ❑ 12-1
Metropolis
New York, NY
1000+

7 ❑ 12-1-T
Old Dutch
Brooklyn, NY
750

8 ❑ 12-2
Pioneer
Minneapolis, MN
25

9 ❑ 12-2
G. Heileman
La Crosse, WI
20

10 ❑ 12-1
Pilsener
Cleveland, OH
140

11 ❑ 12-1
Pilsener
Cleveland, OH
225

12 ❑ 12-2
Pilsener
Cleveland, OH
15

13 ❑ 12-1
Pilsener
Cleveland, OH
15

14 ❑ 12-1
Pilsener
Cleveland, OH
15

15 ❑ 12-1
Pilsener
Columbus, OH
18

16 ❑ 12-2
Pilsener
Pittsburgh, PA
20

17 ❑ 12-2
Stevens Point
Stevens Point, WI
25

18 ❑ 12-2
Stevens Point
Stevens Point, WI
25

19 ❑ 12-2
Stevens Point
Stevens Point, WI
30

20 ❑ 12-2
Stevens Point
Stevens Point, WI
10

21 ❑ 12-1-OT
Poth
Philadelphia, PA
1000+

22 ❑ 12-1-OT
Poth
Philadelphia, PA
1000+

23 ❑ 12-1-OT
Poth
Philadelphia, PA
1000+

24 ❑ 12-1-OT
Poth
Philadelphia, PA
1000+

25 ❑ 12-1
Potosi
Potosi, WI
20

26 ❑ 12-2
Potosi
Potosi, WI
12

27 ❑ 12-1-OT
Manhattan
Chicago, IL
250

28 ❑ 12-1-OT
Prima
Chicago, IL
250

29 ❑ 12-2
Prima
Chicago, IL
45

30 ❑ 12-2
Prima
Chicago, IL
18

31 ❑ 12-2
Prima
Chicago, IL
18

32 ❑ 12-2
Prima
Chicago, IL
8

33 ❑ 12-1
Canadian Ace
Chicago, IL
45

34 ❑ 12-2
Canadian Ace
Chicago, IL
22

35 ❑ 12-2
Canadian Ace
Chicago, IL
12

36 ❑ 12-1
Prima Bismarck
Chicago, IL
45

37 ❑ 12-2-T
Globe
San Francisco, CA
1000+

38 ❑ 11-2
Hawaii
Honolulu, HI
100

39 ❑ 12-1-W
Adam Scheidt
Norristown, PA
75

40 ❑ 12-1-T
Adam Scheidt
Norristown, PA
50

1 ❑ 12-1
Adam Scheidt
Norristown, PA
50

2 ❑ 12-2
Adam Scheidt
Norristown, PA
40

3 ❑ 12-2
Adam Scheidt
Norristown, PA
30

4 ❑ 12-2
Adam Scheidt
Norristown, PA
30

5 ❑ 12-2
Valley Forge
Norristown, PA
30

6 ❑ 12-2
Valley Forge
Norristown, PA
30

7 ❑ 12-2
C. Schmidt & Sons
Philadelphia, PA
35

8 ❑ 12-2
C. Schmidt & Sons
Philadelphia, PA
35

9 ❑ 12-2
D.G. Yuengling & Son
Pottsville, PA
100

10 ❑ 12-2
Old Reading
Reading, PA
125

11 ❑ 12-2
Old Reading
Reading, PA
10

12 ❑ 12-2
Reading
Reading, PA
5

13 ❑ 12-1-OT
Indianapolis
Indianapolis, IN
1000+

14 ❑ 12-2
Progress
Oklahoma City, OK
125

15 ❑ 12-2
Schoenhofen Edelweiss
Chicago, IL
30

16 ❑ 12-2
Drewry's Ltd.
South Bend, IN
30

17 ❑ 12-1
Queen City
Cumberland, MD
150

18 ❑ 12-1
Queen City
Cumberland, MD
50

19 ❑ 12-2
Rahr-Green Bay
Green Bay, WI
65

20 ❑ 12-2
Rahr-Green Bay
Green Bay, WI
20

21 ❑ 12-2
Rahr-Green Bay
Green Bay, WI
20

22 ❑ 12-2
Rahr-Green Bay
Oshkosh, WI
20

23 ❑ 12-1-T
Rainier
San Francisco, CA
50

24 ❑ 12-1-T
Rainier
San Francisco, CA
50

25 ❑ 12-1
Rainier
San Francisco, CA
22

26 ❑ 12-1
Rainier
San Francisco, CA
22

27 ❑ 12-2
Rainier
San Francisco, CA
22

28 ❑ 12-1-T
Rainier
San Francisco, CA
225

29 ❑ 12-1-W
Rainier
San Francisco, CA
225

30 ❑ 12-1-T
Rainier
San Francisco, CA
60

31 ❑ 12-1-W
Rainier
San Francisco, CA
250

32 ❑ 12-1-T
Rainier
San Francisco, CA
45

33 ❑ 12-1-W
Rainier
San Francisco, CA
45

34 ❑ 12-1-T
Rainier
San Francisco, CA
45

35 ❑ 12-1-W
Rainier
San Francisco, CA
45

36 ❑ 12-1-T
Rainier
San Francisco, CA
45

37 ❑ 12-1-T
Rainier
San Francisco, CA
110

38 ❑ 12-1
Rainier
San Francisco, CA
110

39 ❑ 12-2
Rainier
San Francisco, CA
Flesh Stein
125

40 ❑ 12-1
Sicks' Seattle
Seattle, WA
225

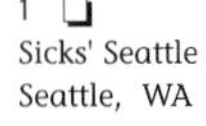

1 ❑ 12-2
Sicks' Seattle
Seattle, WA
15

2 ❑ 12-2
Sicks' Seattle
Seattle, WA
15

3 ❑ 12-2
Sicks' Rainier
Seattle, WA
15

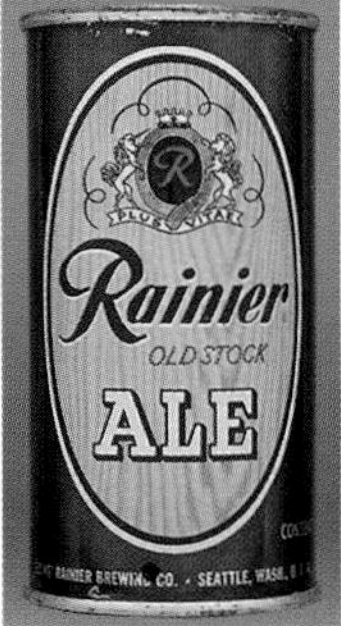

4 ❑ 11-2
Sicks' Rainier
Seattle, WA
12

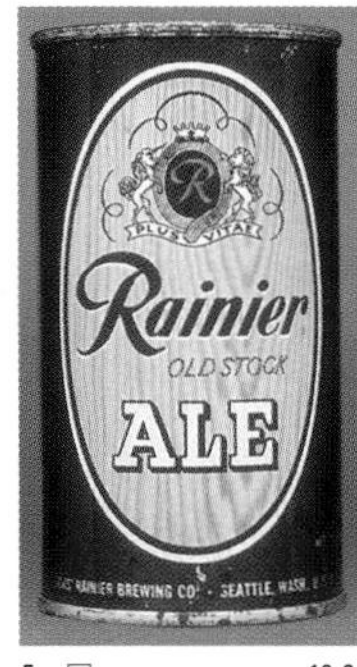

5 ❑ 12-2
Sicks' Rainier
Seattle, WA
12

6 ❑ 11-2
Sicks' Rainier
Seattle, WA
12

7 ❑ 12-1-OT
Century
Seattle, WA
N.O. 4% 1000+

8 ❑ 12-1-OT
Century
Seattle, WA
1000+

9 ❑ 12-1-OT
Seattle Br. & Malt.
Seattle, WA
800

10 ❑ 12-1-OT
Sicks' Seattle Br.&Malt. Co.
Seattle, WA
150

11 ❑ 12-1-OT
Sicks' Seattle Br.&Malt. Co.
Seattle, WA
150

12 ❑ 12-1
Sicks' Seattle
Seattle, WA
40

13 ❑ 12-1
Sicks' Seattle
Seattle, WA
40

14 ❑ 12-1
Sicks' Seattle
Seattle, WA
40

15 ❑ 12-1
Sicks' Seattle Br.&Malt. Co.
Seattle, WA
Set Can 35

16 ❑ 11-1
Sicks' Rainier
Seattle, WA
Set Can 35

17 ❑ 12-1
Sicks' Seattle
Seattle, WA
175

18 ❑ 12-1
Sicks' Seattle
Seattle, WA
175

19 ❑ 12-1
Sicks' Seattle
Seattle, WA
175

20 ❑ 12-1
Sicks' Seattle
Seattle, WA
175

21 ❑ 12-1
Sicks' Seattle
Seattle, WA
175

22 ❑ 12-1
Sicks' Seattle
Seattle, WA
175

23 ❑ 12-2
Sicks' Seattle
Seattle, WA
10

24 ❑ 12-1-OT
Sicks' Spokane
Spokane, WA
110

25 ❑ 12-1-OT
Sicks' Spokane
Spokane, WA
125

26 ❑ 12-1
Sicks' Spokane
Spokane, WA
30

27 ❑ 12-1
Sicks' Spokane
Spokane, WA
30

28 ❑ 12-1
Sicks' Spokane
Spokane, WA
45

29 ❑ 11-1
Sicks' Spokane
Spokane, WA
Set Can 35

30 ❑ 12-1
Sicks' Spokane
Spokane, WA
Set Can 35

31 ❑ 12-1-OT
Adam Scheidt
Norristown, PA
200

32 ❑ 12-1-OT
Adam Scheidt
Norristown, PA
200

33 ❑ 12-1-OT
Adam Scheidt
Norristown, PA
100

34 ❑ 12-1
Adam Scheidt
Norristown, PA
12

35 ❑ 12-1
Valley Forge
Norristown, PA
10

36 ❑ 12-1
C. Schmidt & Sons
Philadelphia, PA
10

37 ❑ 12-2
Pacific
Oakland, CA
1000+

38 ❑ 12-2
Old Reading
Reading, PA
30

39 ❑ 12-2
Old Reading
Reading, PA
20

40 ❑ 12-2
Reading
Reading, PA
5

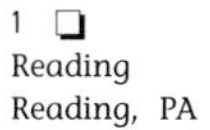

1 ❑ 12-2
Reading
Reading, PA
5

2 ❑ 12-2
Carling
Atlanta, GA
15

3 ❑ 12-2
Carling
Atlanta, GA
10

4 ❑ 12-2
Carling
Belleville, IL
15

5 ❑ 12-2
Carling
Belleville, IL
10

6 ❑ 12-2
Carling
Baltimore, MD
10

7 ❑ 12-2
Carling
Natick, MA
15

8 ❑ 12-2
Carling
Natick, MA
10

9 ❑ 12-2
Carling
Frankenmuth, MI
15

10 ❑ 12-2
Carling
Frankenmuth, MI
10

11 ❑ 12-2
Carling
St. Louis, MO
15

12 ❑ 12-1
Br. Corp. of Amer.
Cleveland, OH
25

Front of 14, 15

14 ❑ 12-1
Br. Corp. of Amer.
Cleveland, OH
25

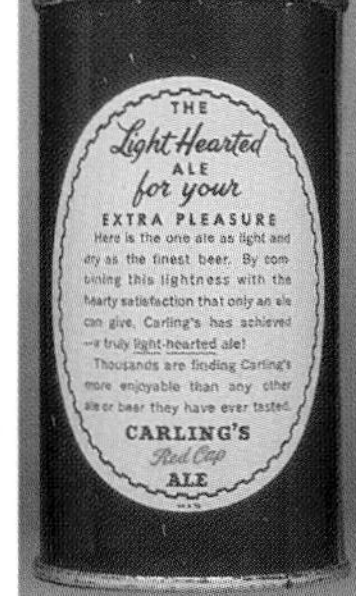

15 ❑ 12-1
Br. Corp. of Amer.
Cleveland, OH
25

16 ❑ 12-1
Carling
Cleveland, OH
25

17 ❑ 12-2
Carling
Cleveland, OH
15

18 ❑ 12-2
Carling
Cleveland, OH
10

19 ❑ 11-2
Carling
Tacoma, WA
15

20 ❑ 12-2
Carling
Tacoma, WA
10

21 ❑ 12-1
Best
Chicago, IL
50

22 ❑ 12-1
Cumberland
Cumberland, MD
60

23 ❑ 12-1-T
Ruppert-Virginia
Norfolk, VA
80

24 ❑ 12-1
Ruppert-Virginia
Norfolk, VA
80

25 ❑ 12-1
Century
Norfolk, VA
60

26 ❑ 12-1
Atlantic
Chicago, IL
30

27 ❑ 12-1
Atlantic
Chicago, IL
22

28 ❑ 12-1
CV
Chicago, IL
22

29 ❑ 12-1
Drewry's Ltd.
South Bend, IN
22

30 ❑ 12-2
Terre Haute
Terre Haute, IN
35

31 ❑ 12-2-OT
Red Top
Cincinnati, OH
Silver 1000+

32 ❑ 12-2-OT
Red Top
Cincinnati, OH
White 1000+

33 ❑ 12-1-OT
Red Top
Cincinnati, OH
1000+

34 ❑ 12-1
Red Top
Cincinnati, OH
75

35 ❑ 12-2-OT
Red Top
Cincinnati, OH
1000+

36 ❑ 12-1-OT
Red Top
Cincinnati, OH
750

37 ❑ 12-1
Red Top
Cincinnati, OH
25

38 ❑ 12-1
Red Top
Cincinnati, OH
Baseball 110

39 ❑ 12-1
Red Top
Cincinnati, OH
Baseball 110

40 ❑ 12-1
Red Top
Cincinnati, OH
Bowling 110

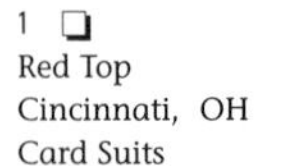
1 ❑ 12-1
Red Top
Cincinnati, OH
Card Suits 110

2 ❑ 12-1
Red Top
Cincinnati, OH
Circus 110

3 ❑ 12-1
Red Top
Cincinnati, OH
Cooking 110

4 ❑ 12-1
Red Top
Cincinnati, OH
Cooking 110

5 ❑ 12-1
Red Top
Cincinnati, OH
Fishing 110

6 ❑ 12-1
Red Top
Cincinnati, OH
Golf 110

7 ❑ 12-1
Red Top
Cincinnati, OH
Haywagon 110

8 ❑ 12-1
Red Top
Cincinnati, OH
Haywagon 110

9 ❑ 12-1
Red Top
Cincinnati, OH
Horse Racing 110

10 ❑ 12-1
Red Top
Cincinnati, OH
Horse Racing 110

11 ❑ 12-1
Red Top
Cincinnati, OH
Music Notes 110

12 ❑ 12-1
Red Top
Cincinnati, OH
Party Flags 110

13 ❑ 12-1
Red Top
Cincinnati, OH
Party Flags 110

14 ❑ 12-1
Red Top
Cincinnati, OH
Radio 110

15 ❑ 12-1
Wunderbrau
Cincinnati, OH
Cooking 110

16 ❑ 12-1
Wunderbrau
Cincinnati, OH
Golf 110

17 ❑ 12-1
Wunderbrau
Cincinnati, OH
Haywagon 110

18 ❑ 12-1
Wunderbrau
Cincinnati, OH
Party Flags 110

19 ❑ 12-1
Wunderbrau
Cincinnati, OH
Music Notes 110

20 ❑ 12-1
Red Top
Cincinnati, OH
75

21 ❑ 12-2
Red Top
Cincinnati, OH
75

22 ❑ 12-2
Red Top
Cincinnati, OH
35

23 ❑ 12-2
Red Top
Cincinnati, OH
35

24 ❑ 12-2
Standard
Cleveland, OH
200

25 ❑ 12-1-0T
Regal Amber
San Francisco, CA
1000+

Front of 27, 28

27 ❑ 12-1-0T
Regal Amber
San Francisco, CA
500

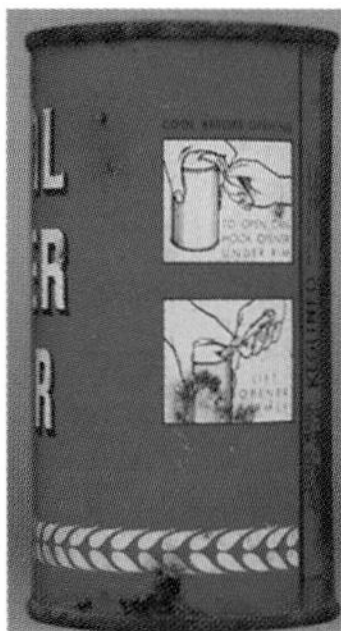
28 ❑ 12-1-0T
Regal Amber
San Francisco, CA
500

29 ❑ 12-1-0W
Regal Amber
San Francisco, CA
600

30 ❑ 12-1-T
Regal Amber
San Francisco, CA
1000+

31 ❑ 12-2-T
Regal Amber
San Francisco, CA
90

32 ❑ 12-2-W
Regal Amber
San Francisco, CA
100

33 ❑ 12-1-T
Regal Amber
San Francisco, CA
300

34 ❑ 12-2-T
Regal Amber
San Francisco, CA
150

35 ❑ 12-2-T
Regal Amber
San Francisco, CA
20

36 ❑ 12-2-T
Regal Amber
San Francisco, CA
20

37 ❑ 12-2
Regal Amber
San Francisco, CA
20

38 ❑ 12-2
Regal Amber
San Francisco, CA
20

39 ❑ 12-2
Regal Amber
San Francisco, CA
20

40 ❑ 12-2
Regal Amber
San Francisco, CA
10

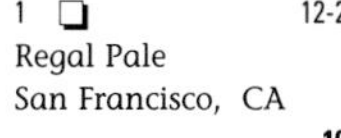

1 ❑ 12-2
Regal Pale
San Francisco, CA
10

2 ❑ 12-2
Regal Pale
San Francisco, CA
10

3 ❑ 12-2
Regal Pale
San Francisco, CA
10

4 ❑ 12-2
Regal Pale
San Francisco, CA
10

5 ❑ 12-2
Regal Pale
San Francisco, CA
12

6 ❑ 11-2
Regal Pale
San Francisco, CA
12

7 ❑ 12-2
Regal Pale
San Francisco, CA
Dull **6**

8 ❑ 11-2
Regal Pale
San Francisco, CA
6

9 ❑ 12-2
Regal Pale
San Francisco, CA
6

10 ❑ 12-2
Regal Amber
San Francisco, CA
400

11 ❑ 12-2
Regal Amber
San Francisco, CA
75

12 ❑ 12-2
Regal Pale
San Francisco, CA
Met. Blue **75**

13 ❑ 12-2
Regal Pale
San Francisco, CA
Enam. Blue **75**

14 ❑ 12-2
Regal Pale
San Francisco, CA
75

15 ❑ 11-2
Regal Pale
San Francisco, CA
Metallic **90**

16 ❑ 11-2
Regal Pale
San Francisco, CA
Enamel **90**

17 ❑ 12-2
Maier
Los Angeles, CA
20

18 ❑ 12-2
Maier
Los Angeles, CA
6

19 ❑ 12-2
Maier
Los Angeles, CA
12

20 ❑ 12-2
Maier
Los Angeles, CA
12

21 ❑ 12-2
Maier
Los Angeles, CA
150

22 ❑ 12-2
American
Miami, FL
200

23 ❑ 12-1
American
Miami, FL
90

24 ❑ 12-1
American
Miami, FL
60

25 ❑ 12-1
American
Miami, FL
60

26 ❑ 12-2
Anheuser-Busch
Miami, FL
300

27 ❑ 12-2
Anheuser-Busch
Miami, FL
225

28 ❑ 12-2
Anheuser-Busch
Miami, FL
100

29 ❑ 12-1
Anheuser-Busch
Miami, FL
125

30 ❑ 12-2
Anheuser-Busch
Miami, FL
60

31 ❑ 12-2
Regal
Miami, FL
75

32 ❑ 12-2
Regal
Miami, FL
35

33 ❑ 12-2
Regal
Miami, FL
35

34 ❑ 12-2
Atlantic
Chicago, IL
10

35 ❑ 12-2
Drewry's Ltd.
South Bend, IN
10

36 ❑ 12-1
American
New Orleans, LA
135

37 ❑ 12-2
American
New Orleans, LA
135

38 ❑ 12-1
American
New Orleans, LA
65

39 ❑ 12-1
American
New Orleans, LA
40

40 ❑ 12-2
American
New Orleans, LA
30

1 ❑ 12-2
American
New Orleans, LA
25

2 ❑ 12-2
American
New Orleans, LA
10

3 ❑ 12-1-OT
Regal
Detroit, MI
1000+

4 ❑ 12-2
Jos. Huber
Monroe, WI
10

5 ❑ 12-1
Jos. Huber
Monroe, WI
10

6 ❑ 12-2
Maier
Los Angeles, CA
45

7 ❑ 12-2
Maier
Los Angeles, CA
65

8 ❑ 12-2
Maier
Los Angeles, CA
30

9 ❑ 12-1
Regent
Pensacola, FL
125

10 ❑ 12-1
Regent
Pensacola, FL
40

11 ❑ 12-1
Spearman
Pensacola, FL
40

12 ❑ 12-1
Metropolis
Trenton, NJ
200

13 ❑ 12-1
Metropolis
Trenton, NJ
200

14 ❑ 12-1
Century
Norfolk, VA
100

15 ❑ 12-1
Century
Norfolk, VA
25

16 ❑ 12-1
Century
Norfolk, VA
25

17 ❑ 12-1
Century
Norfolk, VA
250

18 ❑ 12-2
Atlas
Chicago, IL
25

19 ❑ 12-2
Canadian Ace
Chicago, IL
25

20 ❑ 12-2
Empire
Chicago, IL
25

21 ❑ 12-2
United States
Chicago, IL
25

22 ❑ 12-2
Reisch
Springfield, IL
40

23 ❑ 12-1
Renner
Fort Wayne, IN
12

24 ❑ 12-1
Renner
Youngstown, OH
15

25 ❑ 12-2
Wisconsin
Burlington, WI
125

26 ❑ 12-2
Wisconsin
Waukesha, WI
125

27 ❑ 12-2
Wisconsin
Waukesha, WI
Metallic
40

28 ❑ 12-2
Wisconsin
Waukesha, WI
Enamel
40

29 ❑ 12-2
Wisconsin
Waukesha, WI
40

30 ❑ 12-2
Fox Head
Waukesha, WI
40

31 ❑ 12-2
Maier
Los Angeles, CA
18

32 ❑ 12-2
Maier
Los Angeles, CA
18

33 ❑ 12-2
Maier
Los Angeles, CA
18

34 ❑ 12-1-T
Rubsam & Horrmann
New York, NY
1000+

35 ❑ 12-1-T
Rubsam & Horrmann
New York, NY
700

36 ❑ 12-1
Rubsam & Horrmann
New York, NY
700

37 ❑ 12-1-T
Rubsam & Horrmann
New York, NY
500

38 ❑ 12-1-T
Rubsam & Horrmann
New York, NY
135

39 ❑ 12-1-T
Rubsam & Horrmann
New York, NY
135

40 ❑ 12-1
Rubsam & Horrmann
New York, NY
175

1 ❑ 12-1
Rubsam & Horrmann
New York, NY
100

2 ❑ 12-1
Rubsam & Horrmann
New York, NY
100

3 ❑ 12-2
Jax Ice
Jacksonville, FL
750

4 ❑ 12-1
United States
Chicago, IL
30

5 ❑ 12-1
Rheingold
Los Angeles, CA
10

6 ❑ 12-2
Rheingold
Los Angeles, CA
10

7 ❑ 12-1
Liebmann
Orange, NJ
5

8 ❑ 12-1
Liebmann
Orange, NJ
5

9 ❑ 12-2
Liebmann
Orange, NJ
Diane Baker
275

10 ❑ 12-2
Liebmann
Orange, NJ
Beverly Christensen 275

11 ❑ 12-2
Liebmann
Orange, NJ
Tami Connor
275

12 ❑ 12-2
Liebmann
Orange, NJ
Margie McNally 275

13 ❑ 12-2
Liebmann
Orange, NJ
Suzy Ruel
275

14 ❑ 12-2
Liebmann
Orange, NJ
Kathleen Wallace 275

15 ❑ 12-2
Liebmann
Orange, NJ
Margie McNally
275

16 ❑ 12-1
Liebmann
Orange, NJ
150

17 ❑ 12-2
Liebmann
Orange, NJ
150

18 ❑ 12-2
Liebmann
Orange, NJ
35

19 ❑ 12-2
Rheingold
Orange, NJ
5

20 ❑ 12-2
Rheingold
Orange, NJ
35

21 ❑ 12-1-OT
Liebmann
New York, NY
250

22 ❑ 12-1-T
Liebmann
New York, NY
90

23 ❑ 12-1-T
Liebmann
New York, NY
140

24 ❑ 12-1-T
Liebmann
New York, NY
35

25 ❑ 12-1-T
Liebmann
New York, NY
25

26 ❑ 12-2
Liebmann
New York, NY
12

27 ❑ 12-1-T
Liebmann
New York, NY
1000+

28 ❑ 12-1-T
Liebmann
New York, NY
90

29 ❑ 12-1
Liebmann
New York, NY
90

30 ❑ 12-1
Liebmann
New York, NY
90

31 ❑ 12-2
Liebmann
New York, NY
800

32 ❑ 12-1-OT
Liebmann
New York, NY
700

33 ❑ 12-1-T
Liebmann
New York, NY
400

34 ❑ 12-1-T
Liebmann
New York, NY
400

35 ❑ 12-1-W
Liebmann
New York, NY
225

36 ❑ 12-1-T
Liebmann
New York, NY
200

37 ❑ 12-1-W
Liebmann
New York, NY
700

38 ❑ 12-1-T
Liebmann
New York, NY
250

Front of 40 and 124-1

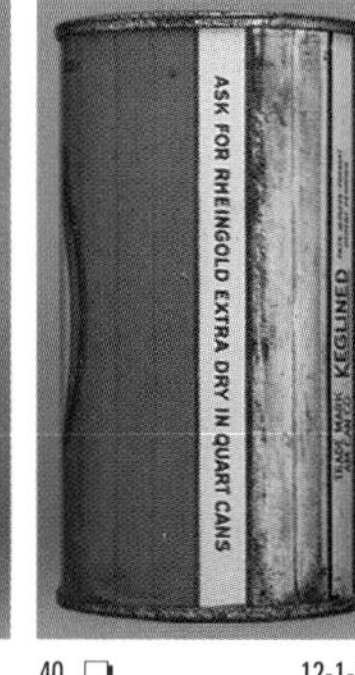

40 ❑ 12-1-T
Liebmann
New York, NY
250

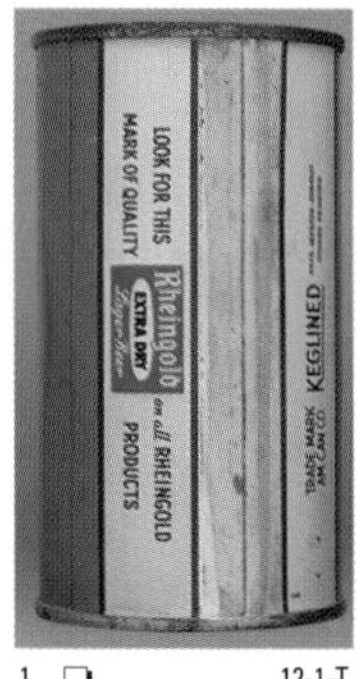
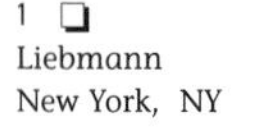
1 ❑ 12-1-T
Liebmann
New York, NY
250

2 ❑ 12-1-T
Liebmann
New York, NY
600

3 ❑ 12-1-T
Liebmann
New York, NY
12

4 ❑ 12-1
Liebmann
New York, NY
7

5 ❑ 12-1
Liebmann
New York, NY
7

6 ❑ 12-1
Liebmann
New York, NY
5

7 ❑ 12-2
Liebmann
New York, NY
Diane Baker 275

8 ❑ 12-2
Liebmann
New York, NY
Beverly Christensen 275

9 ❑ 12-2
Liebmann
New York, NY
Tami Connor 275

10 ❑ 12-2
Liebmann
New York, NY
Margie McNally 275

11 ❑ 12-2
Liebmann
New York, NY
Suzy Ruel 275

12 ❑ 12-2
Liebmann
New York, NY
Kathleen Wallace 275

13 ❑ 12-2
Liebmann
New York, NY
Margie McNally 275

14 ❑ 12-2
Liebmann
New York, NY
5

15 ❑ 12-1-T
Liebmann
New York, NY
1000+

16 ❑ 12-1-T
Liebmann
New York, NY
1000+

17 ❑ 12-1
Liebmann
New York, NY
150

18 ❑ 12-2
Liebmann
New York, NY
150

19 ❑ 12-2
Liebmann
New York, NY
35

20 ❑ 12-2
Rheingold
New York, NY
75

21 ❑ 12-2
Rheingold
New York, NY
5

22 ❑ 12-2
Rheingold
New York, NY
3

23 ❑ 12-2
Rheingold
New York, NY
35

24 ❑ 12-1-OT
Century
Seattle, WA
1000+

25 ❑ 12-1-OT
Century
Seattle, WA
N.O. 4% 1000+

26 ❑ 12-1-OT
Seattle Br. & Malt. Co.
Seattle, WA
750

27 ❑ 11-2
Highlander
Seattle, WA
10

28 ❑ 12-2
Highlander
Seattle, WA
10

29 ❑ 11-2
Rheinlander
Seattle, WA
10

30 ❑ 12-2
Rheinlander
Seattle, WA
10

31 ❑ 11-2
Spokane
Spokane, WA
10

32 ❑ 12-1
Rhinelander
Rhinelander, WI
20

33 ❑ 12-2
Rialto
Trenton, NJ
125

34 ❑ 12-2
Rialto
Trenton, NJ
110

35 ❑ 12-2
Wausau
Wausau, WI
275

36 ❑ 12-2
Home
Richmond, VA
50

37 ❑ 12-2
Home
Richmond, VA
40

38 ❑ 12-2
Home
Richmond, VA
Sm. BEER 40

39 ❑ 12-2
Home
Richmond, VA
Lg. BEER 40

40 ❑ 12-2
Home
Richmond, VA
Met., Sm. BEER 30

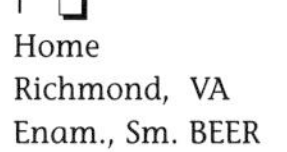
1 ❑ 12-2
Home
Richmond, VA
Enam., Sm. BEER 30

2 ❑ 12-2
Home
Richmond, VA
Met., Lg BEER 30

3 ❑ 12-2
Home
Richmond, VA
Enam., Lg BEER 30

4 ❑ 12-2
Home
Richmond, VA
500

5 ❑ 12-2
Home
Richmond, VA
450

6 ❑ 12-2
Home
Richmond, VA
500

7 ❑ 12-2
Home
Richmond, VA
500

8 ❑ 12-2
Home
Richmond, VA
Met. Blue 400

9 ❑ 12-2
Home
Richmond, VA
Enam. Blue 400

10 ❑ 12-1-T
San Francisco
San Francisco, CA
1000+

11 ❑ 12-2
Atlantic
Chicago, IL
160

12 ❑ 12-2
Roger Wilco
Trenton, NJ
600

13 ❑ 12-1-T
Roger Williams
Cranston, RI
Metallic 1000+

14 ❑ 12-1-T
Roger Williams
Cranston, RI
Enamel 1000+

15 ❑ 12-1
Latrobe
Latrobe, PA
75

16 ❑ 12-1
Latrobe
Latrobe, PA
12

17 ❑ 12-1-OT
No Mandatory
Chicago, IL
1000+

18 ❑ 12-2
Maier
Los Angeles, CA
150

19 ❑ 12-2
Maier
Los Angeles, CA
150

20 ❑ 12-2
Maier
Los Angeles, CA
45

21 ❑ 12-2
Maier
Los Angeles, CA
45

22 ❑ 12-2
Royal
Chicago, IL
30

23 ❑ 12-2
Royal
Chicago, IL
30

24 ❑ 12-1
Geo. Wiedemann
Newport, KY
175

25 ❑ 12-1
Geo. Wiedemann
Newport, KY
100

26 ❑ 12-1
Geo. Wiedemann
Newport, KY
100

27 ❑ 12-2
Maier
Los Angeles, CA
20

28 ❑ 12-2
Maier
Los Angeles, CA
20

29 ❑ 12-2
Sunshine
Reading, PA
200

30 ❑ 12-2
Duluth
Duluth, MN
15

31 ❑ 12-2
Jacob Ruppert
New Bedford, MA
6

32 ❑ 12-2
Jacob Ruppert
Orange, NJ
4

33 ❑ 12-1-T
Jacob Ruppert
New York, NY
Copper Trim 135

34 ❑ 12-1-T
Jacob Ruppert
New York, NY
Silver Trim 135

35 ❑ 12-1-T
Jacob Ruppert
New York, NY
175

36 ❑ 12-1
Jacob Ruppert
New York, NY
175

37 ❑ 12-2
Jacob Ruppert
New York, NY
225

38 ❑ 12-2
Jacob Ruppert
New York, NY
35

39 ❑ 12-1-OT
Jacob Ruppert
New York, NY
250

40 ❑ 12-1-OT
Jacob Ruppert
New York, NY
125

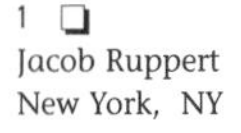

1 ❑ 12-1-T
Jacob Ruppert
New York, NY
75

2 ❑ 12-1-T
Jacob Ruppert
New York, NY
60

3 ❑ 12-1-T
Jacob Ruppert
New York, NY
60

4 ❑ 12-1-W
Jacob Ruppert
New York, NY
65

5 ❑ 12-1-T
Jacob Ruppert
New York, NY
60

6 ❑ 12-1-W
Jacob Ruppert
New York, NY
60

7 ❑ 12-1-T
Jacob Ruppert
New York, NY
35

8 ❑ 12-1-T
Jacob Ruppert
New York, NY
Cream Background 35

9 ❑ 12-1-T
Jacob Ruppert
New York, NY
White Background 35

10 ❑ 12-1
Jacob Ruppert
New York, NY
35

11 ❑ 12-1
Jacob Ruppert
New York, NY
Cream Background 35

12 ❑ 12-1
Jacob Ruppert
New York, NY
White Background 35

13 ❑ 12-2
Jacob Ruppert
New York, NY
18

14 ❑ 12-2
Jacob Ruppert
New York, NY
18

15 ❑ 12-2
Jacob Ruppert
New York, NY
12

16 ❑ 12-2
Jacob Ruppert
New York, NY
12

17 ❑ 12-2
Jacob Ruppert
New York, NY
12

18 ❑ 12-2
Jacob Ruppert
New York, NY
12

19 ❑ 12-2
Jacob Ruppert
New York, NY
12

20 ❑ 12-2
Jacob Ruppert
New York, NY
12

21 ❑ 12-2
Jacob Ruppert
New York, NY
5

22 ❑ 12-2
Jacob Ruppert
New York, NY
5

23 ❑ 12-2
Jacob Ruppert
New York, NY
3

24 ❑ 12-1-OT
Jacob Ruppert
New York, NY
1000+

25 ❑ 12-1-T
Jacob Ruppert
New York, NY
1000+

26 ❑ 12-1-T
Jacob Ruppert
New York, NY
800

Front of 28, 29

28 ❑ 12-1
Jacob Ruppert
New York, NY
175

29 ❑ 12-1
Jacob Ruppert
New York, NY
175

30 ❑ 12-2
Jacob Ruppert
New York, NY
450

31 ❑ 12-2
Jacob Ruppert
New York, NY
550

32 ❑ 12-2
Jacob Ruppert
New York, NY
50

33 ❑ 12-1-OT
Jacob Ruppert
New York, NY
1000+

34 ❑ 12-1
Jacob Ruppert
New York, NY
125

35 ❑ 12-1
Jacob Ruppert
New York, NY
75

36 ❑ 12-2
Jacob Ruppert
New York, NY
90

37 ❑ 12-2
Jacob Ruppert
New York, NY
80

38 ❑ 12-1-T
Ruppert-Virginia
Norfolk, VA
85

39 ❑ 12-1
Ruppert-Virginia
Norfolk, VA
80

40 ❑ 12-1
Ruppert-Virginia
Norfolk, VA
80

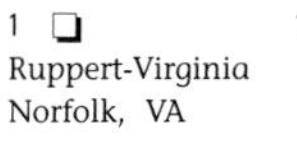

1 ❑ 12-1
Ruppert-Virginia
Norfolk, VA
80

2 ❑ 12-1
Ruppert-Virginia
Norfolk, VA
100

3 ❑ 12-1
Ruppert-Virginia
Norfolk, VA
200

4 ❑ 12-1
Arizona
Phoenix, AZ
175

5 ❑ 12-2
Grace Bros.
Santa Rosa, CA
90

6 ❑ 12-1-OT
Grace Bros.
Santa Rosa, CA
1000+

7 ❑ 12-1-T
North Bay
Santa Rosa, CA
700

8 ❑ 12-2
Schoenhofen Edelweiss
Chicago, IL
60

9 ❑ 12-2
Drewry's Ltd.
South Bend, IN
60

10 ❑ 12-2
Cerveceria San Juan
Hammonton, NJ
150

11 ❑ 12-2
Cerveceria San Juan
Hammonton, NJ
175

12 ❑ 12-1
Geo. Muehlebach
Kansas City, MO
200

Side of 12

14 ❑ 12-1
Geo. Muehlebach
Kansas City, MO
200

Side of 14

16 ❑ 12-2
Jax Ice
Jacksonville, FL
500

17 ❑ 12-2
Maier
Los Angeles, CA
175

18 ❑ 12-2
Savoy
Davenport, IA
150

19 ❑ 12-2
Atlantic
Chicago, IL
125

20 ❑ 12-2
Savoy
Chicago, IL
125

21 ❑ 12-2
Leisy
Cleveland, OH
125

22 ❑ 12-2
Savoy
Cleveland, OH
125

23 ❑ 12-1
F. & M. Schaefer
Baltimore, MD
World's Fair
12

Side of 23

25 ❑ 12-1
F. & M. Schaefer
Albany, NY
35

26 ❑ 12-2
F. & M. Schaefer
Albany, NY
30

27 ❑ 12-1
F. & M. Schaefer
Albany, NY
18

28 ❑ 12-1
F. & M. Schaefer
Albany, NY
15

29 ❑ 12-1
F. & M. Schaefer
Albany, NY
15

30 ❑ 12-1
F. & M. Schaefer
Albany, NY
15

31 ❑ 12-1
F. & M. Schaefer
Albany, NY
15

32 ❑ 12-1
F. & M. Schaefer
Albany, NY
10

33 ❑ 12-1
F. & M. Schaefer
Albany, NY
10

34 ❑ 12-1
F. & M. Schaefer
Albany, NY
10

35 ❑ 12-1
F. & M. Schaefer
Albany, NY
5

36 ❑ 12-1
F. & M. Schaefer
Albany, NY
World's Fair
12

Side of 36

38 ❑ 12-1-T
F. & M. Schaefer
New York, NY
30

39 ❑ 12-1-W
F. & M. Schaefer
New York, NY
40

40 ❑ 12-1-W
F. & M. Schaefer
New York, NY
100

1 ❑ 12-1-T
F. & M. Schaefer
Brooklyn, NY
18

2 ❑ 12-1-T
F. & M. Schaefer
Brooklyn, NY
18

3 ❑ 12-1-T
F. & M. Schaefer
Brooklyn, NY
18

4 ❑ 12-1
F. & M. Schaefer
Brooklyn, NY
15

5 ❑ 12-1
F. & M. Schaefer
Brooklyn, NY
15

6 ❑ 12-1
F. & M. Schaefer
Brooklyn, NY
15

7 ❑ 12-1
F. & M. Schaefer
Brooklyn, NY
15

8 ❑ 12-1
F. & M. Schaefer
Brooklyn, NY
15

9 ❑ 12-1
F. & M. Schaefer
Brooklyn, NY
15

10 ❑ 12-1
F. & M. Schaefer
Brooklyn, NY
15

11 ❑ 12-1
F. & M. Schaefer
New York, NY
12

12 ❑ 12-1
F. & M. Schaefer
New York, NY
12

13 ❑ 12-1
F. & M. Schaefer
New York, NY
12

14 ❑ 12-1
F. & M. Schaefer
New York, NY
12

15 ❑ 12-2
F. & M. Schaefer
New York, NY
12

16 ❑ 12-1
F. & M. Schaefer
New York, NY
5

17 ❑ 12-1
F. & M. Schaefer
New York, NY
5

18 ❑ 12-1
F. & M. Schaefer
New York, NY
World's Fair
5

Side of 18

20 ❑ 12-1-T
F. & M. Schaefer
Brooklyn, NY
1000+

21 ❑ 12-1-T
F. & M. Schaefer
Brooklyn, NY
350

22 ❑ 12-1
F. & M. Schaefer
Brooklyn, NY
350

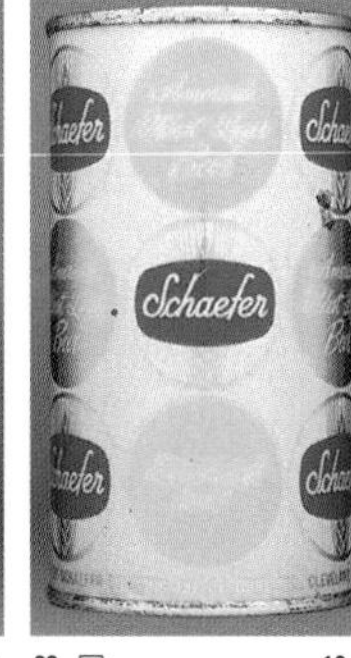

23 ❑ 12-1
Schaefer of Ohio
Cleveland, OH
15

24 ❑ 12-2
August Schell
New Ulm, MN
18

25 ❑ 12-2
Jos. Schlitz
Los Angeles, CA
1954
8

26 ❑ 12-2
Jos. Schlitz
Los Angeles, CA
1954
8

27 ❑ 12-2
Jos. Schlitz
Los Angeles, CA
1960
5

28 ❑ 12-2
Jos. Schlitz
Los Angeles, CA
1960
5

29 ❑ 12-2
Jos. Schlitz
Los Angeles, CA
1960
5

30 ❑ 12-2
Jos. Schlitz
San Francisco, CA
1960
5

31 ❑ 12-2
Jos. Schlitz
San Francisco, CA
1962
5

32 ❑ 12-2
Jos. Schlitz
San Francisco, CA
1962
5

33 ❑ 12-2
Jos. Schlitz
Tampa, FL
1958
18

34 ❑ 12-2
Jos. Schlitz
Tampa, FL
1962
5

35 ❑ 12-2
Jos. Schlitz
Tampa, FL
5

36 ❑ 12-2
Jos. Schlitz
Kansas City, MO
1957
6

37 ❑ 12-2
Jos. Schlitz
Kansas City, MO
1960
3

38 ❑ 12-2
Jos. Schlitz
Kansas City, MO
1960
3

39 ❑ 12-2
Jos. Schlitz
Kansas City, MO
1960
3

40 ❑ 12-2
Jos. Schlitz
Kansas City, MO
1962
3

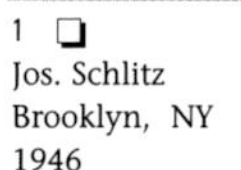

1 ❑ 12-1-T
Jos. Schlitz
Brooklyn, NY
1946 10

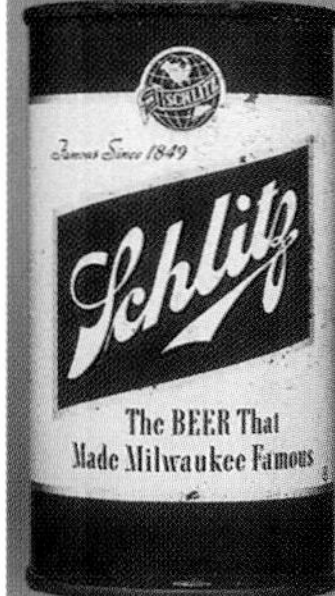

2 ❑ 12-1
Jos. Schlitz
Brooklyn, NY
1949 6

3 ❑ 12-1
Jos. Schlitz
Brooklyn, NY
1949 6

4 ❑ 12-2
Jos. Schlitz
Brooklyn, NY
1954 6

5 ❑ 12-2
Jos. Schlitz
Brooklyn, NY
1954 6

6 ❑ 12-2
Jos. Schlitz
Brooklyn, NY
1957 6

7 ❑ 12-2
Jos. Schlitz
Brooklyn, NY
1960 3

8 ❑ 12-2
Jos. Schlitz
Brooklyn, NY
1960 3

9 ❑ 12-2
Jos. Schlitz
Brooklyn, NY
1962 3

10 ❑ 12-2
Jos. Schlitz
Brooklyn, NY
1962 3

11 ❑ 10-2
Jos. Schlitz
Longview, TX
1960 8

Side of 13

13 ❑ 12-1-W
Jos. Schlitz
Milwaukee, WI
600

Side of 13

15 ❑ 12-1-T
Jos. Schlitz
Milwaukee, WI
350

16 ❑ 12-1-W
Jos. Schlitz
Milwaukee, WI
1937 375

17 ❑ 12-2-W
Jos. Schlitz
Milwaukee, WI
1945 250

18 ❑ 12-1-W
Jos. Schlitz
Milwaukee, WI
1946 65

Side of 18

20 ❑ 12-1-T
Jos. Schlitz
Milwaukee, WI
1946 10

21 ❑ 12-1-T
Jos. Schlitz
Milwaukee, WI
1946 10

Side of 21

23 ❑ 12-1-W
Jos. Schlitz
Milwaukee, WI
1949 65

Side of 23

25 ❑ 12-1-T
Jos. Schlitz
Milwaukee, WI
1949 5

26 ❑ 12-1
Jos. Schlitz
Milwaukee, WI
1949 5

27 ❑ 12-2
Jos. Schlitz
Milwaukee, WI
1954 3

28 ❑ 12-2
Jos. Schlitz
Milwaukee, WI
1954 3

29 ❑ 12-2
Jos. Schlitz
Milwaukee, WI
1957 3

30 ❑ 10-2
Jos. Schlitz
Milwaukee, WI
1958 3

31 ❑ 10-2
Jos. Schlitz
Milwaukee, WI
1960 10

32 ❑ 12-2
Jos. Schlitz
Milwaukee, WI
1960 2

33 ❑ 10-2
Jos. Schlitz
Milwaukee, WI
1960 10

34 ❑ 10-2
Jos. Schlitz
Milwaukee, WI
1960 10

35 ❑ 12-2
Jos. Schlitz
Milwaukee, WI
1960 2

36 ❑ 10-2
Jos. Schlitz
Milwaukee, WI
1960 10

37 ❑ 12-2
Jos. Schlitz
Milwaukee, WI
1960 2

38 ❑ 10-2
Jos. Schlitz
Milwaukee, WI
1962 10

39 ❑ 12-2
Jos. Schlitz
Milwaukee, WI
1962 2

40 ❑ 12-2
Jos. Schlitz
Milwaukee, WI
1966 75

1 ❑ 12-2-OT
Jacob Schmidt
St. Paul, MN
90

2 ❑ 12-2-W
Jacob Schmidt
St. Paul, MN
750

3 ❑ 12-2-T
Jacob Schmidt
St. Paul, MN
25

4 ❑ 12-2
Jacob Schmidt
St. Paul, MN
25

5 ❑ 12-2
Jacob Schmidt
St. Paul, MN
15

6 ❑ 12-1
Jacob Schmidt
St. Paul, MN
12

7 ❑ 12-1
Jacob Schmidt
St. Paul, MN
40

8 ❑ 12-1
Jacob Schmidt
St. Paul, MN
12

9 ❑ 12-1
Jacob Schmidt
St. Paul, MN
15

10 ❑ 12-1
Jacob Schmidt
St. Paul, MN
15

11 ❑ 12-1
Jacob Schmidt
St. Paul, MN
15

12 ❑ 12-1
Jacob Schmidt
St. Paul, MN
15

13 ❑ 12-1
Jacob Schmidt
St. Paul, MN
CLT 1/2 of 1%
60

14 ❑ 12-1
Jacob Schmidt
St. Paul, MN
350

15 ❑ 12-1
Pfeiffer DBA Jacob Sch.
St.Paul, MN
CLT 1/2 of 1%
20

16 ❑ 12-1
Pfeiffer DBA Jacob Sch.
St. Paul, MN
10

17 ❑ 12-1
Pfeiffer DBA Jacob Sch.
St. Paul, MN
10

18 ❑ 12-1
Pfeiffer DBA Jacob Sch.
St. Paul, MN
10

19 ❑ 12-1
Pfeiffer DBA Jacob Sch.
St. Paul, MN
10

20 ❑ 12-1
Pfeiffer DBA Jacob Sch.
St. Paul, MN
10

21 ❑ 12-1
Pfeiffer DBA Jacob Sch.
St. Paul, MN
10

22 ❑ 12-1
Pfeiffer DBA Jacob Sch.
St. Paul, MN
10

23 ❑ 12-1
Pfeiffer DBA Jacob Sch.
St. Paul, MN
10

24 ❑ 12-1
Pfeiffer DBA Jacob Sch.
St. Paul, MN
10

25 ❑ 12-1
Pfeiffer DBA Jacob Sch.
St. Paul, MN
10

26 ❑ 12-1
Pfeiffer DBA Jacob Sch.
St. Paul, MN
10

27 ❑ 12-1
Pfeiffer DBA Jacob Sch.
St. Paul, MN
10

28 ❑ 12-1
Pfeiffer DBA Jacob Sch.
St. Paul, MN
10

29 ❑ 12-1
Pfeiffer DBA Jacob Sch.
St. Paul, MN
15

30 ❑ 12-1
Pfeiffer DBA Jacob Sch.
St. Paul, MN
5

31 ❑ 12-1
Associated DBA Jac. Sch.
St. Paul, MN
10

32 ❑ 12-1
Associated DBA Jac. Sch.
St. Paul, MN
10

33 ❑ 12-1
Associated DBA Jac. Sch.
St. Paul, MN
10

34 ❑ 12-1
Associated DBA Jac. Sch.
St. Paul, MN
10

35 ❑ 12-1
Associated DBA Jac. Sch.
St. Paul, MN
10

36 ❑ 12-1
Associated DBA Jac. Sch.
St. Paul, MN
10

37 ❑ 12-1
Associated DBA Jac. Sch.
St. Paul, MN
10

38 ❑ 12-1
Associated DBA Jac. Sch.
St. Paul, MN
10

39 ❑ 12-1
Associated DBA Jac. Sch.
St. Paul, MN
10

40 ❑ 12-1
Associated DBA Jac. Sch.
St. Paul, MN
10

1 ❑ 12-1
Associated DBA Jac. Sch.
St. Paul, MN
10

2 ❑ 12-1
Associated DBA Jac. Sch.
St. Paul, MN
10

3 ❑ 12-1
Associated DBA Jac. Sch.
St. Paul, MN
10

4 ❑ 12-1
Associated DBA Jac. Sch.
St. Paul, MN
10

5 ❑ 12-1
Associated DBA Jac. Sch.
St. Paul, MN
10

6 ❑ 12-1
Associated DBA Jac. Sch.
St. Paul, MN
15

7 ❑ 12-1
Associated DBA Jac. Sch.
St. Paul, MN
CLT 1/2 of 1% 5

8 ❑ 12-1
Jac. Schmidt Div. of Assoc.
St. Paul, MN
12

9 ❑ 12-1
Jac. Schmidt Div. of Assoc.
St. Paul, MN
12

10 ❑ 12-1
Jac. Schmidt Div. of Assoc.
St. Paul, MN
12

11 ❑ 12-1
Jac. Schmidt Div. of Assoc.
St. Paul, MN
12

12 ❑ 12-1
Jac. Schmidt Div. of Assoc.
St. Paul, MN
12

13 ❑ 12-1
Jac. Schmidt Div. of Assoc.
St. Paul, MN
12

14 ❑ 12-1
Jac. Schmidt Div. of Assoc.
St. Paul, MN
12

15 ❑ 12-1
Jac. Schmidt Div. of Assoc.
St. Paul, MN
15

16 ❑ 12-1
G. Heileman
La Crosse, WI
3

17 ❑ 12-1
Schmidt
Detroit, MI
125

18 ❑ 12-1
E & B
Detroit, MI
50

19 ❑ 12-1
E & B
Detroit, MI
65

20 ❑ 12-1
E & B
Detroit, MI
65

21 ❑ 12-2
E & B
Detroit, MI
125

22 ❑ 12-2
E & B
Detroit, MI
150

23 ❑ 12-1
Schmidt & Sons
Norristown, PA
20

24 ❑ 12-1
Schmidt & Sons
Norristown, PA
20

25 ❑ 12-2
Schmidt & Sons
Norristown, PA
5

26 ❑ 12-1
C. Schmidt & Sons
Philadelphia, PA
40

27 ❑ 12-1
C. Schmidt & Sons
Philadelphia, PA
35

28 ❑ 12-2
C. Schmidt & Sons
Philadelphia, PA
20

29 ❑ 12-1
C. Schmidt & Sons
Philadelphia, PA
20

30 ❑ 12-1
C. Schmidt & Sons
Philadelphia, PA
20

31 ❑ 12-1
C. Schmidt & Sons
Philadelphia, PA
20

32 ❑ 12-2
C. Schmidt & Sons
Philadelphia, PA
5

33 ❑ 12-1
C. Schmidt & Sons
Philadelphia, PA
110

34 ❑ 12-2
C. Schmidt & Sons
Philadelphia, PA
30

35 ❑ 12-2
Rhinelander
Rhinelander, WI
25

36 ❑ 12-2
Wausau
Wausau, WI
25

37 ❑ 12-1
Schoenling
Cincinnati, OH
350

38 ❑ 12-1
Schoenling
Cincinnati, OH
75

Front of 40 and 132-1

40 ❑ 12-1
Schoenling
Cincinnati, OH
25

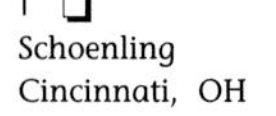
1 ❑ 12-1
Schoenling
Cincinnati, OH
25

2 ❑ 12-1
Schoenling
Cincinnati, OH
15

3 ❑ 12-1
Schoenling
Cincinnati, OH
500

4 ❑ 12-1-T
Schultz
Union City, NJ
1000+

5 ❑ 12-1-T
Schultz
Union City, NJ
1000+

6 ❑ 12-2
Gold Brau
Chicago, IL
35

7 ❑ 12-2
Gold Brau
Chicago, IL
35

8 ❑ 12-2
Gold Brau
Chicago, IL
800

9 ❑ 12-1
Sebewaing
Sebewaing, MI
30

10 ❑ 12-1
Sebewaing
Sebewaing, MI
30

11 ❑ 12-1-T
Chr. Heurich
Washington, DC
550

12 ❑ 12-1-T
Chr. Heurich
Washington, DC
400

Side of 12

14 ❑ 12-1-T
Chr. Heurich
Washington, DC
100

15 ❑ Front Like 14 12-1-T
Chr. Heurich
Washington, DC
135

16 ❑ 12-1-T
Chr. Heurich
Washington, DC
175

17 ❑ 12-1-T
Chr. Heurich
Washington, DC
800

18 ❑ 12-1-T
Chr. Heurich
Washington, DC
900

Side of 18

20 ❑ 12-1-T
Chr. Heurich
Washington, DC
75

21 ❑ 12-1-T
Chr. Heurich
Washington, DC
60

22 ❑ 12-1
Chr. Heurich
Washington, DC
65

23 ❑ 12-1
Columbia
Shenandoah, PA
60

24 ❑ 12-1
Columbia
Shenandoah, PA
60

25 ❑ 12-2
Spearman
Pensacola, FL
135

26 ❑ 12-2
Quality
Pensacola, FL
135

27 ❑ 12-1
Best
Chicago, IL
100

28 ❑ 12-2
Altes
Detroit, MI
50

29 ❑ 12-2
Altes
Detroit, MI
50

30 ❑ 12-2
Eastern Brewing Corp.
Hammonton, NJ
500

31 ❑ 12-2
Century
Norfolk, VA
135

32 ❑ 12-2
Century
Norfolk, VA
Metallic
225

33 ❑ 12-2
Century
Norfolk, VA
Dull
225

34 ❑ 12-2
Schoenhofen Edelweiss
Chicago, IL
40

35 ❑ 12-2
Drewry's Ltd.
South Bend, IN
35

36 ❑ 12-2
Drewry's Ltd.
South Bend, IN
10

37 ❑ 12-1-T
Globe
Baltimore, MD
600

38 ❑ 12-1
Walter
Pueblo, CO
100

39 ❑ 12-2
Walter
Pueblo, CO
50

40 ❑ 12-2
Walter
Pueblo, CO
10

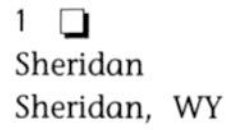

1 ❑ 12-1-T
Sheridan
Sheridan, WY
50

2 ❑ 12-1
Sheridan
Sheridan, WY
40

3 ❑ 12-2
Horlacher
Allentown, PA
10

4 ❑ 12-2
Old Dutch
Allentown, PA
10

5 ❑ 12-2
Colonial
Hammonton, NJ
50

6 ❑ 12-2
Sunshine
Reading, PA
50

7 ❑ 12-1
Eagle
Catasauqua, PA
135

8 ❑ 12-1
Lebanon Valley
Lebanon, PA
135

9 ❑ 12-1-OT
Sicks'
Salem, OR
110

10 ❑ 12-1-OT
Sicks'
Salem, OR
250

11 ❑ 12-1-OW
Sicks'
Salem, OR
45

12 ❑ 12-1-T
Sicks'
Salem, OR
125

13 ❑ 12-1
Sicks'
Salem, OR
110

14 ❑ 12-1
Sicks'
Salem, OR
100

15 ❑ 12-1-OW
Seattle Br. & Malt.
Seattle, WA
600

16 ❑ 12-1-OT
Seattle Br. & Malt.
Seattle, WA
250

17 ❑ 12-1-OT
Seattle Br. & Malt.
Seattle, WA
250

18 ❑ 12-1-OW
Sicks' Seattle
Seattle, WA
250

19 ❑ 12-1-OW
Sicks' Seattle
Seattle, WA
110

20 ❑ 12-1-OT
Sicks' Seattle
Seattle, WA
110

21 ❑ 12-1-OT
Sicks' Seattle
Seattle, WA
110

22 ❑ 12-1-T
Sicks' Seattle
Seattle, WA
Gold
110

23 ❑ 12-1-T
Sicks' Seattle
Seattle, WA
Silver
110

24 ❑ 12-1
Sicks' Seattle
Seattle, WA
110

25 ❑ 12-1
Sicks' Seattle Br.&Malt
Seattle, WA
100

26 ❑ 12-1
Sicks' Seattle
Seattle, WA
100

27 ❑ 12-1
Sicks' Seattle
Seattle, WA
100

28 ❑ 12-2
Pacific
Oakland, CA
400

29 ❑ 11-2
Regal Pale
San Francisco, CA
50

30 ❑ 11-2
Regal Pale
San Francisco, CA
50

31 ❑ 12-2
Reno
Reno, NV
40

32 ❑ 12-2
Pacific
Oakland, CA
650

33 ❑ 12-1
Southern
Tampa, FL
450

34 ❑ 12-1
Southern
Tampa, FL
Enamel
400

35 ❑ 12-1
Southern
Tampa, FL
Metallic
400

36 ❑ 12-2
Southern
Tampa, FL
225

37 ❑ 12-2
Southern
Tampa, FL
225

38 ❑ 12-1
International
Tampa, FL
100

39 ❑ 12-1
International
Tampa, FL
100

40 ❑ 12-1
International
Tampa, FL
100

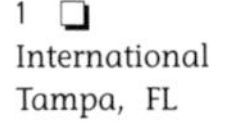
1 ❑ 12-1
International
Tampa, FL
100

2 ❑ 12-1
Southern
Tampa, FL
175

3 ❑ 12-2
Southern
Tampa, FL
200

4 ❑ 12-2
Southern
Tampa, FL
350

5 ❑ 12-2
Southern
Tampa, FL
350

6 ❑ 12-2
Southern
Tampa, FL
350

7 ❑ 12-2
Southern
Tampa, FL
350

8 ❑ 12-2
Southern
Tampa, FL
350

9 ❑ 12-2
Southern
Tampa, FL
350

10 ❑ 12-2
Southern
Tampa, FL
350

11 ❑ 12-1
International
Tampa, FL
90

12 ❑ 12-1
International
Tampa, FL
90

13 ❑ 12-2
Menominee-Marinette
Menominee, MI
45

14 ❑ 12-1-W
Peter Fox
Chicago, IL
900

15 ❑ 12-1-T
Peter Fox
Chicago, IL
400

16 ❑ 12-1
Lebanon Valley
Lebanon, PA
325

17 ❑ 12-1-OT
Star
Vancouver, WA
DNCMT (4%) PC 850

18 ❑ 12-1-OT
Interstate
Vancouver, WA
DNCMT (4%) PC 850

19 ❑ 12-1-T
Duquesne
Pittsburgh, PA
400

20 ❑ 12-2
Duquesne
Pittsburgh, PA
70

21 ❑ 12-2
Duquesne
Pittsburgh, PA
80

22 ❑ 12-2
Duquesne
Pittsburgh, PA
60

23 ❑ 12-1
William Simon
Buffalo, NY
18

24 ❑ 12-1
Jax Ice
Jacksonville, FL
275

25 ❑ 12-2
Drewry's Ltd.
South Bend, IN
35

26 ❑ 12-2
Skooner
St. Charles, MO
175

27 ❑ 12-2
Arizona
Phoenix, AZ
200

28 ❑ 12-2
Grace Bros.
Santa Rosa, CA
175

29 ❑ 12-1
Galveston-Houston
Galveston, TX
50

30 ❑ 12-2
Galveston-Houston
Galveston, TX
35

31 ❑ 12-1
Spearman
Pensacola, FL
110

32 ❑ 12-1
Sewanee
Pensacola, FL
110

33 ❑ 12-1
Spearman
Pensacola, FL
45

34 ❑ 12-2
Spearman
Pensacola, FL
250

35 ❑ 12-2
Spearman
Pensacola, FL
275

36 ❑ 12-2
Sewanee
Pensacola, FL
350

37 ❑ 12-2
Spearman
Pensacola, FL
100

38 ❑ 12-2
Spearman
Pensacola, FL
80

39 ❑ 12-1
Century
Norfolk, VA
45

40 ❑ 12-2
Spearman
Norfolk, VA
75

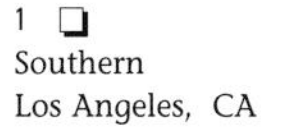

1 ❑ 12-1
Southern
Los Angeles, CA
80

2 ❑ 12-1
Southern
Los Angeles, CA
80

3 ❑ 12-1
Southern
Los Angeles, CA
80

4 ❑ 12-1-OT
Grace Bros LTD.
Los Angeles, CA
175

5 ❑ 12-1-OT
Grace Bros.
Santa Rosa, CA
175

6 ❑ 12-1-OT
Grace Bros.
Santa Rosa, CA
175

7 ❑ 12-2
Maier
Los Angeles, CA
20

8 ❑ 12-2
Maier
Los Angeles, CA
20

9 ❑ 12-2
Grace Bros.
Santa Rosa, CA
30

10 ❑ 12-2
Maier
Los Angeles, CA
12

11 ❑ 12-2
Maier
Los Angeles, CA
12

12 ❑ 12-2
Maier
Los Angeles, CA
12

13 ❑ 12-2-T
St. Claire
San Jose, CA
1000+

14 ❑ 12-2-T
St. Claire
San Jose, CA
1000+

15 ❑ 12-2-T
St. Claire
San Jose, CA
850

Front of 17, 18, 19

17 ❑ 12-1
Griesedieck Western
Belleville, IL
10

18 ❑ 12-1
Griesedieck Western
Belleville, IL
10

19 ❑ 12-1
Griesedieck Western
Belleville, IL
10

20 ❑ 12-1
Carling
Belleville, IL
10

21 ❑ 12-1
Carling
Belleville, IL
8

22 ❑ 12-1
Carling
Belleville, IL
5

23 ❑ 12-2
Carling
Belleville, IL
5

24 ❑ 12-1
Griesedieck Western
St. Louis, MO
15

Front of 26, 27

26 ❑ 12-1
Griesedieck Western
St. Louis, MO
10

27 ❑ 12-1
Griesedieck Western
St. Louis, MO
10

28 ❑ 12-1
Carling
St. Louis, MO
10

29 ❑ 12-1
Standard
Rochester, NY
20

30 ❑ 12-1
Standard
Rochester, NY
20

31 ❑ 12-1
Standard
Rochester, NY
60

32 ❑ 12-1
Standard Rochester
Rochester, NY
20

33 ❑ 12-2
Standard Rochester
Rochester, NY
10

34 ❑ 12-1
Standard Rochester
Rochester, NY
60

35 ❑ 12-2
Standard Rochester
Rochester, NY
150

36 ❑ 12-1
Standard
Cleveland, OH
Dull Gold
40

37 ❑ 12-1
Standard
Cleveland, OH
Metalic Gold
40

38 ❑ 12-1
Standard
Cleveland, OH
50

39 ❑ 12-2
Star Union
Chicago, IL
12

40 ❑ 12-2
Star Union
Peru, IL
12

1 ❑ 12-2
Fuhrmann & Schmidt
Shamokin, PA
60

2 ❑ 12-1
Stegmaier
Wilkes-Barre, PA
70

3 ❑ 12-1
Stegmaier
Wilkes-Barre, PA
40

4 ❑ 12-1
Stegmaier
Wilkes-Barre, PA
22

5 ❑ 12-2
Stegmaier
Wilkes-Barre, PA
Metallic
15

6 ❑ 12-2
Stegmaier
Wilkes-Barre, PA
Dull
15

7 ❑ 12-2
Stegmaier
Wilkes-Barre, PA
8

8 ❑ 12-2
Maier
Los Angeles, CA
12

9 ❑ 12-2
Maier
Los Angeles, CA
35

10 ❑ 12-2
Grace Bros.
Santa Rosa, CA
15

11 ❑ 12-2
Grace Bros.
Santa Rosa, CA
15

12 ❑ 12-2
Grace Bros.
Santa Rosa, CA
15

13 ❑ 12-2
Grace Bros.
Santa Rosa, CA
15

14 ❑ 12-2
Grace Bros.
Santa Rosa, CA
15

15 ❑ 12-2
ABC
Los Angeles, CA
18

16 ❑ 12-2
Maier
Los Angeles, CA
5

17 ❑ 12-2
Maier
Los Angeles, CA
5

18 ❑ 12-2-T
St. Claire
San Jose, CA
1000+

19 ❑ 12-2
Haberle Congress
Rochester, NY
75

20 ❑ 12-2
Haberle Congress
Syracuse, NY
75

21 ❑ 12-1
Geo. F. Stein
Buffalo, NY
1000+

22 ❑ 12-2
Geo. F. Stein
Buffalo, NY
300

23 ❑ 12-1
Geo. F. Stein
Buffalo, NY
115

24 ❑ 12-2
Geo. F. Stein
Buffalo, NY
135

25 ❑ 12-2
Geo. F. Stein
Buffalo, NY
125

26 ❑ 12-1
Geo. F. Stein
Buffalo, NY
55

27 ❑ 12-1
Leisy
Cleveland, OH
65

28 ❑ 12-1-OT
Sterling
Evansville, IN
750

29 ❑ 12-2
Sterling
Evansville, IN
225

30 ❑ 12-2
Sterling
Evansville, IN
125

31 ❑ 12-1-OT
Sterling
Evansville, IN
250

32 ❑ 12-1-OT
Sterling
Evansville, IN
25

33 ❑ 12-1-O
Sterling
Evansville, IN
25

34 ❑ 12-2
Sterling
Evansville, IN
10

35 ❑ 12-2
Sterling
Evansville, IN
10

36 ❑ 12-2
Sterling
Evansville, IN
25

37 ❑ 12-2
Sterling
Evansville, IN
10

38 ❑ 12-2
Sterling
Evansville, IN
10

39 ❑ 12-2
Sterling
Evansville, IN
5

40 ❑ 12-1
Rice Lake
Rice Lake, WI
175

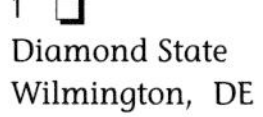

1 ❑ 12-1
Diamond State
Wilmington, DE
400

2 ❑ 12-2
Stolz
Tampa, FL
60

3 ❑ 12-2
International
Covington, KY
60

4 ❑ 12-2
International
Buffalo, NY
Metallic 45

5 ❑ 12-2
International
Buffalo, NY
Enamel 45

6 ❑ 12-1
Jones
Smithton, PA
150

7 ❑ 12-1-T
Storz
Omaha, NE
225

Front of 9, 10

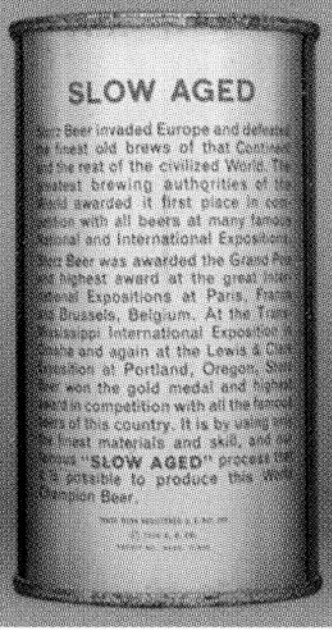

9 ❑ 12-1-T
Storz
Omaha, NE
225

10 ❑ 12-1-T
Storz
Omaha, NE
225

11 ❑ 12-1-T
Storz
Omaha, NE
1000+

12 ❑ 12-1-T
Storz
Omaha, NE
1000+

13 ❑ 12-1-T
Storz
Omaha, NE
1000+

14 ❑ 12-1-T
Storz
Omaha, NE
250

15 ❑ 12-1-T
Storz
Omaha, NE
1000+

16 ❑ 12-1-T
Storz
Omaha, NE
200

17 ❑ 12-1-T
Storz
Omaha, NE
40

18 ❑ 12-1
Storz
Omaha, NE
Enamel Red 20

19 ❑ 12-1
Storz
Omaha, NE
Metallic Red 20

20 ❑ 12-1
Storz
Omaha, NE
12

21 ❑ 12-2
Storz
Omaha, NE
10

22 ❑ 12-2
Storz
Omaha, NE
10

23 ❑ 12-2
Storz
Omaha, NE
15

24 ❑ 12-2
Storz
Omaha, NE
8

25 ❑ 12-2
Storz
Omaha, NE
8

26 ❑ 12-2
Storz
Omaha, NE
8

27 ❑ 12-2
Storz
Omaha, NE
6

28 ❑ 12-2
Storz
Omaha, NE
6

29 ❑ 12-1
Stroh
Detroit, MI
7

30 ❑ 12-1
Stroh
Detroit, MI
Metallic 7

31 ❑ 12-1
Stroh
Detroit, MI
Enamel 7

32 ❑ 12-2-OT
Stroudsburg
Stroudsburg, PA
1000+

33 ❑ 12-2-OT
Stroudsburg
Stroudsburg, PA
1000+

34 ❑ 12-2-OT
Stroudsburg
Stroudsburg, PA
NY Dist. bf 1000+

35 ❑ 12-1
Sunshine
Reading, PA
20

36 ❑ 12-2
Sunshine
Reading, PA
35

37 ❑ 12-2
Sunshine
Reading, PA
22

38 ❑ 12-2
Sunshine
Reading, PA
1000+

39 ❑ 12-2
Maier
Los Angeles, CA
110

40 ❑ 12-2
Pacific
Oakland, CA
500

FLATS

1 ❑ 11-2
Fisher
Salt Lake City, UT
150

2 ❑ 12-2
Lucky Lager
Salt Lake City, UT
150

3 ❑ 12-2
General
Salt Lake City, UT
150

4 ❑ 12-2
Superior
Chicago, IL
40

5 ❑ 12-2
Supreme
Chicago, IL
100

6 ❑ 12-2
Cumberland
Cumberland, MD
175

7 ❑ 12-1-T
Rainier
San Francisco, CA
400

8 ❑ 12-2
Maier
Los Angeles, CA
150

9 ❑ 12-2
Pacific
Oakland, CA
160

10 ❑ 12-2
Grace Bros.
Santa Rosa, CA
110

11 ❑ 12-2
Grace Bros.
Santa Rosa, CA
110

12 ❑ 12-1-OT
American
Rochester, NY
DULL GRAY
65

13 ❑ 12-1-OT
American
Rochester, NY
60

14 ❑ 12-1-OT
American
Rochester, NY
400

15 ❑ 12-1-T
Atlantic
Chicago, IL
20

Front of 17, 18

17 ❑ 12-1
Atlantic
Chicago, IL
18

18 ❑ 12-1
Atlantic
Chicago, IL
15

19 ❑ 12-1
Atlantic
Chicago, IL
15

20 ❑ 12-1
Atlantic
Chicago, IL
15

21 ❑ 12-1
Atlantic
Chicago, IL
15

22 ❑ 12-2
Atlantic
Chicago, IL
30

23 ❑ 12-2
Atlantic
Chicago, IL
45

24 ❑ 12-2
Atlantic
Chicago, IL
6

25 ❑ 12-2
Atlantic
South Bend, IN
7

26 ❑ 12-1
Pittsburgh
Pittsburgh, PA
100

27 ❑ 12-1
Pittsburgh
Pittsburgh, PA
250

28 ❑ 12-2
Blatz
Milwaukee, WI
175

29 ❑ 12-2
Blatz
Milwaukee, WI
175

30 ❑ 12-1
Gold Brau
Chicago, IL
20

31 ❑ 12-1
Gold Brau
Chicago, IL
20

32 ❑ 12-1-OT
Manhattan
Chicago, IL
1000+

33 ❑ 12-1-T
Tivoli Union
Denver, CO
135

34 ❑ 12-1
Tivoli Union
Denver, CO
80

35 ❑ 12-1
Tivoli Union
Denver, CO
60

36 ❑ 12-1
Tivoli
Denver, CO
50

37 ❑ 12-2
Tivoli
Denver, CO
30

38 ❑ 12-2
Tivoli
Denver, CO

39 ❑ 12-2
Tivoli
Denver, CO
25

40 ❑ 12-2
Tivoli
Denver, CO
25

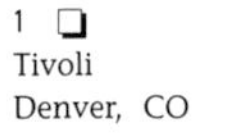
1 ❑ 12-1
Tivoli
Denver, CO
50

Back of 1

3 ❑ 12-2
Tivoli
Denver, CO
20

4 ❑ 12-2
Tivoli
Denver, CO
85

5 ❑ 12-2
Tivoli
Denver, CO
35

6 ❑ 12-2
Tivoli
Denver, CO
200

7 ❑ 12-2
Rochester
Rochester, NY
70

8 ❑ 12-2
Rochester
Rochester, NY
40

9 ❑ 12-2
Standard Rochester
Rochester, NY
70

10 ❑ 12-2
Standard Rochester
Rochester, NY
175

11 ❑ 12-2
Standard Rochester
Rochester, NY
175

12 ❑ 12-2
Standard Rochester
Rochester, NY
35

13 ❑ 12-2
Standard Rochester
Rochester, NY
35

14 ❑ 12-2
Standard Rochester
Rochester, NY
200

15 ❑ 12-2
Standard Rochester
Rochester, NY
10

16 ❑ 12-2
Pacific
Oakland, CA
500

17 ❑ 12-1
Sunshine
Reading, PA
125

18 ❑ 12-1-OT
No Mandatory
1000+

19 ❑ 12-1-OT
Star
Vancouver, WA
1000+

20 ❑ 12-1-OT
Interstate
Vancouver, WA
DULL GRAY, 4% 1000+

21 ❑ 12-1-OT
Interstate
Vancouver, WA
DNCMT (4%) PC 1000+

22 ❑ 12-1-T
Monarch
Los Angeles, CA
1000+

23 ❑ 12-1-OT
ABC
St. Louis, MO
1000+

24 ❑ 12-1-OT
John F. Trommer, Inc.
Orange, NJ
750

25 ❑ 12-2-T
John F. Trommer, Inc.
Orange, NJ
475

26 ❑ 12-1-W
John F. Trommer, Inc.
Orange, NJ
110

27 ❑ 12-2-T
John F. Trommer, Inc.
Orange, NJ
50

28 ❑ 12-2-T
John F. Trommer, Inc.
Orange, NJ
50

29 ❑ 12-2
John F. Trommer, Inc.
Orange, NJ
50

30 ❑ 12-1-OT
John F. Trommer, Inc.
Orange, NJ
450

31 ❑ 12-2-T
John F. Trommer, Inc.
Orange, NJ
200

32 ❑ 12-2-T
John F. Trommer, Inc.
Orange, NJ
125

33 ❑ 12-2-T
John F. Trommer, Inc.
Orange, NJ
100

34 ❑ 12-1-OT
John F. Trommer, Inc.
Orange, NJ
1000+

35 ❑ 12-1-OT
John F. Trommer, Inc.
Orange, NJ
1000+

36 ❑ 12-1
John F. Trommer, Inc.
Orange, NJ
1000+

37 ❑ 12-2
Piel Bros.
Brooklyn, NY
60

38 ❑ 12-2
Piel Bros.
Brooklyn, NY
60

39 ❑ 12-3
Trommer
Brooklyn, NY
30

40 ❑ 12-1
Trophy
Chicago, IL
60

1 ❑ 12-2
B.B.
Chicago, IL
25

2 ❑ 12-2
Schoenhofen Edelweiss
Chicago, IL
25

3 ❑ 12-2
Drewry's Ltd.
South Bend, IN
20

4 ❑ 12-1
Tampa Florida
Tampa, FL
200

5 ❑ 12-2
Tampa Florida
Tampa, FL
30

6 ❑ 12-1
Tampa Florida
Tampa, FL
500

7 ❑ 12-2
Tampa Florida
Tampa, FL
80

8 ❑ 12-2
Tampa Florida
Tampa, FL
50

9 ❑ 12-1-OT
Northampton
Northampton, PA
700

10 ❑ 12-2-OT
Northampton
Northampton, PA
300

11 ❑ 12-2-OT
Northampton
Northampton, PA
750

12 ❑ 12-2-OT
Northampton
Northampton, PA
300

13 ❑ 12-2-OT
Northampton
Northampton, PA
375

14 ❑ 12-2-OW
Northampton
Northampton, PA
500

15 ❑ 12-2-OT
Northampton
Northampton, PA
375

16 ❑ 12-2-OT
Northampton
Northampton, PA
300

17 ❑ 12-2-OT
Northampton
Northampton, PA
325

18 ❑ 12-1-OT
No Mandatory

American Liquor 1000+

19 ❑ 12-1-OT
Wehle
West Haven, CT
1000+

20 ❑ 12-2
Spearman
Pensacola, FL
25

21 ❑ 12-2
Spearman
Pensacola, FL
22

22 ❑ 12-1
Best
Chicago, IL
10

23 ❑ 12-1
Best
Chicago, IL
7

24 ❑ 12-2
Canadian Ace
Chicago, IL
20

25 ❑ 12-2
Tudor
Chicago, IL
18

26 ❑ 12-2
Tudor
Chicago, IL
15

27 ❑ 12-2
Tudor
Chicago, IL
15

28 ❑ 12-2
Tudor
Chicago, IL
10

29 ❑ 12-2
Tudor
Chicago, IL
10

30 ❑ 12-2
Tudor
Chicago, IL
10

31 ❑ 12-1
Cumberland
Cumberland, MD
Green Can 20

32 ❑ 12-2
Cumberland
Cumberland, MD
18

33 ❑ 12-2
Cumberland
Cumberland, MD
15

34 ❑ 12-2
Cumberland
Cumberland, MD
12

35 ❑ 12-2
Cumberland
Cumberland, MD
10

36 ❑ 12-1
Metropolis
Trenton, NJ
20

37 ❑ 12-2
Metropolis
Trenton, NJ
18

38 ❑ 12-2
Metropolis
Trenton, NJ
15

39 ❑ 12-1
Metropolis
Trenton, NJ
15

40 ❑ 12-1
Metropolis
Trenton, NJ
12

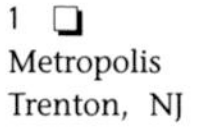
1 ❑ 12-2
Metropolis
Trenton, NJ
25

2 ❑ 12-2
Metropolis
Trenton, NJ
10

3 ❑ 12-2
Metropolis
Trenton, NJ
10

4 ❑ 12-1
Metropolis
Trenton, NJ
110

5 ❑ 12-2
Metropolis
Trenton, NJ
125

6 ❑ 12-2
Metropolis
Trenton, NJ
135

7 ❑ 12-2
Tudor
Trenton, NJ
12

8 ❑ 12-2
Tudor
Trenton, NJ
12

9 ❑ 12-2
Tudor
Trenton, NJ
70

10 ❑ 12-1
Geo. F. Stein
Buffalo, NY
30

11 ❑ 12-1
Geo. F. Stein
Buffalo, NY
20

12 ❑ 12-1
Hornell
Hornell, NY
25

13 ❑ 12-2
Hornell
Hornell, NY
20

14 ❑ 12-1
Hornell
Hornell, NY
18

15 ❑ 12-2
Hornell
Hornell, NY
15

16 ❑ 12-2
Hornell
Hornell, NY
12

17 ❑ 12-2
Hornell
Hornell, NY
150

18 ❑ 12-2
Five Star
New York, NY
200

19 ❑ 12-1-T
Metropolis
New York, NY
25

20 ❑ 12-1-T
Metropolis
New York, NY
50

21 ❑ 12-1-T
Metropolis
New York, NY
30

22 ❑ 12-1
Burkhardt
Akron, OH
30

23 ❑ 12-1
Burkhardt
Akron, OH
20

24 ❑ 12-1
Century
Norfolk, VA
20

25 ❑ 12-1
Century
Norfolk, VA
15

26 ❑ 12-1
Tudor
Norfolk, VA
20

27 ❑ 12-2
Tudor
Norfolk, VA
18

28 ❑ 12-2
Tudor
Norfolk, VA
15

29 ❑ 12-1
Tudor
Norfolk, VA
12

30 ❑ 12-2
Tudor
Norfolk, VA
10

31 ❑ 12-2
Tudor
Norfolk, VA
12

32 ❑ 12-2
Tuxedo
Chicago, IL
110

33 ❑ 12-2
Tuxedo
Chicago, IL
110

34 ❑ 12-2
Tuxedo
Norfolk, VA
125

35 ❑ 11-2
Atlantic
Spokane, WA
110

36 ❑ 12-2
Atlantic
Spokane, WA
110

37 ❑ 12-2
Tuxedo
Spokane, WA
125

38 ❑ 12-2
Tuxedo
Spokane, WA
110

39 ❑ 12-2
Atlantic
Chicago, IL
20

40 ❑ 12-2
Terre Haute
Terre Haute, IN
60

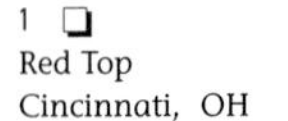

1 ❑ 12-2
Red Top
Cincinnati, OH
60

2 ❑ 12-1
Red Top
Cincinnati, OH
30

3 ❑ 12-2
Canadian Ace
Chicago, IL
300

4 ❑ 12-2
Colonial
Hammonton, NJ
250

5 ❑ 12-2
Uchtorff
Davenport, IA
125

6 ❑ 12-1-OT
Becker Products
Ogden, UT
175

7 ❑ 12-1-OT
Becker Products
Ogden, UT
110

8 ❑ 12-2
Becker Products
Ogden, UT
100

9 ❑ 11-2
Becker Products
Ogden, UT
100

10 ❑ 12-2
Becker Products
Ogden, UT
100

11 ❑ 12-1-T
Roger Williams
Cranston, RI
325

12 ❑ 12-2
Fuhrmann & Schmidt
Shamokin, PA
75

13 ❑ 12-2
Fuhrmann & Schmidt
Shamokin, PA
350

14 ❑ 12-1
Gettelman, Div. of Miller
Milwaukee, WI
12

15 ❑ 12-1
Gettelman, Div. of Miller
Milwaukee, WI
10

16 ❑ 12-1
Gettelman, Div. of Miller
Milwaukee, WI
12

17 ❑ 12-1-T
West End
Utica, NY
35

18 ❑ 12-1
West End
Utica, NY
30

19 ❑ 12-1
West End
Utica, NY
20

20 ❑ 12-2
West End
Utica, NY
20

21 ❑ 12-1-T
West End
Utica, NY
25

22 ❑ 12-1
West End
Utica, NY
18

23 ❑ 12-1
West End
Utica, NY
18

24 ❑ 12-2
West End
Utica, NY
15

25 ❑ 12-2
West End
Utica, NY
10

26 ❑ 12-2
West End
Utica, NY
10

27 ❑ 12-2
West End
Utica, NY
8

28 ❑ 12-2
West End
Utica, NY
175

29 ❑ 12-2
West End
Utica, NY
110

30 ❑ 12-1
El Dorado
Stockton, CA
50

31 ❑ 12-1
El Dorado
Stockton, CA
350

Front of 33, 34

33 ❑ 12-1-OT
Adam Scheidt
Norristown, PA
200

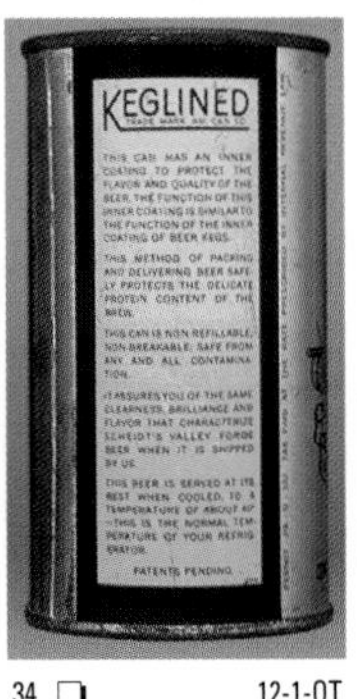

34 ❑ 12-1-OT
Adam Scheidt
Norristown, PA
200

35 ❑ 12-1-OW
Adam Scheidt
Norristown, PA
200

36 ❑ 12-1-OW
Adam Scheidt
Norristown, PA
400

37 ❑ 12-1-OW
Adam Scheidt
Norristown, PA
400

38 ❑ 12-1-OT
Adam Scheidt
Norristown, PA
75

39 ❑ 12-1-OT
Adam Scheidt
Norristown, PA
75

40 ❑ 12-1
Adam Scheidt
Norristown, PA
30

1 ❑ 12-1
Adam Scheidt
Norristown, PA
20

2 ❑ 12-1
Adam Scheidt
Norristown, PA
22

3 ❑ 12-1
Adam Scheidt
Norristown, PA
10

4 ❑ 12-1-OT
Adam Scheidt
Norristown, PA
700

Side of 4

6 ❑ 12-1-OT
Adam Scheidt
Norristown, PA
800

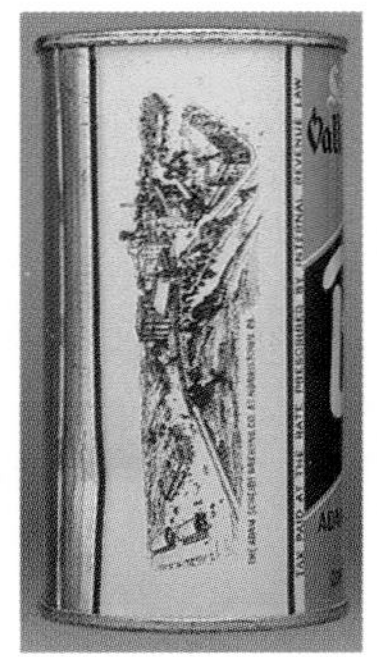
Side of 6

8 ❑ 12-1
Adam Scheidt
Norristown, PA
80

9 ❑ 12-1
Adam Scheidt
Norristown, PA
75

10 ❑ 12-2
Valley Forge
Norristown, PA
10

11 ❑ 12-1
Valley Forge
Norristown, PA
75

12 ❑ 12-2
C. Schmidt & Sons
Norristown, PA
10

13 ❑ 12-2
C. Schmidt & Sons
Norristown, PA
6

14 ❑ 12-2
Maier
Los Angeles, CA
10

15 ❑ 12-2
Maier
Los Angeles, CA
10

16 ❑ 12-2
Maier
Los Angeles, CA
10

17 ❑ 12-2
Maier
Los Angeles, CA
8

18 ❑ 12-2
Pacific
Oakland, CA
45

19 ❑ 12-2
Van Merritt
Chicago, IL
10

20 ❑ 12-1
Van Merritt
Joliet, IL
15

21 ❑ 12-2
Van Merritt
Joliet, IL
10

22 ❑ 12-2
Old Crown
Fort Wayne, IN
6

23 ❑ 12-1
Van Merritt
Burlington, WI
75

24 ❑ 12-2
Van Merritt
Oconto, WI
12

25 ❑ 12-2
Maier
Los Angeles, CA
10

26 ❑ 12-2
Pacific
Oakland, CA
45

27 ❑ 12-2
Pacific
Oakland, CA
50

28 ❑ 12-2
Grace Bros.
Santa Rosa, CA
12

29 ❑ 12-2
Grace Bros.
Santa Rosa, CA
12

30 ❑ 12-2
Grace Bros.
Santa Rosa, CA
12

31 ❑ 12-2
Sewanee
Pensacola, FL
Metallic
150

32 ❑ 12-2
Sewanee
Pensacola, FL
Enamel
85

33 ❑ 12-2
Spearman
Pensacola, FL
85

34 ❑ 12-2
Sewanee
Pensacola, FL
1000+

35 ❑ 12-2
Century
Norfolk, VA
85

36 ❑ 11-1
Atlantic
Spokane, WA
350

37 ❑ 12-1
Atlantic
Spokane, WA
350

38 ❑ 12-2
Atlas
Chicago, IL
60

39 ❑ 12-2
Drewry's Ltd.
South Bend, IN
50

40 ❑ 12-1-OT
Forest City
Cleveland, OH
375

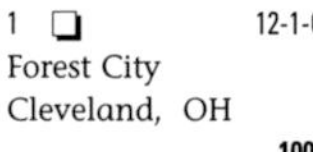

1 ❑ 12-1-OT
Forest City
Cleveland, OH
1000+

2 ❑ 12-1-T
Forest City
Cleveland, OH
900

3 ❑ 12-1-OT
Forest City
Cleveland, OH
375

4 ❑ 12-1-OT
Forest City
Cleveland, OH
110

5 ❑ 12-1-T
Forest City
Cleveland, OH
Brewed Right... 300

6 ❑ 12-1-T
Forest City
Cleveland, OH
Brewed With... 300

7 ❑ 12-1-T
Forest City
Cleveland, OH
Every Resource... 300

8 ❑ 12-1-T
Forest City
Cleveland, OH
For the Keenest... 300

9 ❑ 12-1-T
Forest City
Cleveland, OH
Waldorf Beer... 300

10 ❑ 12-1-OT
Forest City
Cleveland, OH
1000+

11 ❑ 12-1
Walter
Pueblo, CO
900

Side of 11

13 ❑ 12-1
Walter
Pueblo, CO
900

Side of 13

15 ❑ 12-1-T
Walter
Pueblo, CO
30

16 ❑ 12-1
Walter
Pueblo, CO
25

17 ❑ 12-2
Walter
Pueblo, CO
250

18 ❑ 12-2
Walter
Pueblo, CO
20

19 ❑ 12-2
Walter
Pueblo, CO
4

20 ❑ 12-2
Walter
Pueblo, CO
18

21 ❑ 12-1
Walter
Eau Claire, WI
20

22 ❑ 12-1
Walter
Eau Claire, WI
30

23 ❑ 12-1
Walter
Eau Claire, WI
12

24 ❑ 12-2
Walter
Eau Claire, WI
8

25 ❑ 12-2
Weber Waukesha
La Crosse, WI
30

26 ❑ 12-2
Weber Wauk. Div. G. Heil.
La Crosse, WI
30

27 ❑ 12-2
G. Heileman
La Crosse, WI
30

28 ❑ 12-2
Weber Waukesha
Sheboygan, WI
40

29 ❑ 12-2
Weber Waukesha
Waukesha, WI
Metallic 15

30 ❑ 12-2
Weber Waukesha
Waukesha, WI
Enamel 15

31 ❑ 12-2
Weber Waukesha
Waukesha, WI
Metallic 20

32 ❑ 12-2
Weber Waukesha
Waukesha, WI
Enamel 20

33 ❑ 12-2
Weber Waukesha
Waukesha, WI
30

34 ❑ 12-2
Weber Waukesha
Waukesha, WI
35

35 ❑ 12-1-OT
Wehle
West Haven, CT
Long Opener 350

36 ❑ 12-1-OT
Wehle
West Haven, CT
250

37 ❑ 12-1-OT
Wehle
West Haven, CT
900

38 ❑ 12-1-OT
Wehle
West Haven, CT
1000+

39 ❑ 12-1
Hampden
Willimansett, MA
150

40 ❑ 12-2
Walter
Pueblo, CO
85

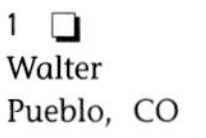

1 ❑ 12-2
Walter
Pueblo, CO
45

2 ❑ 12-2
Walter
Pueblo, CO
75

3 ❑ 12-2
Walter
Pueblo, CO
35

4 ❑ 12-2
Fesenmeier
Huntington, WV
65

5 ❑ 12-2
Fesenmeier
Huntington, WV
50

6 ❑ 12-2
Sioux City
Sioux City, IA
550

7 ❑ 12-2
Grace Bros.
Santa Rosa, CA
65

8 ❑ 12-2
Grace Bros.
Santa Rosa, CA
65

9 ❑ 11-2
Silver Springs
Tacoma, WA
Enamel **30**

10 ❑ 11-2
Silver Springs
Tacoma, WA
Metallic **30**

11 ❑ 12-1
Silver Springs
Tacoma, WA
30

12 ❑ 12-2
Eastern Brewing Corp.
Hammonton, NJ
85

13 ❑ 12-2
Walter
Eau Claire, WI
175

14 ❑ 12-1-OT
Manhattan
Chicago, IL
1000+

15 ❑ 12-1-OT
Westminster
Chicago, IL
1000+

16 ❑ 12-1-OT
Lubeck
Toledo, OH
1000+

17 ❑ 12-2
Minneapolis
Minneapolis, MN
8

18 ❑ 12-2
White Label
Minneapolis, MN
8

19 ❑ 12-2
Columbia
Shenandoah, PA
60

20 ❑ 12-2
Columbia
Shenandoah, PA
65

21 ❑ 12-1
Geo. Wiedemann
Newport, KY
15

22 ❑ 12-1
Geo. Wiedemann
Newport, KY
8

23 ❑ 12-1
Geo. Wiedemann
Newport, KY
8

24 ❑ 12-1
Geo. Wiedemann
Newport, KY
8

25 ❑ 12-1
Geo. Wiedemann
Newport, KY
8

26 ❑ 12-1
Geo. Wiedemann
Newport, KY
8

27 ❑ 12-1
Geo. Wiedemann
Newport, KY
8

28 ❑ 12-1
Geo. Wiedemann
Newport, KY
8

29 ❑ 12-1
Geo. Wiedemann
Newport, KY
8

30 ❑ 12-1
Geo. Wiedemann
Newport, KY
8

31 ❑ 12-1
Geo. Wiedemann
Newport, KY
8

32 ❑ 12-1
Geo. Wiedemann
Newport, KY
8

33 ❑ 12-1
Geo. Wiedemann
Newport, KY
8

34 ❑ 12-1
Geo. Wiedemann
Newport, KY
8

35 ❑ 12-1
Geo. Wiedemann
Newport, KY
8

36 ❑ 12-1
Geo. Wiedemann
Newport, KY
8

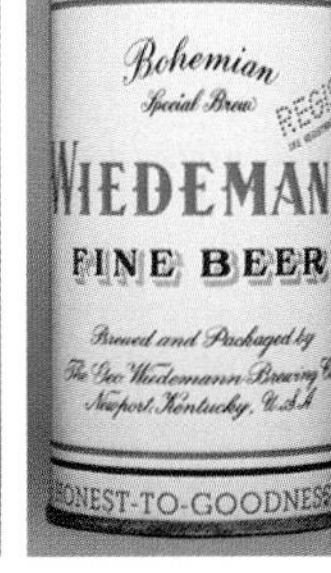

37 ❑ 12-1
Geo. Wiedemann
Newport, KY
8

38 ❑ 12-1
Geo. Wiedemann
Newport, KY
8

39 ❑ 12-1
Geo. Wiedemann
Newport, KY
8

40 ❑ 12-2
Geo. Wiedemann
Newport, KY
8

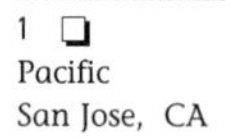

1 ❑ 12-1
Pacific
San Jose, CA
56

2 ❑ 12-2
Wieland's
San Jose, CA
60

3 ❑ 12-1
Wiessner
Baltimore, MD
175

4 ❑ 12-1
Wiessner
Baltimore, MD
110

5 ❑ 12-2
Hofbrau
Allentown, PA
25

6 ❑ 12-1
Fitzgerald Bros.
Troy, NY
500

7 ❑ 12-1-OT
San Francisco
San Francisco, CA
600

8 ❑ 12-1-OT
San Francisco
San Francisco, CA
325

9 ❑ 12-1-OT
San Francisco
San Francisco, CA
325

10 ❑ 12-1-OT
San Francisco
San Francisco, CA
325

11 ❑ 12-1-OT
San Francisco
San Francisco, CA
225

12 ❑ 12-2
Walter
Pueblo, CO
50

13 ❑ 12-2
Windsor
Chicago, IL
400

14 ❑ 12-1-OT
Whitewater
Whitewater, WI
110

15 ❑ 12-2
Swiss
Monroe, WI
5

16 ❑ 12-2
Jos. Huber
Monroe, WI
5

17 ❑ 12-2
Jos. Huber
Monroe, WI
5

18 ❑ 12-2
Swiss
Monroe, WI
60

19 ❑ 12-2
Jos. Huber
Monroe, WI
100

20 ❑ 12-2
Jos. Huber
Monroe, WI
5

21 ❑ 12-1
Wisconsin
Burlington, WI
12

22 ❑ 12-2
Wisconsin
La Crosse, WI
8

23 ❑ 12-2
Wisconsin
La Crosse, WI
5

24 ❑ 12-2
G. Heileman
La Crosse, WI
8

25 ❑ 12-1
Fox Head
Waukesha, WI
22

26 ❑ 12-2
Wisconsin
Waukesha, WI
8

27 ❑ 12-2
Wisconsin
Waukesha, WI
8

28 ❑ 12-2
Wisconsin
Waukesha, WI
10

29 ❑ 12-2
Fox Head
Waukesha, WI
8

30 ❑ 12-1
Fox Head
Waukesha, WI
50

31 ❑ 12-1
Fox Head
Waukesha, WI
45

32 ❑ 12-1
Fox Head
Waukesha, WI
45

33 ❑ 12-1
Spring City
Waukesha, WI
40

34 ❑ 12-1
Spring City
Waukesha, WI
40

35 ❑ 12-2
Wunderbrau
Cincinnati, OH
60

36 ❑ 12-2
Wunderbrau
Cincinnati, OH
3.2-7% b.f.
75

37 ❑ 12-2
Wunderbrau
Cincinnati, OH
40

38 ❑ 12-2
Wunderbrau
Cincinnati, OH
100

39 ❑ 12-2
Red Top
Cincinnati, OH
55

40 ❑ 12-2
Yankee
New York, NY
45

1 ❑ 12-2
Enterprise
Fall River, MA
600

2 ❑ 12-2
Colonial
Hammonton, NJ
65

3 ❑ 12-2
Wm. Gretz
Philadelphia, PA
500

4 ❑ 12-1
Old Reading
Reading, PA
100

5 ❑ 12-1
Old Reading
Reading, PA
12

6 ❑ 12-1
Reading
Reading, PA
8

7 ❑ 12-1
D.G. Yuengling & Son
Pottsville, PA
15

8 ❑ 12-2
D.G. Yuengling & Son
Pottsville, PA
15

9 ❑ 12-1
Pilsen
Chicago, IL
65

10 ❑ 12-1
Pilsen
Chicago, IL
65

11 ❑ 12-1
Pilsen
Chicago, IL
20

12 ❑ 12-1
Pilsen
Chicago, IL
15

13 ❑ 12-1
Pilsen
Chicago, IL
10

14 ❑ 12-2-OT
Grace Bros.
Santa Rosa, CA
1000+

15 ❑ 12-2
Kingsbury
Sheboygan, WI
15

16 ❑ 12-2
Kingsbury, Div. G. Heil.
Sheboygan, WI
15

17 ❑ 12-2
Kingsbury, Div. G. Heil.
Sheboygan, WI
15

18 ❑ 12-2
Kingsbury
Sheboygan, WI
10

"You're telling me? I know a good mouthpiece when I see one—and boy, those Cap-Sealed cans have it. You can drink right out of the can, from a clean, cap-protected surface. And that's not all, either . . ."

"Note this, Sousa! No special opener needed with *that* top. Any opener will do. No bother about deposits or returns, and the beer chills *fast!* I'll join the chorus for Cap-Sealed cans any time!" *(Two sizes: 12 oz. and qt.)*

USE ANY OPENER

CAP-SEALED CAN FOR BEER

CONTINENTAL © CAN COMPANY

ASK FOR BEER OR ALE IN "EASY TO OPEN" CAP-SEALED CANS

Before the advent of television and million-dollar Super Bowl commercials, brewers and can makers had to depend heavily on popular magazines like Collier's *and the* Saturday Evening Post, *as well as on local newspapers, to communicate the virtues of canned beer. Here are a pair of examples, courtesy of Continental Can Company and the Iroquois Beverage Corporation.*

Iroquois

LEADS AGAIN!

TEN REASONS WHY THE *Cap Sealed Can* IS BETTER FOR BEER

1. Opens and pours like a bottle.
2. Special lining, applied after can is made, guards taste.
3. Beer flows from a clean, cap-protected opening.
4. Small opening keeps out flavor-robbing oxygen.
5. Can shields beer from harmful light.
6. Cools faster.
7. No deposits; no returns.
8. Saves half the weight.
9. Takes less room in refrigerator.
10. Drink right from can.

First* BUFFALO BREWERY TO OFFER BEER and ALE *in Cans

Now you can enjoy the old fashioned flavor of Iroquois Beer and Ale *plus* the convenience of buying these great brews in cap-sealed cans.

First Buffalo brewery to offer beer and ale in this type of container, Iroquois has selected the cap-sealed can for its many advantages—believes you will appreciate them just as Iroquois sales leadership proves you appreciate the old-time taste of Iroquois Beer and Ale.

Whenever and wherever limited refrigeration space, quicker cooling or freedom from bottle returns is an important consideration, always ask for Iroquois Beer, Ale or Half & Half in cans.

BUFFALO'S LARGEST SELLING BEER and ALE

In 12-Ounce and 32-Ounce Bottles — in Cans — on Tap

IROQUOIS BEVERAGE CORPORATION, BUFFALO, N. Y.

Cones Section Explanatory Notes

Assume that unless stated otherwise, all cones have concave bottoms and no alcohol statement.

Blackhawk 152-29, 30, 31. Catalog defines the difference between these cans as one or two lips on the Indian silhouette. This is easier to differentiate by the length of the feathers.

Bub's 154-29 through 155-1. The first three cans have "over 80 years" at the top of the label. The following two cans have "over 90 years" at the top of the label.

Fort Pitt 163-10 and 163-11 have different wording in the ribbon on each side of the shield as captioned.

Frankenmuth 163-28 has the "e" in Ale filled in with orange and 163-29 is black.

Frankenmuth 164-2 has the original mandatory blacked out.

Stegmaier's Gold Medal 165-31 is listed with other Gold Medal cans from Stegmaier for brand continuity.

Indian Queen 168-32 is listed with Hohenadel Indian Queen for brand continuity.

National Ale 174-24 lists the New York Office at 5 Bedford St. N.Y.C.

National Ale 174-25 lists the New York Office at 67 West 44th Street. N.Y.C.

National Ale 174-26 lists the New York Office at 45 E. 43rd St. N.Y.C.

Old Bohemian Brand 175-6 listed with National Bohemian.

Old German 176-17 has a Union label lower left while 176-18 does not.

All **Heileman's Old Style** and **Old Style Lager** cans are shown together.

The first five **Heileman's,** 177-3 through 177-7, are prior to IRTP statement cans. Note that 177-6 and 177-7 have concave bottoms and cones without inverted ribs.

Regal 181-13 through 181-15 have light blue shading in shield, while 181-16 and 181-17 do not.

Schmidt Select 184-24 This can does have a 4% alcohol statement.

Little Dutch 188-20 and 188-21 are listed with Wacker.

CONES

10, 12 & 16 Oz.

1 ❑ 12-1-T
Aztec
San Diego, CA
400

2 ❑ 12-1-T
Aztec
San Diego, CA
275

3 ❑ 12-1-T
Aztec
San Diego, CA
225

4 ❑ 12-1-T
East Idaho
Pocatello, ID
150

5 ❑ 12-1
East Idaho
Pocatello, ID
135

6 ❑ 12-1
East Idaho
Pocatello, ID
DNCMT 3.2%
135

7 ❑ 12-1
East Idaho
Pocatello, ID
DNCMT 4%
135

8 ❑ 12-1-T
Altes
San Diego, CA
100

9 ❑ 12-1
Altes
San Diego, CA
100

10 ❑ 12-2
Altes
San Diego, CA
110

11 ❑ 12-2
Altes
San Diego, CA
140

12 ❑ 12-1-T
Altes
Detroit, MI
85

13 ❑ 12-1
Altes
Detroit, MI
90

14 ❑ 12-2
Altes
Detroit, MI
100

15 ❑ 12-1-T
Ambassador
Los Angeles, CA
1000+

16 ❑ 12-1-T
American
Baltimore, MD
150

17 ❑ 12-1
American
Baltimore, MD
150

18 ❑ 12-1-T
Arizona
Phoenix, AZ
1000+

19 ❑ 12-1-T
Apex
Seattle, WA
4%
1000+

20 ❑ 12-1-T
Rochester
Rochester, NY
1000+

21 ❑ 12-1
Louis Ziegler
Beaver Dam, WI
1000+

22 ❑ 12-1-T
Aztec
San Diego, CA
1000+

23 ❑ 12-1-T
Atlantic
Atlanta, GA
275

24 ❑ 12-1
Atlantic
Atlanta, GA
250

25 ❑ 12-1-T
Atlantic
Atlanta, GA
200

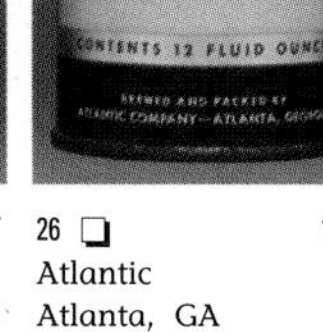

26 ❑ 12-1
Atlantic
Atlanta, GA
200

27 ❑ 12-1
August Wagner
Columbus, OH
225

28 ❑ 12-1-T
Red Top
Cincinnati, OH
125

29 ❑ 12-1-T
Red Top
Cincinnati, OH
3.2-7%
125

30 ❑ 12-1
Red Top
Cincinnati, OH
125

31 ❑ 12-1-T
Greater New York
New York, NY
800

32 ❑ 12-1-T
Mount Carbon
Pottsville, PA
135

1 ❑ 12-1
Mount Carbon
Pottsville, PA
135

2 ❑ 12-1
Mount Carbon
Pottsville, PA
135

3 ❑ 12-1
Bavarian
Covington, KY
125

Front of 5-8

5 ❑ 12-1-T
No Mandatory
"Better Buy", N.O. 4 % 450

6 ❑ 12-1-T
Rainier
San Francisco, CA
"Better Buy" 450

7 ❑ 12-1-T
Rainier
San Francisco, CA
"Better Buy", N.O. 4 % 450

8 ❑ 12-1-T
Rainier
San Francisco, CA
"Bert McDowell" 450

9 ❑ 12-1-T
Rainier
San Francisco, CA
"Bert McDowell", N.O. 4% 450

10 ❑ 12-2
Becker Br. and Malt.
Evanston, WY
135

11 ❑ 12-1-W
Largay
Waterbury, CT
900

12 ❑ 12-1-W
American
New Orleans, LA
900

13 ❑ 12-1-T
Jos. Schlitz
Milwaukee, WI
900

14 ❑ 12-1-T
Jos. Schlitz
Milwaukee, WI
900

15 ❑ 12-1-T
Franklin
Columbus, OH
NMT 3.2% 450

16 ❑ 12-1-T
Franklin
Columbus, OH
3.2-7% 450

17 ❑ 12-1-T
Franklin
Columbus, OH
225

18 ❑ 12-1
Franklin
Columbus, OH
225

19 ❑ 12-1-T
Berghoff
Ft. Wayne, IN
Flat Bottom 250

20 ❑ 12-1-T
Berghoff
Ft. Wayne, IN
100

21 ❑ 12-1-T
Berghoff
Ft. Wayne, IN
65

22 ❑ 12-1-T
Berghoff
Ft. Wayne, IN
50

23 ❑ 12-1-W
Berghoff
Ft. Wayne, IN
750

24 ❑ 12-1-T
Berghoff
Ft. Wayne, IN
50

25 ❑ 12-1-T
Berghoff
Ft. Wayne, IN
4% 50

26 ❑ 12-1
Berghoff
Ft. Wayne, IN
50

27 ❑ 12-1
Berghoff
Ft. Wayne, IN
650

28 ❑ 12-1-T
Beverwyck
Albany, NY
Flat Bottom 300

29 ❑ 12-1-T
Beverwyck
Albany, NY
Flat Bottom 300

30 ❑ 12-1-T
Beverwyck
Albany, NY
100

31 ❑ 12-1-T
Beverwyck
Albany, NY
100

32 ❑ 12-1-T
Beverwyck
Albany, NY
150

1 ❑ 12-1-T
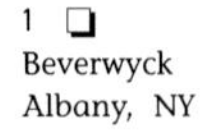
Beverwyck
Albany, NY
275

2 ❑ 12-1-T
Beverwyck
Albany, NY
90

3 ❑ 12-1-T
Beverwyck
Albany, NY
85

4 ❑ 12-1-T
Beverwyck
Albany, NY
85

5 ❑ 12-1-T
Beverwyck
Albany, NY
75

6 ❑ 12-1-T
Beverwyck
Albany, NY
75

7 ❑ 12-1-T
Beverwyck
Albany, NY
75

8 ❑ 12-1-T
Beverwyck
Albany, NY
75

9 ❑ 12-1-T
Beverwyck
Albany, NY
Flat Bottom
300

10 ❑ 12-1-T
Beverwyck
Albany, NY
100

11 ❑ 12-1-T
Beverwyck
Albany, NY
100

12 ❑ 12-1-T
Beverwyck
Albany, NY
250

13 ❑ 12-1-T
Beverwyck
Albany, NY
250

14 ❑ 12-1-T
Beverwyck
Albany, NY
250

15 ❑ 12-1-W
Beverwyck
Albany, NY
700

16 ❑ 12-1-W
Beverwyck
Albany, NY
375

17 ❑ 12-1-T
Beverwyck
Albany, NY
375

18 ❑ 12-1-T
Beverwyck
Albany, NY
250

19 ❑ 12-1-T
Beverwyck
Albany, NY
225

20 ❑ 12-1
Billings
Billings, MT
DNCMT 4%
600

21 ❑ 12-1
Billings
Billings, MT
CNMT 4%
450

22 ❑ 12-2
Cleveland Home
Cleveland, OH
350

23 ❑ 12-2
Cleveland Home
Cleveland, OH
DNCMT 3.2%
350

24 ❑ 12-2
Cleveland Home
Cleveland, OH
3.2-7%
350

25 ❑ 12-1-T
Blackhawk
Davenport, IA
200

26 ❑ 12-1-T
Blackhawk
Davenport, IA
DNCMT 4%
200

27 ❑ 12-1-T
Blackhawk
Davenport, IA
175

28 ❑ 12-1-T
Blackhawk
Davenport, IA
DNCMT 4%
175

29 ❑ 12-1
Blackhawk
Davenport, IA
Long Feathers
200

30 ❑ 12-1
Blackhawk
Davenport, IA
Long Feathers, 4%
200

31 ❑ 12-1
Blackhawk
Davenport, IA
Short Feathers, 4%
200

32 ❑ 12-1
Blackhawk
Davenport, IA
225

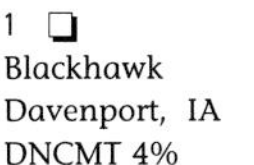

1 ❑ 12-1
Blackhawk
Davenport, IA
DNCMT 4% 225

2 ❑ 12-1-T
Blatz
Milwaukee, WI
Flat Bottom 450

3 ❑ 12-1-T
Blatz
Milwaukee, WI
350

4 ❑ 12-1-T
Blatz
Milwaukee, WI
300

5 ❑ 12-1-T
Blatz
Milwaukee, WI
Flat Bottom 175

6 ❑ 12-1-T
Blatz
Milwaukee, WI
Flat Bot. MT 4-1/2% 175

7 ❑ 12-1-T
Blatz
Milwaukee, WI
175

8 ❑ 12-1-T
Blatz
Milwaukee, WI
MT 4-1/2% 65

CONES

9 ❑ 12-1-T
Blatz
Milwaukee, WI
DNCMT 4% 65

10 ❑ 12-1-T
Blatz
Milwaukee, WI
65

11 ❑ 12-1-W
Blatz
Milwaukee, WI
100

12 ❑ 12-1-T
Blatz
Milwaukee, WI
MT 4-1/2% 65

13 ❑ 12-1-T
Blatz
Milwaukee, WI
65

14 ❑ 12-1-T
Blatz
Milwaukee, WI
55

15 ❑ 12-1-T
Blatz
Milwaukee, WI
65

16 ❑ 12-1-T
Blatz
Milwaukee, WI
MT 4-1/2% 65

17 ❑ 12-1-T
Blatz
Milwaukee, WI
65

18 ❑ 12-1-T
Blatz
Milwaukee, WI
DNCMT 4% 65

19 ❑ 12-1-T
Blatz
Milwaukee, WI
MT 4-1/2% 65

20 ❑ 12-1-T
Blatz
Milwaukee, WI
65

21 ❑ 12-1-T
Blatz
Milwaukee, WI
DNCMT 4% 65

22 ❑ 12-1-T
Blatz
Milwaukee, WI
MT 4-1/2% 65

23 ❑ 12-1-W
Blatz
Milwaukee, WI
250

24 ❑ 12-1-W
Blatz
Milwaukee, WI
250

25 ❑ 12-1-T
Blatz
Milwaukee, WI
55

26 ❑ 12-1-T
Blatz
Milwaukee, WI
DNCMT 4% 55

27 ❑ 12-1-T
Blatz
Milwaukee, WI
55

28 ❑ 12-1-T
Blatz
Milwaukee, WI
3.2-7% 55

29 ❑ 12-1-T
Blitz-Weinhard
Portland, OR
750

30 ❑ 12-1-T
North Bay
Santa Rosa, CA
350

31 ❑ 12-1-T
Dallas-Ft. Worth
Dallas, TX
400

32 ❑ 12-1
Dallas-Ft. Worth
Dallas, TX
400

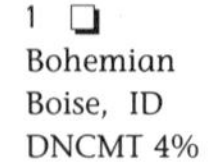

1 ❑ 12-1-T
Bohemian
Boise, ID
DNCMT 4% 65

2 ❑ 12-1
Bohemian
Boise, ID
DNCMT 4% 65

3 ❑ 12-2
Bohemian
Boise, ID
150

4 ❑ 12-2
Bohemian
Boise, ID
250

5 ❑ 12-1-T
Bohemian
Spokane, WA
DNCMT 4% 250

6 ❑ 12-1-T
Bohemian
Spokane, WA
DNCMT 4% 90

7 ❑ 12-1-T
Bohemian
Spokane, WA
DNCMT 4% 65

8 ❑ 12-1
Bohemian
Spokane, WA
DNCMT 4% 65

9 ❑ 12-1
Spearman
Pensacola, FL
300

10 ❑ 12-1-T
Boston Beer
Boston, MA
1000+

11 ❑ 12-1-T
Ind. Milwaukee
Milwaukee, WI
65

12 ❑ 12-1-T
Ind. Milwaukee
Milwaukee, WI
4% 65

13 ❑ 12-1
Ind. Milwaukee
Milwaukee, WI
65

14 ❑ 12-1
Ind. Milwaukee
Milwaukee, WI
3.2-7% 65

15 ❑ 12-1
Ind. Milwaukee
Milwaukee, WI
4% 65

16 ❑ 12-1-T
Peter Breidt
Elizabeth, NJ
1000+

17 ❑ 12-1-W
Peter Breidt
Elizabeth, NJ
800

18 ❑ 12-1-T
Peter Breidt
Elizabeth, NJ
125

19 ❑ 12-1
Peter Breidt
Elizabeth, NJ
125

20 ❑ 12-1-T
Peter Breidt
Elizabeth, NJ
1000+

21 ❑ 12-1
Rice Lake
Rice Lake, WI
110

22 ❑ 12-1
Rice Lake
Rice Lake, WI
3.2% 110

23 ❑ 12-1
Rice Lake
Rice Lake, WI
STRONG 200

24 ❑ 12-1-T
Brockert
Worcester, MA
225

25 ❑ 12-1-T
Brockert
Worcester, MA
1000+

26 ❑ 12-1-T
Bruckmann
Cincinnati, OH
600

27 ❑ 12-2-T
Bruckmann
Cincinnati, OH
DNCMT 3.2% 200

28 ❑ 12-2-T
Bruckmann
Cincinnati, OH
135

29 ❑ 12-1
Peter Bub
Winona, MN
90

30 ❑ 12-1
Peter Bub
Winona, MN
DNCMT 3.2% 90

31 ❑ 12-1
Peter Bub
Winona, MN
STRONG 90

32 ❑ 12-1
Peter Bub
Winona, MN
90

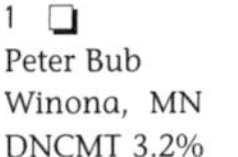
1 ❑ 12-1
Peter Bub
Winona, MN
DNCMT 3.2% **90**

2 ❑ 12-1
Peter Bub
Winona, MN
DNCMT 3.2% **75**

3 ❑ 12-1
Peter Bub
Winona, MN
STRONG **75**

4 ❑ 12-2
Buckeye
Toledo, OH
275

5 ❑ 12-1-T
Buckeye
Toledo, OH
800

6 ❑ 12-1-T
Buckeye
Toledo, OH
800

7 ❑ 12-1-T
Buckeye
Toledo, OH
CNMT 3.2% **300**

8 ❑ 12-1-T
Buckeye
Toledo, OH
3.2-7% **300**

9 ❑ 12-1-T
Buckeye
Toledo, OH
3.2-7% **150**

10 ❑ 12-1-T
Buckeye
Toledo, OH
125

11 ❑ 12-1
Buckeye
Toledo, OH
125

12 ❑ 12-2
Buckeye
Toledo, OH
85

13 ❑ 12-2
Buckeye
Toledo, OH
175

14 ❑ 12-1-T
Buffalo
Sacramento, CA
325

15 ❑ 12-1-T
Buffalo
Sacramento, CA
400

16 ❑ 12-1-T
Warsaw
Warsaw, IL
125

17 ❑ 12-1
Warsaw
Warsaw, IL
125

18 ❑ 12-1
Warsaw
Warsaw, IL
4% bf **125**

19 ❑ 12-1
Warsaw
Warsaw, IL
4% At Seam **125**

20 ❑ 12-1-T
Burger
Cincinnati, OH
400

21 ❑ 12-1
Burger
Cincinnati, OH
400

22 ❑ 12-1-T
Burger
Cincinnati, OH
Flat Bottom **700**

23 ❑ 12-1-T
Burger
Cincinnati, OH
600

24 ❑ 12-1-T
Burger
Cincinnati, OH
325

25 ❑ 12-1-T
Burger
Cincinnati, OH
175

26 ❑ 12-1-T
Burger
Cincinnati, OH
175

27 ❑ 12-1
Burger
Cincinnati, OH
60

28 ❑ 12-1
Burger
Cincinnati, OH
60

29 ❑ 12-1
Burger
Cincinnati, OH
225

30 ❑ 12-1-W
Burger
Cincinnati, OH
950

31 ❑ 12-1-T
Burger
Cincinnati, OH
250

32 ❑ 12-1-T
Burger
Cincinnati, OH
3.2-7% **250**

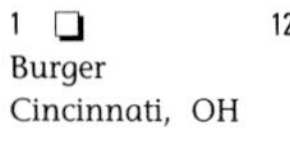

1 ❑ 12-1-T
Burger
Cincinnati, OH
250

2 ❑ 12-1-T
Burkhardt
Akron, OH
3.2-7% 120

3 ❑ 12-1-T
Burkhardt
Akron, OH
110

4 ❑ 12-1
Burkhardt
Akron, OH
110

5 ❑ 12-1
Burkhardt
Akron, OH
200

6 ❑ 12-1-T
Bushkill Products
Easton, PA
1000+

7 ❑ 12-1-T
Butte
Butte, MT
900

8 ❑ 12-1
Butte
Butte, MT
CNMT 4% 100

9 ❑ 12-1
Canadian Ace
Chicago, IL
85

10 ❑ 12-1
Canadian Ace
Chicago, IL
125

11 ❑ 12-1
Canadian Ace
Chicago, IL
CMT 3.2% 125

12 ❑ 12-1-T
Canadian Ace
Chicago, IL
85

13 ❑ 12-1
Canadian Ace
Chicago, IL
50

14 ❑ 12-1
Canadian Ace
Chicago, IL
CMT 3.2% 50

15 ❑ 12-1
Canadian Ace
Chicago, IL
DNCMT 4PC 50

16 ❑ 12-1
Canadian Ace
Chicago, IL
100

17 ❑ 12-1
Canadian Ace
Chicago, IL
CMT 3.2% 100

18 ❑ 12-1
Canadian Ace
Chicago, IL
STRONG 100

19 ❑ 12-1
Standard
Scranton, PA
75

20 ❑ 12-1
Standard
Scranton, PA
150

21 ❑ 12-1-T
Brewing Corp. of America
Cleveland, OH
100

22 ❑ 12-1-T
Brewing Corp. of America
Cleveland, OH
100

23 ❑ 12-1-W
Brewing Corp. of America
Cleveland, OH
375

24 ❑ 12-1-W
Carling's, Inc.
Cleveland, OH
375

25 ❑ 12-1-W
Brewing Corp. of America
Cleveland, OH
Like 26 175

26 ❑ 12-1-T
Brewing Corp. of America
Cleveland, OH
110

27 ❑ 12-1-T
Brewing Corp. of America
Cleveland, OH
110

28 ❑ 12-1
Brewing Corp. of America
Cleveland, OH
110

29 ❑ 12-1-T
Brewing Corp. of America
Cleveland, OH
85

30 ❑ 12-1
Brewing Corp. of America
Cleveland, OH
90

31 ❑ 12-1-T
Duquesne Plant #2
Stowe Twp., PA
1000+

32 ❑ 12-1-T
Duquesne Plant #2
Stowe Twp., PA
1000+

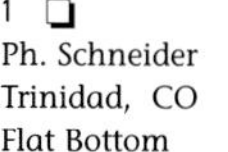

1 ❑ 12-1-T
Ph. Schneider
Trinidad, CO
Flat Bottom **400**

2 ❑ 12-1-T
Ph. Schneider
Trinidad, CO **500**

3 ❑ 12-1
Harold C. Johnson
Lomira, WI **100**

4 ❑ 12-1-T
Terre Haute
Terre Haute, IN **450**

5 ❑ 12-1-W
Terre Haute
Terre Haute, IN **600**

6 ❑ 12-1-T
Terre Haute
Terre Haute, IN
"1944" On Policy **50**

7 ❑ 12-1-T
Terre Haute
Terre Haute, IN
"1949" On Policy **50**

8 ❑ 12-1
Terre Haute
Terre Haute, IN
"1954" On Policy **50**

9 ❑ 12-1
Terre Haute
Terre Haute, IN **45**

10 ❑ 12-1
Terre Haute
Terre Haute, IN **40**

11 ❑ 12-1
Terre Haute
Terre Haute, IN **75**

12 ❑ 12-1
Terre Haute
Terre Haute, IN **75**

13 ❑ 12-1
Terre Haute
Terre Haute, IN **75**

14 ❑ 12-1
Terre Haute
Terre Haute, IN **450**

15 ❑ 12-1-T
Eagle
Chester, PA **1000+**

16 ❑ 12-1-T
White Eagle
Chicago, IL **600**

17 ❑ 12-1-T
White Eagle
Chicago, IL
DNCMT 4% **600**

18 ❑ 12-1-T
Hudepohl
Cincinnati, OH **1000+**

19 ❑ 12-2
Oshkosh
Oshkosh, WI **120**

20 ❑ 12-1-T
Renner Co.
Youngstown, OH **1000+**

21 ❑ 12-1-T
Enterprise
Fall River, MA **250**

22 ❑ 12-1-T
Enterprise
Fall River, MA **200**

23 ❑ 12-1-T
Enterprise
Fall River, MA **175**

24 ❑ 12-1-T
Cold Spring
Cold Spring, MN
MT 3.2% **350**

25 ❑ 12-1-T
Cold Spring
Cold Spring, MN
NMT 3.2% **75**

26 ❑ 12-1
Cold Spring
Cold Spring, MN
NMT 3.2% **75**

27 ❑ 12-1
Cold Spring
Cold Spring, MN
Contains 5% **75**

28 ❑ 12-1
Cold Spring
Cold Spring, MN
STRONG **75**

29 ❑ 12-1
Cold Spring
Cold Spring, MN
NMT 3.2% **200**

30 ❑ 12-1
Cold Spring
Cold Spring, MN
STRONG **200**

31 ❑ 12-1
Cold Spring
Cold Spring, MN **125**

32 ❑ 12-1
Cold Spring
Cold Spring, MN
NMT 3.2% **125**

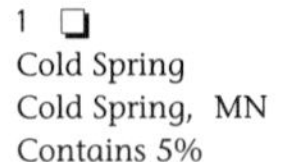

1 ❑ 12-1
Cold Spring
Cold Spring, MN
Contains 5% 125

2 ❑ 12-1
Cold Spring
Cold Spring, MN
STRONG 125

3 ❑ 12-1
F. W. Cook
Evansville, IN
500

4 ❑ 12-1-T
F. W. Cook
Evansville, IN
Cherokee 300

5 ❑ 12-1
F. W. Cook
Evansville, IN
Cherokee 300

6 ❑ 12-1
F. W. Cook
Evansville, IN
Cherokee NIEO 4% 300

7 ❑ 12-1
F. W. Cook
Evansville, IN
Robert E Lee 300

8 ❑ 12-1-T
F. W. Cook
Evansville, IN
225

9 ❑ 12-1
F. W. Cook
Evansville, IN
200

10 ❑ 12-1-T
Liebert & Obert
Philadelphia, PA
1000+

11 ❑ 12-2
Copper Country
Hancock, MI
100

12 ❑ 12-2
Copper Country
Hancock, MI
DNCMT 3.2% 100

13 ❑ 12-2
A. Haas
Hancock, MI
100

14 ❑ 12-2
A. Haas
Hancock, MI
DNCMT 3.2% 100

15 ❑ 12-2
A. Haas
Hancock, MI
STRONG 110

16 ❑ 12-1-T
Cremo
New Britain, CT
800

17 ❑ 12-1-T
Cremo
New Britain, CT
800

18 ❑ 12-1-T
Cremo
New Britain, CT
800

19 ❑ 12-1-T
Cremo
New Britain, CT
1000+

20 ❑ 12-1-T
Croft
Boston, MA
Three Products 300

21 ❑ 12-1-T
Croft
Boston, MA
400

22 ❑ 12-1-T
Croft
Boston, MA
200

23 ❑ 12-2-T
Dawson's
New Bedford, MA
550

24 ❑ 12-2-T
Dawson's
New Bedford, MA
450

25 ❑ 12-2-T
Dawson's
New Bedford, MA
500

26 ❑ 12-2-T
Dawson's
New Bedford, MA
100

27 ❑ 12-2-T
Dawson's
New Bedford, MA
100

28 ❑ 12-2-T
Dawson's
New Bedford, MA
100

29 ❑ 12-2-T
Dawson's
New Bedford, MA
100

30 ❑ 12-2-T
Dawson's
New Bedford, MA
100

31 ❑ 12-2-T
Dawson's
New Bedford, MA
350

32 ❑ 12-2-T
Dawson's
New Bedford, MA
300

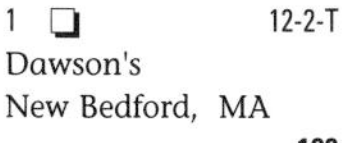

1 ❑ 12-2-T
Dawson's
New Bedford, MA
100

2 ❑ 12-2-T
Dawson's
New Bedford, MA
250

3 ❑ 12-2-T
Dawson's
New Bedford, MA
1000+

4 ❑ 12-2-T
Dawson's
New Bedford, MA
500

5 ❑ 12-2-T
Dawson's
New Bedford, MA
400

6 ❑ 12-2-T
Dawson's
New Bedford, MA
400

7 ❑ 12-2-T
Dawson's
New Bedford, MA
500

8 ❑ 12-2-T
Dawson's
New Bedford, MA
600

9 ❑ 12-1
Cleveland Home
Cleveland, OH
650

10 ❑ 12-1
Corona
San Juan, PR
300

11 ❑ 12-1
Diamond State
Wilmington, DE
400

12 ❑ 12-1-T
Christian Diehl
Defiance, OH
70 Years
85

13 ❑ 12-1-T
Christian Diehl
Defiance, OH
70 Years 3.2%
85

14 ❑ 12-1-T
Christian Diehl
Defiance, OH
70 Years 3.2-7%
85

15 ❑ 12-1-T
Christian Diehl
Defiance, OH
75 Years
85

16 ❑ 12-1
Christian Diehl
Defiance, OH
85

17 ❑ 12-1
Christian Diehl
Defiance, OH
125

18 ❑ 12-1-T
Dixie
New Orleans, LA
250

19 ❑ 12-1
Dixie
New Orleans, LA
200

20 ❑ 12-1-T
Chas. Schaefer
New York, NY
750

21 ❑ 12-1-T
Du Bois
Du Bois, PA
1000+

22 ❑ 12-1
Du Bois
Du Bois, PA
110

23 ❑ 12-1-T
Duquesne Plant #2
Stowe Twp., PA
Flat Bottom
250

24 ❑ 12-1-T
Duquesne Plant #2
Stowe Twp., PA
150

25 ❑ 12-1-T
Duquesne Plant #2
Stowe Twp., PA
90

26 ❑ 12-1-W
Duquesne Plant #2
Stowe Twp., PA
125

27 ❑ 12-1-T
Duquesne Plant #2
Stowe Twp., PA
75

28 ❑ 12-1-T
Duquesne Plant #2
Stowe Twp., PA
75

29 ❑ 12-1
Duquesne Plant #2
Stowe Twp., PA
75

30 ❑ 12-1-T
Duquesne Plant #2
Stowe Twp., PA
60

31 ❑ 12-1-T
Duquesne Plant #2
Stowe Twp., PA
60

32 ❑ 12-1
Duquesne Plant #2
Stowe Twp., PA
60

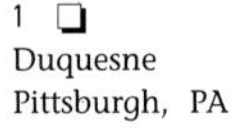

1 ❏ 12-2
Duquesne
Pittsburgh, PA
60

2 ❏ 12-1
Duquesne
Pittsburgh, PA
100

3 ❏ 12-1
Duquesne
Pittsburgh, PA
70

4 ❏ 12-1-T
Pittsburgh
Pittsburgh, PA
1000+

5 ❏ 12-1-T
Pittsburgh
Pittsburgh, PA
1000+

6 ❏ 12-1-T
Pittsburgh
Pittsburgh, PA
750

7 ❏ 12-1-T
Pittsburgh
Pittsburgh, PA
100

8 ❏ 12-1
Pittsburgh
Pittsburgh, PA
100

9 ❏ 12-1-T
Albion
San Francisco, CA
1000+

10 ❏ 12-1-T
El Rey
San Francisco, CA
1000+

11 ❏ 12-1-T
Los Angeles
Los Angeles, CA
55

12 ❏ 12-1-T
Los Angeles
Los Angeles, CA
55

13 ❏ 12-1-T
Ekhardt & Becker
Detroit, MI
750

14 ❏ 12-1-T
Ekhardt & Becker
Detroit, MI
3.2-7%
750

15 ❏ 12-1-T
Ekhardt & Becker
Detroit, MI
50

16 ❏ 12-1-T
Ekhardt & Becker
Detroit, MI
3.2-7%
60

17 ❏ 12-1-T
Ekhardt & Becker
Detroit, MI
75

18 ❏ 12-1-T
Ekhardt & Becker
Detroit, MI
3.2-7%
75

19 ❏ 12-1-T
Ekhardt & Becker
Detroit, MI
75

20 ❏ 12-1
Ekhardt & Becker
Detroit, MI
75

21 ❏ 12-1-T
Ebling
New York, NY
Flat Bottom
500

22 ❏ 12-1-T
Ebling
New York, NY
500

23 ❏ 12-1-T
Ebling
New York, NY
Paper Label
600

24 ❏ 12-1-T
Ebling
New York, NY
Flat Bottom
550

25 ❏ 12-1-T
Ebling
New York, NY
550

26 ❏ 12-1-T
Ebling
New York, NY
Paper Label
600

27 ❏ 12-1-T
Schoenhofen Edelweiss
Chicago, IL
150

28 ❏ 12-1-T
Schoenhofen Edelweiss
Chicago, IL
50

29 ❏ 12-1-T
Schoenhofen Edelweiss
Chicago, IL
65

30 ❏ 12-1-T
Schoenhofen Edelweiss
Chicago, IL
DNCMT 4 PC
65

31 ❏ 12-1
Schoenhofen Edelweiss
Chicago, IL
65

32 ❏ 12-1
Schoenhofen Edelweiss
Chicago, IL
750

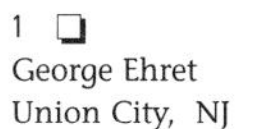

1 ❑ 12-1-T
George Ehret
Union City, NJ
350

2 ❑ 12-1-T
George Ehret
Brooklyn, NY
400

3 ❑ 12-1-T
El Rey
San Francisco, CA
Flat Bot. Exc. 4% 450

4 ❑ 12-1-T
El Rey
San Francisco, CA
N.O. 4% 300

5 ❑ 12-1-T
El Rey
San Francisco, CA
Exceeds 4% 300

6 ❑ 12-1-T
Prima
Chicago, IL
1000+

7 ❑ 12-1-T
Westminster
Chicago, IL
1000+

8 ❑ 12-1-T
Otto Erlanger
Philadelphia, PA
1000+

9 ❑ 12-1-T
Otto Erlanger
Philadelphia, PA
750

10 ❑ 12-1-T
Otto Erlanger
Philadelphia, PA
350

11 ❑ 12-1
Otto Erlanger
Philadelphia, PA
350

12 ❑ 12-1-T
Esslinger's
Philadelphia, PA
Flat Bottom 300

13 ❑ 12-1-T
Esslinger's
Philadelphia, PA
Flat Bottom 300

14 ❑ 12-1-T
Esslinger's
Philadelphia, PA
Flat Bottom 375

15 ❑ 12-1-T
Esslinger's
Philadelphia, PA
Flat Bottom 375

16 ❑ 12-1-T
Esslinger's
Philadelphia, PA
Long Text 375

17 ❑ 12-1
Esslinger's
Philadelphia, PA
Flat Bottom 375

18 ❑ 12-2-T
Eagle
San Francisco, CA
1000+

19 ❑ 12-1
Falls
Fergus Falls, MN
350

20 ❑ 12-1
Falls
Fergus Falls, MN
N.O. 3.2% 350

21 ❑ 12-1
Falls
Fergus Falls, MN
STRONG 350

22 ❑ 12-1
Falls
Fergus Falls, MN
N.O. 3.2% 325

23 ❑ 12-1-T
Falls
Fergus Falls, MN
500

24 ❑ 12-1
Falstaff
New Orleans, LA
60

25 ❑ 12-1-T
Falstaff
St. Louis, MO
40

26 ❑ 12-1-T
Falstaff
St. Louis, MO
DNCMT 5 PC (5%) 40

27 ❑ 12-1-T
Falstaff
St. Louis, MO
40

28 ❑ 12-1-T
Falstaff
St. Louis, MO
40

29 ❑ 12-1
Falstaff
St. Louis, MO
40

30 ❑ 12-1
Falstaff
St. Louis, MO
DNCMT 4 PC (4%) 40

31 ❑ 12-1-T
Falstaff
Omaha, NE
40

32 ❑ 12-1-T
Falstaff
Omaha, NE
Over 3.2% 40

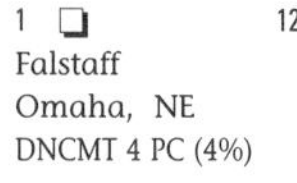

1 ❑ 12-1-T
Falstaff
Omaha, NE
DNCMT 4 PC (4%) **40**

2 ❑ 12-1
Falstaff
Omaha, NE
40

3 ❑ 12-1
Falstaff
Omaha, NE
DNCMT 4 PC (4%) **40**

4 ❑ 12-1
Fauerbach
Madison, WI
90

5 ❑ 12-1-T
Frank Fehr
Louisville, KY
150

6 ❑ 12-1-T
Frank Fehr
Louisville, KY
150

7 ❑ 12-1
Frank Fehr
Louisville, KY
150

8 ❑ 12-1-T
Fitger
Duluth, MN
Flat Bottom **750**

9 ❑ 12-1-T
Fitger
Duluth, MN
600

10 ❑ 12-1-T
Fitger
Duluth, MN
4% **600**

11 ❑ 12-1-T
Fitger
Duluth, MN
STRONG **600**

12 ❑ 12-1-T
Fitger
Duluth, MN
275

13 ❑ 12-1-T
Fitger
Duluth, MN
DNCMT 4% **275**

14 ❑ 12-1-T
Fitger
Duluth, MN
Contains 4.9 PC **275**

15 ❑ 12-1-T
Fitger
Duluth, MN
STRONG **275**

16 ❑ 12-1-T
Fitger
Duluth, MN
200

17 ❑ 12-1-T
Fitger
Duluth, MN
STRONG **200**

18 ❑ 12-1
Fitger
Duluth, MN
90

19 ❑ 12-1
Fitger
Duluth, MN
CNMT 3.2% **90**

20 ❑ 12-1
Fitger
Duluth, MN
STRONG **90**

21 ❑ 12-1
Fitger
Duluth, MN
90

22 ❑ 12-1
Fitger
Duluth, MN
CNMT 3.2% **90**

23 ❑ 12-1
Fitger
Duluth, MN
STRONG **90**

24 ❑ 12-1-T
Fitger
Duluth, MN
70

25 ❑ 12-1-T
Fitger
Duluth, MN
CNMT 3.2% **70**

26 ❑ 12-1-T
Fitger
Duluth, MN
NMT 4.9% **70**

27 ❑ 12-1-T
Fitger
Duluth, MN
STRONG **70**

28 ❑ 12-1
Fitger
Duluth, MN
70

29 ❑ 12-1
Fitger
Duluth, MN
CNMT 3.2% **70**

30 ❑ 12-1
Fitger
Duluth, MN
STRONG **70**

31 ❑ 12-1-T
Fitzgerald Bros.
Troy, NY
Flat Bottom **400**

32 ❑ 12-1-T
Fitzgerald Bros.
Troy, NY
275

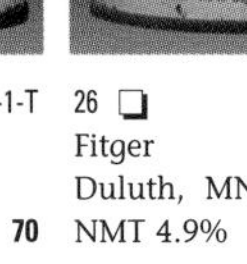

CONES

1 ❑ 12-1-T
Fitzgerald Bros.
Troy, NY
Flat Bottom **600**

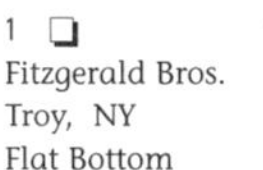

2 ❑ 12-1-T
Fitzgerald Bros.
Troy, NY **275**

3 ❑ 12-1-T
Fitzgerald Bros.
Troy, NY
Flat Bottom **750**

4 ❑ 12-1-T
Fitzgerald Bros.
Troy, NY **600**

5 ❑ 12-1-T
Fitzgerald Bros.
Troy, NY **75**

6 ❑ 12-1-T
Fitzgerald Bros.
Troy, NY
DNCMT 4% **75**

7 ❑ 12-1-T
Fort Pitt
Pittsburgh, PA **200**

8 ❑ 12-1
Fort Pitt
Pittsburgh, PA **200**

9 ❑ 12-2-T
Fort Pitt
Jeanette, PA **80**

10 ❑ 12-2
Fort Pitt
Jeanette, PA
Extra Quality **80**

11 ❑ 12-2
Fort Pitt
Jeanette, PA
Choicest Malt **80**

12 ❑ 12-2-T
Fort Pitt
Pittsburgh, PA **80**

13 ❑ 12-2-T
Fort Pitt
Pittsburgh, PA **80**

14 ❑ 12-2
Fort Pitt
Pittsburgh, PA **80**

15 ❑ 12-1
Fort Pitt
Pittsburgh, PA **125**

16 ❑ 12-1
Utica
Utica, NY **125**

17 ❑ 12-1
Utica
Utica, NY **175**

18 ❑ 12-1
Utica
Utica, NY **125**

19 ❑ 12-1
Utica
Utica, NY **175**

20 ❑ 12-2
Fountain City
Fountain City, WI **150**

21 ❑ 12-2
Fountain City
Fountain City, WI
N.O. 3.2% **200**

22 ❑ 12-2
Fountain City
Fountain City, WI
STRONG **200**

23 ❑ 12-1-T
Fox Deluxe
Marion, IN **225**

24 ❑ 12-1-T
Fox Deluxe
Marion, IN
DNCMT 3.2% **225**

25 ❑ 12-1-T
Fox Deluxe
Marion, IN
5% **225**

26 ❑ 12-1-T
Fox Deluxe
Grand Rapids, MI
3.2% **250**

27 ❑ 12-1-T
Frankenmuth-KY
Louisville, KY **200**

28 ❑ 12-1-T
Frankenmuth
Frankenmuth, MI **250**

29 ❑ 12-1-T
Frankenmuth
Frankenmuth, MI **300**

30 ❑ 12-1-T
Frankenmuth
Frankenmuth, MI **200**

31 ❑ 12-1-T
Frankenmuth
Frankenmuth, MI **200**

32 ❑ 12-1-T
Frankenmuth
Frankenmuth, MI
3.2% **200**

1 ❑ 12-1-T
Frankenmuth
Frankenmuth, MI
3.2-7% 200

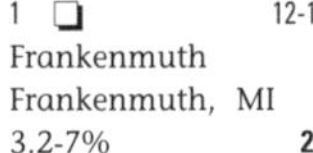

2 ❑ 12-1
Frankenmuth
Frankenmuth, MI
200

3 ❑ 12-1-T
Frederick's
Thornton, IL
1000+

4 ❑ 12-1-T
Free State
Baltimore, MD
225

5 ❑ 12-1-T
Fritz
Freeport, IL
Flat Bottom 900

6 ❑ 12-1-T
Fritz
Freeport, IL
Flat Bot. DNCMT 4% 900

7 ❑ 12-1-T
Fritz
Freeport, IL
800

8 ❑ 12-1-T
Fritz
Freeport, IL
800

9 ❑ 12-1-T
Fuhrmann & Schmidt
Shamokin, PA
275

10 ❑ 12-1-W
Fuhrmann & Schmidt
Shamokin, PA
600

11 ❑ 12-1-T
Fuhrmann & Schmidt
Shamokin, PA
135

12 ❑ 12-1
Fuhrmann & Schmidt
Shamokin, PA
135

13 ❑ 12-1-T
Fuhrmann & Schmidt
Shamokin, PA
600

14 ❑ 12-1-T
Fuhrmann & Schmidt
Shamokin, PA
600

15 ❑ 12-1-T
Fuhrmann & Schmidt
Shamokin, PA
600

16 ❑ 12-1-T
Fuhrmann & Schmidt
Shamokin, PA
400

17 ❑ 12-1
August Wagner
Columbus, OH
90

18 ❑ 12-1
August Wagner
Columbus, OH
400

19 ❑ 12-1-T
Grace Bros.
Santa Rosa, CA
125

20 ❑ 12-1-T
North Bay
Santa Rosa, CA
125

21 ❑ 12-2
William Gerst
Nashville, TN
750

22 ❑ 12-1-T
Gettelman
Milwaukee, WI
100

23 ❑ 12-1
Gettelman
Milwaukee, WI
100

24 ❑ 12-1
Gettelman
Milwaukee, WI
80

25 ❑ 12-1-T
Lion
Wilkes-Barre, PA
150

26 ❑ 12-1-T
Lion
Wilkes-Barre, PA
125

27 ❑ 12-1
Lion
Wilkes-Barre, PA
75

28 ❑ 12-1
Lion
Wilkes-Barre, PA
90

29 ❑ 12-1-T
Lion
Wilkes-Barre, PA
1000+

30 ❑ 12-1-T
Gipps
Peoria, IL
150

31 ❑ 12-1
Gipps
Peoria, IL
110

32 ❑ 12-1
Gipps
Peoria, IL
110

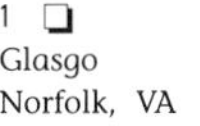

1 ❑ 12-1-T
Glasgo
Norfolk, VA
500

2 ❑ 12-1
Glasgo
Norfolk, VA
500

3 ❑ 12-1-T
Gluek
Minneapolis, MN
Flat Bot. NMT 4 PC **750**

4 ❑ 12-1-T
Gluek
Minneapolis, MN
Flat Bot. CMT 4 PC **750**

5 ❑ 12-1-T
Gluek
Minneapolis, MN
NMT 4 PC **650**

6 ❑ 12-1-T
Gluek
Minneapolis, MN
CMT 4 PC **650**

7 ❑ 12-1-T
Gluek
Minneapolis, MN
CNMT 4 PC **650**

8 ❑ 12-1-T
Gluek
Minneapolis, MN
NMT 4 PC **650**

9 ❑ 12-1-T
M. K. Goetz
St. Joseph, MO
Flat Bottom 5-6% **400**

10 ❑ 12-1-T
M. K. Goetz
St. Joseph, MO
90

11 ❑ 12-1-T
M. K. Goetz
St. Joseph, MO
DNCMT 4 PC **90**

12 ❑ 12-1-T
M. K. Goetz
St. Joseph, MO
5-6% **90**

13 ❑ 12-1-T
M. K. Goetz
St. Joseph, MO
50

14 ❑ 12-1-T
M. K. Goetz
St. Joseph, MO
DNCMT 4 PC **50**

15 ❑ 12-1
M. K. Goetz
St. Joseph, MO
50

16 ❑ 12-1-T
M. K. Goetz
St. Joseph, MO
1000+

17 ❑ 12-1-T
M. K. Goetz
St. Joseph, MO
100

18 ❑ 12-1-T
M. K. Goetz
St. Joseph, MO
DNCMT 4 PC **100**

19 ❑ 12-1
M. K. Goetz
St. Joseph, MO
40

20 ❑ 12-1
M. K. Goetz
St. Joseph, MO
3.2% **40**

21 ❑ 12-1
M. K. Goetz
St. Joseph, MO
DNCMT 4 PC **40**

22 ❑ 12-1
M. K. Goetz
St. Joseph, MO
150

23 ❑ 12-1-T
El Rey
San Francisco, CA
NU 4% **1000+**

24 ❑ 12-1-T
El Rey
San Francisco, CA
N.O. 4% **1000+**

25 ❑ 12-1-W
El Rey
San Francisco, CA
N.O. 4% **1000+**

26 ❑ 12-1-T
Stegmaier
Wilkes-Barre, PA
150

27 ❑ 12-1-T
Stegmaier
Wilkes-Barre, PA
110

28 ❑ 12-1-W
Stegmaier
Wilkes-Barre, PA
600

29 ❑ 12-1-T
Stegmaier
Wilkes-Barre, PA
80

30 ❑ 12-1
Stegmaier
Wilkes-Barre, PA
80

31 ❑ 12-1
Stegmaier
Wilkes-Barre, PA
80

32 ❑ 12-1
Stegmaier
Wilkes-Barre, PA
200

1 ❑ 12-1
Stegmaier
Wilkes-Barre, PA

225

2 ❑ 12-1-T
Southwestern
Oklahoma City, OK
1000+

3 ❑ 12-2-T
Mutual
Ellensburg, WA
DNCMT 4% 500

4 ❑ 12-2-T
Mutual
Ellensburg, WA
DNCMT 4% 350

5 ❑ 12-2-T
Silver Springs
Port Orchard, WA
DNCMT 4% 600

6 ❑ 12-1-T
Tennessee
Memphis, TN
275

7 ❑ 12-1
Tennessee
Memphis, TN
275

8 ❑ 12-2
Tennessee
Memphis, TN
250

9 ❑ 12-1-T
Fernwood
Lansdowne, PA
1000+

10 ❑ 12-1-T
Golden Age
Spokane, WA
NMT 4% 750

11 ❑ 12-1-T
Golden Age
Spokane, WA
NMT 4% 425

12 ❑ 12-1-T
Golden Age
Spokane, WA
NMT 4% 375

13 ❑ 12-1-T
Golden Age
Spokane, WA
325

14 ❑ 12-1-T
Golden Age
Spokane, WA
N.O. 4% 325

15 ❑ 12-1-W
Golden Age
Spokane, WA
900

16 ❑ 12-1-W
Golden Age
Spokane, WA
1000+

17 ❑ 12-1-T
Golden Age
Spokane, WA
200

18 ❑ 12-1-T
Golden Age
Spokane, WA
CNMT 4% 200

19 ❑ 12-1-T
Vernon
Vernon, CA
1000+

20 ❑ 12-1-T
Vernon
Vernon, CA
Exceeds 4% 1000+

21 ❑ 12-1-T
Cincinnati
Cincinnati, OH
400

22 ❑ 12-1-T
Cincinnati
Cincinnati, OH
3.2-7% 400

23 ❑ 12-1
Cincinnati
Cincinnati, OH
400

24 ❑ 12-1
Minneapolis
Minneapolis, MN
5.5% 450

25 ❑ 12-1
Minneapolis
Minneapolis, MN
STRONG 450

26 ❑ 12-1-T
Minneapolis
Minneapolis, MN
Flat Bot. 4 PC 325

27 ❑ 12-1-T
Minneapolis
Minneapolis, MN
NMT 4% 250

28 ❑ 12-1-T
Minneapolis
Minneapolis, MN
4 PC 250

29 ❑ 12-1-T
Minneapolis
Minneapolis, MN
STRONG 250

30 ❑ 12-1-T
Minneapolis
Minneapolis, MN
250

31 ❑ 12-1-T
Minneapolis
Minneapolis, MN
DNCMT 4% 250

32 ❑ 12-1-T
Minneapolis
Minneapolis, MN
250

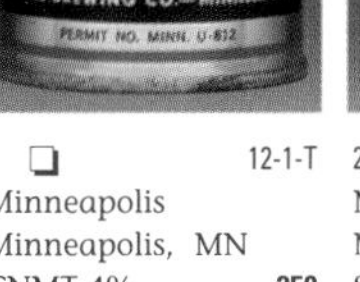

1 ❑ 12-1-T
Minneapolis
Minneapolis, MN
CNMT 4% **250**

2 ❑ 12-1-T
Minneapolis
Minneapolis, MN
STRONG **250**

3 ❑ 12-1-T
Minneapolis
Minneapolis, MN **225**

4 ❑ 12-1-T
Minneapolis
Minneapolis, MN
NMT 4 PC **225**

5 ❑ 12-1-T
Minneapolis
Minneapolis, MN
STRONG **225**

6 ❑ 12-1-T
Minneapolis
Minneapolis, MN **45**

7 ❑ 12-1-T
Minneapolis
Minneapolis, MN
DNCMT 4 PC **45**

8 ❑ 12-1-T
Minneapolis
Minneapolis, MN
5 PC **45**

9 ❑ 12-1-T
Minneapolis
Minneapolis, MN
STRONG **45**

10 ❑ 12-1-T
Minneapolis
Minneapolis, MN
DNCMT 4 PC **40**

11 ❑ 12-1-T
Minneapolis
Minneapolis, MN
5 PC **40**

12 ❑ 12-1-T
Minneapolis
Minneapolis, MN
STRONG **40**

13 ❑ 12-1
Minneapolis
Minneapolis, MN **40**

14 ❑ 12-1
Minneapolis
Minneapolis, MN
DNCMT 4 PC **40**

15 ❑ 12-1
Minneapolis
Minneapolis, MN
5 PC **40**

16 ❑ 12-1
Minneapolis
Minneapolis, MN
STRONG **40**

17 ❑ 12-1-T
Minneapolis
Minneapolis, MN
N.O. 3.2% **45**

18 ❑ 12-1
Minneapolis
Minneapolis, MN
N.O. 3.2% **45**

19 ❑ 12-1
Minneapolis
Minneapolis, MN
N.O. 3.2% **50**

20 ❑ 12-1
Minneapolis
Minneapolis, MN **60**

21 ❑ 12-1
Minneapolis
Minneapolis, MN
DNCMT 4% **60**

22 ❑ 12-1
Minneapolis
Minneapolis, MN
5 PC **60**

23 ❑ 12-1
Minneapolis
Minneapolis, MN
STRONG **60**

24 ❑ 12-1
Minneapolis
Minneapolis, MN
N.O. 3.2% **60**

25 ❑ 12-1
Kiewel
Little Falls, MN
NMT 3.2% **400**

26 ❑ 12-1
Kiewel
Little Falls, MN
5 PC **400**

27 ❑ 12-1-T
R. H. Graupner
Harrisburg, PA **1000+**

28 ❑ 12-1-T
R. H. Graupner
Harrisburg, PA **900**

29 ❑ 12-1-T
Wm. Gretz
Philadelphia, PA **1000+**

30 ❑ 12-1
Wm. Gretz
Philadelphia, PA **300**

31 ❑ 12-1
Wm. Gretz
Philadelphia, PA **300**

32 ❑ 12-1
Wm. Gretz
Philadelphia, PA **375**

CONES

1 ❑ 12-1
Wm. Gretz
Philadelphia, PA
375

2 ❑ 12-1-T
Geo. J. Renner
Akron, OH
600

3 ❑ 12-1
Geo. J. Renner
Akron, OH
250

4 ❑ 12-1-T
Gunther
Baltimore, MD
Flat Bottom
1000+

5 ❑ 12-1-T
Gunther
Baltimore, MD
Flat Bottom
500

6 ❑ 12-1-T
Gunther
Baltimore, MD
Flat Bottom
500

7 ❑ 12-1
Kalispell
Kalispell, MT
CNO 4%
200

8 ❑ 12-1-T
A. Haas
Houghton, MI
150

9 ❑ 12-1-T
A. Haas
Houghton, MI
150

10 ❑ 12-1-T
A. Haas
Houghton, MI
150

11 ❑ 12-1-T
Haberle-Congress
Syracuse, NY
400

12 ❑ 12-1-T
Haberle-Congress
Syracuse, NY
350

13 ❑ 12-1-T
Haberle-Congress
Syracuse, NY
1000+

14 ❑ 12-1-T
James Hanley
Providence, RI
Flat Bottom
150

15 ❑ 12-1-T
James Hanley
Providence, RI
150

16 ❑ 12-1
John Hauenstein
New Ulm, MN
NMT 3.2%
150

17 ❑ 12-1
John Hauenstein
New Ulm, MN
STRONG
150

18 ❑ 12-1
John Hauenstein
New Ulm, MN
NMT 3.2%
100

19 ❑ 12-1
John Hauenstein
New Ulm, MN
STRONG
100

20 ❑ 12-1-T
Valley
Flint, MI
225

21 ❑ 12-1
Valley
Flint, MI
200

22 ❑ 12-1-T
Sioux City
Sioux City, IA
DNCMT 4%
55

23 ❑ 12-1
Sioux City
Sioux City, IA
55

24 ❑ 12-1
Sioux City
Sioux City, IA
N.O. 3.2%
55

25 ❑ 12-1
Sioux City
Sioux City, IA
DNCMT 4%
55

26 ❑ 12-1-T
Missoula
Missoula, MT
350

27 ❑ 12-1-T
Missoula
Missoula, MT
N.O. 4%
350

28 ❑ 12-1-T
Hoff-Brau
Ft. Wayne, IN
325

29 ❑ 12-1-T
Hoff-Brau
Ft. Wayne, IN
325

30 ❑ 12-1
Hoff-Brau
Ft. Wayne, IN
300

31 ❑ 12-1-T
Hoff-Brau
Ft. Wayne, IN
175

32 ❑ 12-1-T
Hohenadel
Philadelphia, PA
1000+

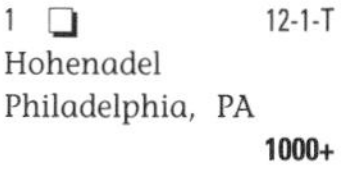

1 ❑ 12-1-T
Hohenadel
Philadelphia, PA
1000+

2 ❑ 12-1-T
Hohenadel
Philadelphia, PA
175

3 ❑ 12-1-T
Hohenadel
Philadelphia, PA
225

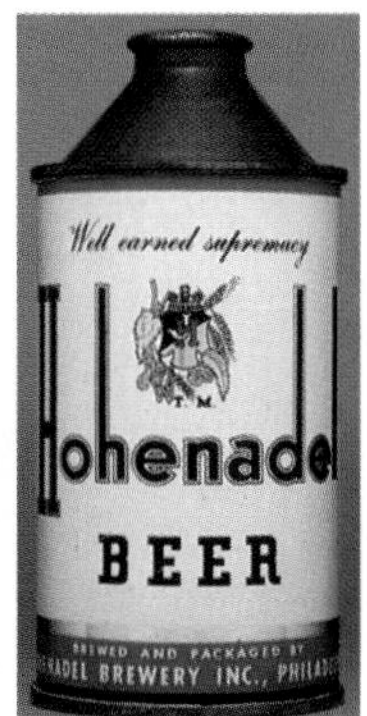

4 ❑ 12-1
Hohenadel
Philadelphia, PA
225

5 ❑ 12-1-T
Honer
Battle Creek, MI
1000+

6 ❑ 12-1-T
Honer
Battle Creek, MI
900

7 ❑ 12-1-T
South Bend
South Bend, IN
275

8 ❑ 12-1
South Bend
South Bend, IN
275

9 ❑ 12-1-T
Jacob Hornung
Philadelphia, PA
350

Front of 11, 12

11 ❑ 12-1-T
Horton Pilsener
New York, NY
175

12 ❑ 12-1-T
Horton Pilsener
New York, NY
175

13 ❑ 12-1-T
Horton Pilsener
New York, NY
200

14 ❑ 12-1-T
City
New York, NY
175

15 ❑ 12-1-T
Horton Pilsener
New York, NY
175

16 ❑ 12-1-T
Horton Pilsener
New York, NY
175

17 ❑ 12-1-T
Horton Pilsener
New York, NY
175

18 ❑ 12-1-W
City
New York, NY
500

19 ❑ 12-1-T
Greater New York
New York, NY
200

20 ❑ 12-1
Jos Huber
Monroe, WI
65

21 ❑ 12-1
Jos Huber
Monroe, WI
DNCMT 4 PC
65

22 ❑ 12-1
Jos Huber
Monroe, WI
STRONG
65

23 ❑ 12-1-T
Hudepohl
Cincinnati, OH
600

24 ❑ 12-1
Hudepohl
Cincinnati, OH
600

25 ❑ 12-1-T
Hudepohl
Cincinnati, OH
1000+

26 ❑ 12-1-T
Hudepohl
Cincinnati, OH
3.2%
1000+

27 ❑ 12-1-T
Hudepohl
Cincinnati, OH
100

28 ❑ 12-1
Hudepohl
Cincinnati, OH
100

29 ❑ 12-1-T
Chas. Schaefer
New York, NY
1000+

30 ❑ 12-1-T
Pittsburgh
Pittsburgh, PA
250

31 ❑ 12-1-T
Pittsburgh
Pittsburgh, PA
225

32 ❑ 12-1-T
Pittsburgh
Pittsburgh, PA
150

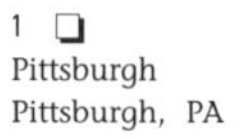

1 ❑ 12-1
Pittsburgh
Pittsburgh, PA
150

2 ❑ 12-1
Pittsburgh
Pittsburgh, PA
150

3 ❑ 12-1
Pittsburgh
Pittsburgh, PA
150

4 ❑ 12-1
Pittsburgh
Pittsburgh, PA
150

5 ❑ 12-1
Pittsburgh
Pittsburgh, PA
175

6 ❑ 12-1-T
Iroquois Bev.
Buffalo, NY
650

7 ❑ 12-1-T
Iroquois Bev.
Buffalo, NY
750

8 ❑ 12-1-W
Iroquois Bev.
Buffalo, NY
350

9 ❑ 12-1-T
Iroquois Bev.
Buffalo, NY
450

10 ❑ 12-1-T
Iroquois Bev.
Buffalo, NY
550

11 ❑ 12-1
Iroquois Bev.
Buffalo, NY
200

12 ❑ 12-1
Iroquois Bev.
Buffalo, NY
300

13 ❑ 12-1-T
Iroquois Bev.
Buffalo, NY
1000+

14 ❑ 12-1-T
Jax
Jacksonville, FL
1000+

15 ❑ 12-1
Jax
Jacksonville, FL
1000+

16 ❑ 12-1-T
New Philadelphia
New Philadelphia, OH
1000+

17 ❑ 12-1-T
New Philadelphia
New Philadelphia, OH
1000+

18 ❑ 12-1-T
R. H. Graupner
Harrisburg, PA
1000+

19 ❑ 12-1-T
Chas. D. Kaier
Mahanoy City, PA
1000+

20 ❑ 12-1
Chas. D. Kaier
Mahanoy City, PA
100

21 ❑ 12-1-T
Kamm & Schellinger
Mishawaka, IN
300

22 ❑ 12-1-T
Kamm & Schellinger
Mishawaka, IN
300

23 ❑ 12-1
Duluth
Duluth, MN
150

24 ❑ 12-1
Duluth
Duluth, MN
3.2%
150

25 ❑ 12-1
Duluth
Duluth, MN
STRONG
150

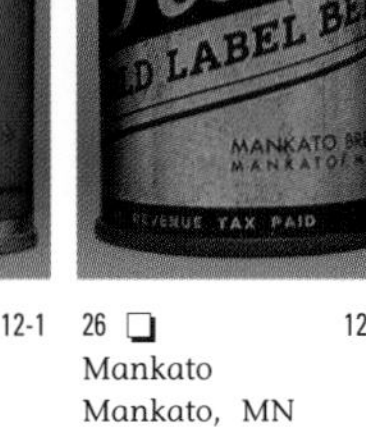

26 ❑ 12-1-T
Mankato
Mankato, MN
Flat Bottom
900

27 ❑ 12-1-T
Mankato
Mankato, MN
Flat Bottom
400

28 ❑ 12-1-T
Mankato
Mankato, MN
Flat Bot. DNCMT 4 PC
400

29 ❑ 12-1-T
Mankato
Mankato, MN
Flat Bot. STRONG
400

30 ❑ 12-1-T
Mankato
Mankato, MN
350

31 ❑ 12-1-T
Mankato
Mankato, MN
STRONG
350

32 ❑ 12-1-T
Mankato
Mankato, MN
Flat Bottom
450

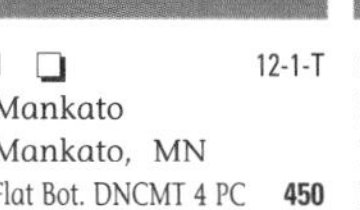

1 ❑ 12-1-T
Mankato
Mankato, MN
Flat Bot. DNCMT 4 PC **450**

2 ❑ 12-1-T
Mankato
Mankato, MN
Flat Bot. DNCMT 4 PC **425**

3 ❑ 12-1-T
Mankato
Mankato, MN
Flat Bot. DNCMT 4 PC **800**

4 ❑ 12-1
Mankato
Mankato, MN
150

5 ❑ 12-1
Mankato
Mankato, MN
3.2% **150**

6 ❑ 12-1
Mankato
Mankato, MN
4% **150**

7 ❑ 12-1
Mankato
Mankato, MN
STRONG **150**

8 ❑ 12-1
Cold Spring
Mankato, MN
3.2% **150**

9 ❑ 12-1
Cold Spring
Mankato, MN
STRONG **150**

10 ❑ 12-1-T
Keeley
Chicago, IL
325

11 ❑ 12-2
Keeley
Chicago, IL
175

12 ❑ 12-1-T
Keeley
Chicago, IL
85

13 ❑ 12-1-T
Keeley
Chicago, IL
DNCMT 4% **85**

14 ❑ 12-1
Keeley
Chicago, IL
85

15 ❑ 12-1
Keeley
Chicago, IL
85

16 ❑ 12-1
Kessler
Helena, MT
N.O. 4% **100**

17 ❑ 12-1-T
Kiewel
Little Falls, MN
4 PC **1000+**

18 ❑ 12-1-T
Kiewel
Little Falls, MN
STRONG **1000+**

19 ❑ 12-1
Franklin
Columbus, OH
1000+

20 ❑ 12-1-T
Rainier
San Francisco, CA
N.O. 4% **1000+**

21 ❑ 12-1-T
Rainier
San Francisco, CA
N.O. 4% **1000+**

22 ❑ 12-1-T
Koch Bev.
Wapakoneta, OH
DNCMT 3.2% **300**

23 ❑ 12-1-T
Koch Bev.
Wapakoneta, OH
3.2-7% **300**

24 ❑ 12-1-T
Erie
Erie, PA
140

25 ❑ 12-1-T
Erie
Erie, PA
140

26 ❑ 12-1
Erie
Erie, PA
150

27 ❑ 12-1
Erie
Erie, PA
100

28 ❑ 12-1-T
Prima-Bismarck
Chicago, IL
150

29 ❑ 12-1-T
Prima-Bismarck
Chicago, IL
110

30 ❑ 12-1-T
Prima-Bismarck
Chicago, IL
DNCMT 4% **110**

31 ❑ 12-1-T
Prima-Bismarck
Chicago, IL
65

32 ❑ 12-1-T
Prima-Bismarck
Chicago, IL
DNCMT 3.2% (PC) **65**

1 ❑ 12-1-T
Prima-Bismarck
Chicago, IL
DNCMT 4% 65

2 ❑ 12-1
Prima-Bismarck
Chicago, IL
DNCMT 3.2% (PC) 65

3 ❑ 12-1-T
Koller
Chicago, IL
450

4 ❑ 12-1-T
LaFayette
LaFayette, IN
500

5 ❑ 12-1-T
LaFayette
LaFayette, IN
500

6 ❑ 12-1-T
G. Krueger
Wilmington, DE
90

7 ❑ 12-1-T
G. Krueger
Wilmington, DE
90

8 ❑ 12-2
G. Krueger
Wilmington, DE
135

9 ❑ 12-1
G. Krueger
Wilmington, DE
350

10 ❑ 12-1-T
G. Krueger
Newark, NJ
125

11 ❑ 12-1-T
G. Krueger
Newark, NJ
90

12 ❑ 12-1-T
Rainier
San Francisco, CA
1000+

13 ❑ 12-1-T
Rainier
San Francisco, CA
1000+

14 ❑ 12-1-T
Kuebler
Easton, PA
500

15 ❑ 12-1-T
Kuebler
Easton, PA
750

16 ❑ 12-1-T
Kuebler
Easton, PA
175

17 ❑ 12-1-T
Kuebler
Easton, PA
200

18 ❑ 12-1-T
Kuebler
Easton, PA
650

19 ❑ 12-1-T
Kuebler
Easton, PA
200

20 ❑ 12-1
Kuebler
Easton, PA
200

21 ❑ 12-1-T
Kuebler
Easton, PA
1000+

22 ❑ 12-1-T
Lebanon Valley
Lebanon, PA
350

23 ❑ 12-1-T
Lebanon Valley
Lebanon, PA
135

24 ❑ 12-1
Lebanon Valley
Lebanon, PA
135

25 ❑ 12-1-T
El Rey
San Francisco, CA
1000+

26 ❑ 12-1-T
Rainier
San Francisco, CA
1000+

27 ❑ 12-2-T
Leisy
Cleveland, OH
350

28 ❑ 12-1-T
Leisy
Cleveland, OH
140

29 ❑ 12-1
Leisy
Cleveland, OH
140

30 ❑ 12-1-T
Rainier
San Francisco, CA
400

31 ❑ 12-1-T
Rainier
San Francisco, CA
N.O. 4% 400

32 ❑ 12-1-T
Rainier
San Francisco, CA
400

1 ❑ 12-1-T
Rainier
San Francisco, CA
450

2 ❑ 12-1-T
Rainier
San Francisco, CA
N.O. 4%
900

3 ❑ 12-1
Cincinnati
Cincinnati, OH
600

4 ❑ 12-1-T
El Dorado
Stockton, CA
900

5 ❑ 12-1-T
Maier
Los Angeles, CA
1000+

6 ❑ 12-1-T
Maier
Los Angeles, CA
325

7 ❑ 12-1-T
Maier
Los Angeles, CA
250

8 ❑ 12-1-T
Maier
Los Angeles, CA
1000+

9 ❑ 12-1-W
Maier
Los Angeles, CA
1000+

Front of 11, 12

11 ❑ 12-1-T
Maier
Los Angeles, CA
125

12 ❑ 12-1-T
Maier
Los Angeles, CA
125

13 ❑ 12-1-T
Yakima Valley
Selah, WA
N.O. 4%
600

14 ❑ 12-1-T
Yakima Valley
Selah, WA
N.O. 4%
400

15 ❑ 12-1
Yakima Valley
Selah, WA
N.O. 4%
375

16 ❑ 12-1-T
Mount Carbon
Pottsville, PA
1000+

17 ❑ 12-1-T
Mount Carbon
Pottsville, PA
500

18 ❑ 12-1-T
Menominee-Marinette
Menominee, MI
100

19 ❑ 12-1
Menominee-Marinette
Menominee, MI
115

20 ❑ 12-1-T
Metz
Omaha, NE
90

21 ❑ 12-1-T
Metz
Omaha, NE
3.2%
90

22 ❑ 12-1-T
Metz
Omaha, NE
DNCMT 4 PC
90

23 ❑ 12-1
Metz
Omaha, NE
90

24 ❑ 12-1
Metz
Omaha, NE
DNCMT 4 PC
90

25 ❑ 12-1-T
Jos. Schlitz
Milwaukee, WI
Flat Bot. 4-3/4%
550

26 ❑ 12-1-T
Jos. Schlitz
Milwaukee, WI
350

27 ❑ 12-1-T
Jos. Schlitz
Milwaukee, WI
Contains 4-3/4%
350

28 ❑ 12-1-T
Jos. Schlitz
Milwaukee, WI
90

29 ❑ 12-1-T
Jos. Schlitz
Milwaukee, WI
3.2-5%
90

30 ❑ 12-1-T
Jos. Schlitz
Milwaukee, WI
4%
90

31 ❑ 12-1-T
Jos. Schlitz
Milwaukee, WI
Contains 4-3/4%
90

32 ❑ 12-1-T
Jos. Schlitz
Milwaukee, WI
90

1 ❑ 12-1-T
Jos. Schlitz
Milwaukee, WI
90

2 ❑ 12-1-T
Jos. Schlitz
Milwaukee, WI
90

3 ❑ 12-1
Jos. Schlitz
Milwaukee, WI
90

4 ❑ 12-1
Mineral Spring
Mineral Point, WI
135

5 ❑ 12-1
Mineral Spring
Mineral Point, WI
135

6 ❑ 12-2-T
Monarch
Chicago, IL
200

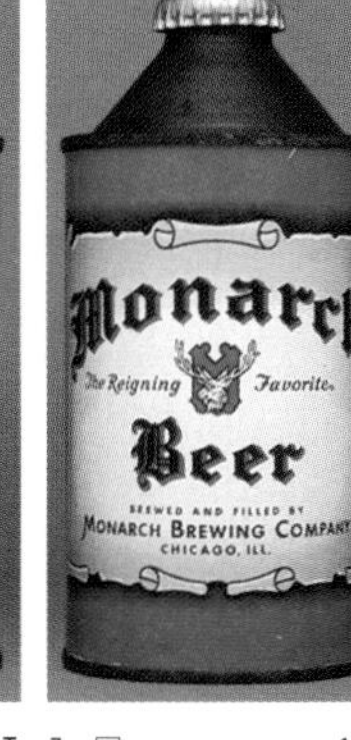

7 ❑ 12-2
Monarch
Chicago, IL
200

8 ❑ 12-2
Monarch
Chicago, IL
125

9 ❑ 12-1-T
Monterey
Salinas, CA
Blue
1000+

10 ❑ 12-1-T
Monterey
Salinas, CA
Black
750

11 ❑ 12-1-T
Burton Products
Patterson, NJ
700

12 ❑ 12-1-T
Geo. Muehlebach
Kansas City, MO
100

13 ❑ 12-1
Geo. Muehlebach
Kansas City, MO
100

14 ❑ 12-1
Geo. Muehlebach
Kansas City, MO
135

15 ❑ 12-1
Geo. Muehlebach
Kansas City, MO
CNMT 4%
150

16 ❑ 12-1
Burkhardt
Akron, OH
400

17 ❑ 12-1-T
Star
Boston, MA
600

18 ❑ 12-1-T
Star
Boston, MA
500

19 ❑ 12-1-T
Cooper
Philadelphia, PA
200

20 ❑ 12-1-T
Cooper
Philadelphia, PA
200

21 ❑ 12-1-T
Cooper
Philadelphia, PA
200

22 ❑ 12-1-T
National
Baltimore, MD
Flat Bottom
750

23 ❑ 12-1-T
National
Baltimore, MD
650

24 ❑ 12-1-T
National
Baltimore, MD
650

25 ❑ 12-1-T
National
Baltimore, MD
650

26 ❑ 12-1-T
National
Baltimore, MD
650

27 ❑ 12-1-T
National
Baltimore, MD
750

28 ❑ 12-1-T
National
Baltimore, MD
750

29 ❑ 12-1-T
National
Baltimore, MD
1000+

30 ❑ 12-1-T
National
Baltimore, MD
1000+

31 ❑ 12-1-T
National
Baltimore, MD
90

32 ❑ 12-1-T
National
Baltimore, MD
N.O. 4%
90

1 ❑ 12-1-T
National
Baltimore, MD
90

2 ❑ 12-1-T
National
Baltimore, MD
3.2-7% 90

3 ❑ 12-1
National
Baltimore, MD
90

4 ❑ 12-1-T
National
Baltimore, MD
150

5 ❑ 12-1-T
National
Baltimore, MD
135

6 ❑ 12-1-T
National
Baltimore, MD
500

7 ❑ 12-1-T
National
Baltimore, MD
75

8 ❑ 12-1
National
Baltimore, MD
75

CONES

9 ❑ 12-1-T
National
Baltimore, MD
1000+

10 ❑ 12-1-T
National
Baltimore, MD
1000+

11 ❑ 12-1-T
L. F. Neuweiler's Sons
Allentown, PA
160

12 ❑ 12-1-T
L. F. Neuweiler's Sons
Allentown, PA
400

13 ❑ 12-1-T
Washington
Columbus, OH
3.2-7% 375

14 ❑ 12-1-T
Washington
Columbus, OH
DNCMT 3.2 PC 375

15 ❑ 12-1-T
Washington
Columbus, OH
3.2-7% 375

16 ❑ 12-1-T
Northern
Superior, WI
CNMT 3.2% 175

17 ❑ 12-1
Northern
Superior, WI
175

18 ❑ 12-1
Northern
Superior, WI
CNMT 3.2% 175

19 ❑ 12-1
Northern
Superior, WI
STRONG 175

20 ❑ 12-2
Northern
Superior, WI
125

21 ❑ 12-2
Northern
Superior, WI
N.O. 3.2% 125

22 ❑ 12-2
Northern
Superior, WI
STRONG 135

23 ❑ 12-2
Oertel
Louisville, KY
85

24 ❑ 12-1-T
Brackenridge
Brackenridge, PA
1000+

25 ❑ 12-1-T
New Philadelphia
New Philadelphia, OH
300

26 ❑ 12-1-T
New Philadelphia
New Philadelphia, OH
300

27 ❑ 12-1-T
New Philadelphia
New Philadelphia, OH
250

28 ❑ 12-1-T
New Philadelphia
New Philadelphia, OH
300

29 ❑ 12-1-T
New Philadelphia
New Philadelphia, OH
250

30 ❑ 12-1-T
New Philadelphia
New Philadelphia, OH
250

31 ❑ 12-1-T
New Philadelphia
New Philadelphia, OH
N.O. 3.2% 250

32 ❑ 12-1-T
Aztec
San Diego, CA
1000+

1 ❑ 12-1-T
Aztec
San Diego, CA
1000+

2 ❑ 12-1-T
Old Dutch
Brooklyn, NY
1000+

3 ❑ 12-1-T
Old Dutch
Brooklyn, NY
900

4 ❑ 12-1-T
Metropolis
New York, NY
125

5 ❑ 12-1-T
Eagle
Catasauqua, PA
1000+

6 ❑ 12-1-T
Old England
Derby, CT
500

7 ❑ 12-1-T
Old England
Derby, CT
400

8 ❑ 12-1-T
Old England
Derby, CT
1000+

9 ❑ 12-1-T
Old England
Derby, CT
1000+

10 ❑ 12-1-T
Cumberland
Cumberland, MD
125

11 ❑ 12-1-T
Cumberland
Cumberland, MD
3.2% 125

12 ❑ 12-1
Cumberland
Cumberland, MD
125

13 ❑ 12-1
Cumberland
Cumberland, MD
50

14 ❑ 12-1
Cumberland
Cumberland, MD
80

15 ❑ 12-1
Billings
Billings, MT
DNCMT 4% 400

16 ❑ 12-1-T
Queen City
Cumberland, MD
45

17 ❑ 12-1-T
Queen City
Cumberland, MD
DNCMT 3.2% 55

18 ❑ 12-1-T
Queen City
Cumberland, MD
DNCMT 3.2% 55

19 ❑ 12-1
Queen City
Cumberland, MD
45

20 ❑ 12-1
Queen City
Cumberland, MD
50

21 ❑ 12-2
Queen City
Cumberland, MD
60

22 ❑ 12-1
Queen City
Cumberland, MD
Paper Label 50

23 ❑ 12-2
Queen City
Cumberland, MD
50

24 ❑ 12-2
Queen City
Cumberland, MD
50

25 ❑ 12-1
Geo. J. Renner
Akron, OH
375

26 ❑ 12-1
Renner Co.
Youngstown, OH
125

27 ❑ 12-1
Lebanon Valley
Lebanon, PA
225

28 ❑ 12-1-T
Cremo
New Britain, CT
500

29 ❑ 12-2-T
Standard
Rochester, NY
1000+

30 ❑ 12-1-T
Old Reading
Reading, PA
1000+

31 ❑ 12-1
Old Reading
Reading, PA
325

32 ❑ 12-1
Old Reading
Reading, PA
250

1 ❑ 12-2
Old Reading
Reading, PA
300

2 ❑ 12-1
Fort Pitt
Jeanette, PA
150

3 ❑ 12-1-PRE
G. Heileman
La Crosse, WI
Flat Bot. N.O. 4% 175

4 ❑ 12-1-PRE
G. Heileman
La Crosse, WI
Flat Bottom 5% 175

5 ❑ 12-1-PRE
G. Heileman
La Crosse, WI
Flat Bot. STRONG 175

6 ❑ 12-1-PRE
G. Heileman
La Crosse, WI
N.O. 4% 150

7 ❑ 12-1-PRE
G. Heileman
La Crosse, WI
N.O. 5% 150

8 ❑ 12-1-T
G. Heileman
La Crosse, WI
150

9 ❑ 12-1-T
G. Heileman
La Crosse, WI
DNCMT 4% 150

10 ❑ 12-1-T
G. Heileman
La Crosse, WI
N.O. 5% STRONG 150

11 ❑ 12-1-T
G. Heileman
La Crosse, WI
STRONG 150

12 ❑ 12-1-T
G. Heileman
La Crosse, WI
% Blacked Out 150

13 ❑ 12-1
G. Heileman
La Crosse, WI
200

14 ❑ 12-1-T
G. Heileman
La Crosse, WI
175

15 ❑ 12-1-T
G. Heileman
La Crosse, WI
110

16 ❑ 12-1-T
G. Heileman
La Crosse, WI
DNCMT 4% 110

17 ❑ 12-1-T
G. Heileman
La Crosse, WI
100

18 ❑ 12-1-T
G. Heileman
La Crosse, WI
DNCMT 4% 100

19 ❑ 12-1-T
G. Heileman
La Crosse, WI
300

20 ❑ 12-1-T
G. Heileman
La Crosse, WI
DNCMT 4% 275

21 ❑ 12-1-T
G. Heileman
La Crosse, WI
70

22 ❑ 12-1-T
G. Heileman
La Crosse, WI
DNCMT 4% 70

23 ❑ 12-1-T
G. Heileman
La Crosse, WI
70

24 ❑ 12-1-T
G. Heileman
La Crosse, WI
% Blacked Out 80

25 ❑ 12-1-T
G. Heileman
La Crosse, WI
CNMT 3.2% 70

26 ❑ 12-1-T
G. Heileman
La Crosse, WI
DNCMT 4 PC 70

27 ❑ 12-1-T
G. Heileman
La Crosse, WI
STRONG 80

28 ❑ 12-1
G. Heileman
La Crosse, WI
70

29 ❑ 12-1
G. Heileman
La Crosse, WI
CNMT 3.2% 70

30 ❑ 12-1
G. Heileman
La Crosse, WI
DNCMT 4 PC 70

31 ❑ 12-1
G. Heileman
La Crosse, WI
STRONG 80

32 ❑ 12-1
G. Heileman
La Crosse, WI
BEER under LAGER 70

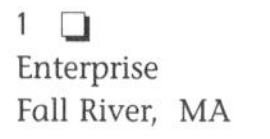
1 ❑ 12-1-T
Enterprise
Fall River, MA
700

2 ❑ 12-1-T
Enterprise
Fall River, MA
700

3 ❑ 12-1-T
Enterprise
Fall River, MA
325

4 ❑ 12-1-T
Enterprise
Fall River, MA
600

5 ❑ 12-1-T
Enterprise
Fall River, MA
400

6 ❑ 12-1-T
Rochester
Rochester, NY
40

7 ❑ 12-1-T
Rochester
Rochester, NY
40

8 ❑ 12-2
Rochester
Rochester, NY
475

9 ❑ 12-1-T
Rochester
Rochester, NY
600

10 ❑ 12-2
Rochester
Rochester, NY
475

11 ❑ 12-1
Koch Bev. and Ice
Wapakoneta, OH
225

12 ❑ 12-1
Peoples
Duluth, MN
8%
450

13 ❑ 12-1-T
New Philadelphia
New Philadelphia, OH
300

14 ❑ 12-1
Virginia
Roanoke, VA
150

15 ❑ 12-1
Virginia
Roanoke, VA
300

16 ❑ 12-1-T
Star Peerless
Belleville, IL
100

17 ❑ 12-1
Star Peerless
Belleville, IL
100

18 ❑ 12-1
Star Peerless
Belleville, IL
NIEO 4%
100

19 ❑ 12-1-T
Henry F. Ortlieb
Philadelphia, PA
500

20 ❑ 12-1-T
Henry F. Ortlieb
Philadelphia, PA
95

21 ❑ 12-1
Henry F. Ortlieb
Philadelphia, PA
95

22 ❑ 12-1
Henry F. Ortlieb
Philadelphia, PA
95

23 ❑ 12-1
Henry F. Ortlieb
Philadelphia, PA
95

24 ❑ 12-1
Henry F. Ortlieb
Philadelphia, PA
95

25 ❑ 12-1-T
Henry F. Ortlieb
Philadelphia, PA
1000+

26 ❑ 12-1-T
Rainier
San Francisco, CA
400

27 ❑ 12-1-T
Rainier
San Francisco, CA
N.O. 4%
400

28 ❑ 12-1-T
Rainier
San Francisco, CA
400

29 ❑ 12-1-T
Rainier
San Francisco, CA
650

30 ❑ 12-1-T
La Crosse
La Crosse, WI
85

31 ❑ 12-1
La Crosse
La Crosse, WI
85

32 ❑ 12-1-T
La Crosse
La Crosse, WI
65

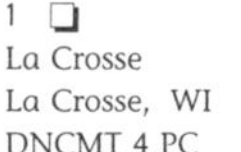

1 ❑ 12-1-T
La Crosse
La Crosse, WI
DNCMT 4 PC **65**

2 ❑ 12-1
La Crosse
La Crosse, WI
65

3 ❑ 12-1
La Crosse
La Crosse, WI
N.O. 3.2% **65**

4 ❑ 12-1
La Crosse
La Crosse, WI
DNCMT 4 PC **65**

5 ❑ 12-1-T
Phoenix
Bay City, MI
900

6 ❑ 12-1-T
Haffenreffer
Boston, MA
150

7 ❑ 12-1-T
Haffenreffer
Boston, MA
150

8 ❑ 12-1-T
Piel Bros.
New York, NY
200

9 ❑ 12-1-T
Piel Bros.
New York, NY
225

10 ❑ 12-1-T
Frederick's
Thornton, IL
400

11 ❑ 12-1-T
Illinois
Thornton, IL
250

12 ❑ 12-1-T
Metropolis
New York, NY
700

13 ❑ 12-1-T
Metropolis
New York, NY
150

14 ❑ 12-1-T
Pilser
New York, NY
1000+

15 ❑ 12-1-T
Pilser
New York, NY
1000+

16 ❑ 12-1-T
Pilser
New York, NY
1000+

17 ❑ 12-1-T
Pilsener
Cleveland, OH
175

18 ❑ 12-1
Pilsener
Cleveland, OH
175

19 ❑ 12-1
Pilsener
Cleveland, OH
250

20 ❑ 12-2-T
Philadelphia
Philadelphia, PA
750

21 ❑ 12-1-T
Potosi
Potosi, WI
DNCMT 4 PC **225**

22 ❑ 12-1
Potosi
Potosi, WI
225

23 ❑ 12-1
Potosi
Potosi, WI
DNCMT 4 PC **225**

24 ❑ 12-1
Potosi
Potosi, WI
70

25 ❑ 12-1
Potosi
Potosi, WI
DNCMT 4 P.C. **70**

26 ❑ 12-1-T
Prima
Chicago, IL
275

27 ❑ 12-1-T
Prima-Bismarck
Chicago, IL
225

28 ❑ 12-1
Prima-Bismarck
Chicago, IL
225

29 ❑ 12-1-T
Progress
Oklahoma City, OK
225

30 ❑ 12-1
Progress
Oklahoma City, OK
200

31 ❑ 12-2
Progress
Oklahoma City, OK
275

32 ❑ 12-1
Rahr-Green Bay
Green Bay, WI
150

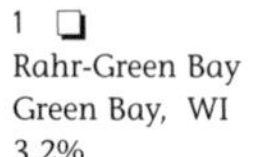

1 ❑ 12-1
Rahr-Green Bay
Green Bay, WI
3.2% **150**

2 ❑ 12-1-T
Rainier
San Francisco, CA
N.O. 6% **65**

3 ❑ 12-1-T
Rainier
San Francisco, CA **65**

4 ❑ 12-1-T
Rainier
San Francisco, CA **65**

5 ❑ 12-1-T
Rainier
San Francisco, CA **70**

6 ❑ 12-1-T
Rainier
San Francisco, CA
N.O. 6% **70**

7 ❑ 12-1-T
Rainier
San Francisco, CA
N.O. 6% **70**

8 ❑ 12-1-T
Rainier
San Francisco, CA
Flat Bot. N.O. 4% **175**

9 ❑ 12-1-T
Rainier
San Francisco, CA
N.O. 4% **100**

10 ❑ 12-1-T
Rainier
San Francisco, CA **100**

11 ❑ 12-1-T
Rainier
San Francisco, CA
N.O. 4% **100**

12 ❑ 12-1-T
Rainier
San Francisco, CA **70**

13 ❑ 12-1-T
Rainier
San Francisco, CA
N.O. 4% **70**

14 ❑ 12-1-T
Rainier
San Francisco, CA **125**

15 ❑ Front Like 14 12-1-T
Rainier
San Francisco, CA
% Blacked Out **135**

16 ❑ 12-1-T
Rainier
San Francisco, CA **100**

17 ❑ 12-1-T
Rainier
San Francisco, CA **70**

18 ❑ 12-1-W
Rainier
San Francisco, CA **900**

19 ❑ 12-1-T
Rainier
San Francisco, CA **65**

20 ❑ 12-1-T
Rainier
San Francisco, CA **90**

21 ❑ 12-1-W
Rainier
San Francisco, CA **90**

22 ❑ 12-1-T
Rainier
San Francisco, CA **90**

23 ❑ 12-1
Rainier
San Francisco, CA **500**

24 ❑ 12-1-T
Largay
Waterbury, CT **900**

25 ❑ 12-1-T
Largay
Waterbury, CT **700**

26 ❑ 12-1-W
Largay
Waterbury, CT **600**

27 ❑ 12-1-T
Largay
Waterbury, CT **600**

28 ❑ 12-1
Largay
Waterbury, CT **600**

29 ❑ 12-1-T
Largay
Waterbury, CT
IRTP tf **750**

30 ❑ 12-1-T
Largay
Waterbury, CT **750**

31 ❑ 12-1-T
Burger
Cincinnati, OH
Flat Bottom **1000+**

32 ❑ 12-1-T
Burger
Cincinnati, OH **1000+**

1 ❑ 12-1
Mathie-Ruder
Wausau, WI
300

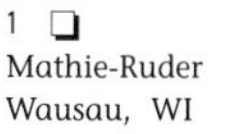

2 ❑ 12-1-T
Red Top
Cincinnati, OH
125

3 ❑ 12-1
Red Top
Cincinnati, OH
125

4 ❑ 12-1-T
Red Top
Cincinnati, OH
125

5 ❑ 12-1
Red Top
Cincinnati, OH
125

6 ❑ 12-1
Red Top
Cincinnati, OH
75

7 ❑ 12-1
Goodhue
Red Wing, MN
3.2%
800

8 ❑ 12-1
Goodhue
Red Wing, MN
STRONG
800

CONES

9 ❑ 12-1-T
American
Miami, FL
275

10 ❑ 12-1
American
Miami, FL
275

11 ❑ 12-1-W
American
New Orleans, LA
250

12 ❑ 12-1-T
American
New Orleans, LA
225

13 ❑ 12-1
Peoples
Duluth, MN
100

14 ❑ 12-1
Peoples
Duluth, MN
CNMT 3.2%
100

15 ❑ 12-1
Peoples
Duluth, MN
STRONG
100

16 ❑ 12-1
Peoples
Duluth, MN
100

17 ❑ 12-1
Peoples
Duluth, MN
CNMT 3.2%
100

18 ❑ 12-2
Reisch
Springfield, IL
70

19 ❑ 12-2
Reisch
Springfield, IL
70

20 ❑ 12-1-T
Renner Co.
Youngstown, OH
225

21 ❑ 12-1-T
Renner Co.
Youngstown, OH
250

22 ❑ 12-1
Renner Co.
Youngstown, OH
250

23 ❑ 12-1-T
Renner Co.
Youngstown, OH
300

24 ❑ 12-1-T
Renner Co.
Youngstown, OH
325

25 ❑ 12-1-T
Renner Co.
Youngstown, OH
200

26 ❑ 12-1-T
Renner Co.
Youngstown, OH
175

27 ❑ 12-1
Renner Co.
Youngstown, OH
175

28 ❑ 12-1
Renner Co.
Youngstown, OH
150

29 ❑ 12-1
Renner Co.
Youngstown, OH
135

30 ❑ 12-1-T
Rhinelander
Rhinelander, WI
150

31 ❑ 12-1
Rhinelander
Rhinelander, WI
150

32 ❑ 12-1
Rhinelander
Rhinelander, WI
135

1 ❏ 12-1
Rhinelander
Rhinelander, WI
110

2 ❏ 12-1
Rhinelander
Rhinelander, WI
100

3 ❏ 12-2
Home
Richmond, VA
90

4 ❏ 12-2
Home
Richmond, VA
90

5 ❏ 12-2
Home
Richmond, VA
90

6 ❏ 12-2
Home
Richmond, VA
125

7 ❏ 12-1
Anaconda
Anaconda, MT
CNMT 4%
125

8 ❏ 12-1
Latrobe
Latrobe, PA
90

9 ❏ 12-1
Columbus
Columbus, NE
600

10 ❏ 12-1-W
Rainier
San Francisco, CA
N.O. 4%
1000+

11 ❏ 12-1-T
Koller
Chicago, IL
375

12 ❏ 12-1-T
Reno
Reno, NV
450

13 ❏ 12-1-T
Reno
Reno, NV
4%
450

14 ❏ 12-1-T
Geo. Wiedemann
Louisville, KY
200

15 ❏ 12-1
Geo. Wiedemann
Louisville, KY
200

16 ❏ 12-1-T
Duluth
Duluth, MN
75

17 ❏ 12-1-T
Duluth
Duluth, MN
CNMT 3.2%
75

18 ❏ 12-1-T
Duluth
Duluth, MN
STRONG
75

19 ❏ 12-1
Duluth
Duluth, MN
75

20 ❏ 12-1
Duluth
Duluth, MN
CNMT 3.2%
75

21 ❏ 12-1
Duluth
Duluth, MN
Contains 5 PC
75

22 ❏ 12-1
Duluth
Duluth, MN
STRONG
75

23 ❏ 12-1
Duluth
Duluth, MN
75

24 ❏ 12-1
Duluth
Duluth, MN
CNMT 3.2%
75

25 ❏ 12-1
Duluth
Duluth, MN
Contains 5 PC
75

26 ❏ 12-1
Duluth
Duluth, MN
STRONG
75

27 ❏ 12-1-T
Union
New Castle, PA
450

28 ❏ 12-1-T
Union
New Castle, PA
500

29 ❏ 12-1-T
Union
New Castle, PA
500

30 ❏ 12-1-T
Union
New Castle, PA
600

31 ❏ 12-1-T
Rainier
San Francisco, CA
N.O. 4%
900

32 ❏ 12-1-T
Rainier
San Francisco, CA
900

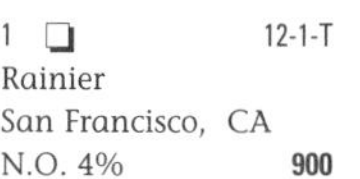

1 ❑ 12-1-T
Rainier
San Francisco, CA
N.O. 4% 900

2 ❑ 12-1-T
Rainier
San Francisco, CA
900

3 ❑ 12-1-T
Southern
Tampa, FL
500

4 ❑ 12-1-T
Southern
Tampa, FL
500

5 ❑ 12-1-T
Southern
Tampa, FL
600

6 ❑ 12-1
Southern
Tampa, FL
600

7 ❑ 12-1
August Schell
New Ulm, MN
CNMT 3.2% 150

8 ❑ 12-1
August Schell
New Ulm, MN
STRONG 150

9 ❑ 12-1-T
Jos. Schlitz
Milwaukee, WI
Flat Bot. 4-3/4% 1000+

10 ❑ 12-1-T
Jos. Schlitz
Milwaukee, WI
Flat Bot. 3.2% to 5% 175

11 ❑ 12-1-T
Jos. Schlitz
Milwaukee, WI
Flat Bot. 4% 175

12 ❑ 12-1-T
Jos. Schlitz
Milwaukee, WI
Flat Bot. 4-3/4% 175

13 ❑ 12-1-T
Jos. Schlitz
Milwaukee, WI
Exceeds 3.2% 135

14 ❑ 12-1-T
Jos. Schlitz
Milwaukee, WI
4-3/4% 135

15 ❑ 12-1-T
Jos. Schlitz
Milwaukee, WI
Flat Bottom 400

16 ❑ 12-1-T
Jos. Schlitz
Milwaukee, WI
Flat Bot. 4-3/4% 400

17 ❑ 12-1-T
Jos. Schlitz
Milwaukee, WI
135

18 ❑ 12-1-T
Jos. Schlitz
Milwaukee, WI
DNCMT 4% 135

19 ❑ 12-1-T
Jos. Schlitz
Milwaukee, WI
Less Than 4% 135

20 ❑ 12-1-T
Jos. Schlitz
Milwaukee, WI
4-3/4% 135

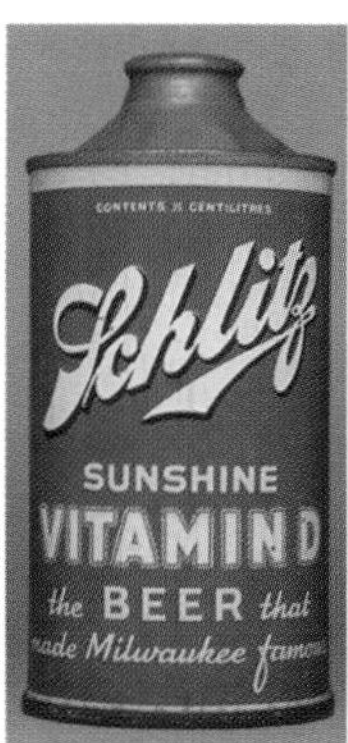

21 ❑ 12-1-W
Jos. Schlitz
Milwaukee, WI
4-3/4% 135

22 ❑ 12-1-T
Jos. Schlitz
Milwaukee, WI
90

23 ❑ 12-1-T
Jos. Schlitz
Milwaukee, WI
DNCMT 4% 90

24 ❑ 12-1-T
Jos. Schlitz
Milwaukee, WI
4-3/4% 90

25 ❑ 12-1-T
Jos. Schlitz
Milwaukee, WI
DNCMT 4% 70

26 ❑ 12-1-T
Jos. Schlitz
Milwaukee, WI
4-3/4% 70

27 ❑ 12-1-W
Jos. Schlitz
Milwaukee, WI
250

28 ❑ 12-1-T
Jos. Schlitz
Milwaukee, WI
45

29 ❑ 12-1-T
Jos. Schlitz
Milwaukee, WI
DNCMT4% 45

30 ❑ 12-1-T
Jos. Schlitz
Milwaukee, WI
3.2% 1000+

31 ❑ 12-1-W
Jos. Schlitz
Milwaukee, WI
325

32 ❑ 12-1-T
K. G. Schmidt
Logansport, IN
140

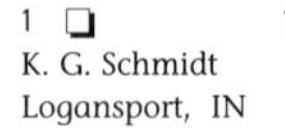

1 ❑ 12-1-T
K. G. Schmidt
Logansport, IN
140

2 ❑ 12-1-T
Schmidt
Detroit, MI
150

3 ❑ 12-1-T
Schmidt
Detroit, MI
275

4 ❑ 12-1-T
Schmidt
Detroit, MI
275

5 ❑ 12-1
Schmidt
Detroit, MI
275

Front of 7, 8

7 ❑ 12-1-T
Schmidt
Detroit, MI
100

8 ❑ 12-1-T
Schmidt
Detroit, MI
100

9 ❑ 12-1
Schmidt
Detroit, MI
90

10 ❑ 12-1
Schmidt
Detroit, MI
250

11 ❑ 12-2
Jacob Schmidt
St. Paul, MN
90

12 ❑ 12-2
Jacob Schmidt
St. Paul, MN
DNCMT 3.2% **90**

13 ❑ 12-2
Jacob Schmidt
St. Paul, MN
DNCMT 4 PC **90**

14 ❑ 12-2
Jacob Schmidt
St. Paul, MN
Contains 5% **90**

15 ❑ 12-2
Jacob Schmidt
St. Paul, MN
STRONG **90**

16 ❑ 12-2
Jacob Schmidt
St. Paul, MN
60

17 ❑ 12-2
Jacob Schmidt
St. Paul, MN
N.O. 3.2% **60**

18 ❑ 12-2
Jacob Schmidt
St. Paul, MN
4% **60**

19 ❑ 12-2
Jacob Schmidt
St. Paul, MN
STRONG **60**

20 ❑ 12-1
Jacob Schmidt
St. Paul, MN
110

21 ❑ 12-1
Jacob Schmidt
St. Paul, MN
4% **110**

22 ❑ 12-1
Jacob Schmidt
St. Paul, MN
STRONG **110**

23 ❑ 12-1
Jacob Schmidt
St. Paul, MN
LT "1/2 of one %", **50**

24 ❑ 12-1
Jacob Schmidt
St. Paul, MN
4% **60**

25 ❑ 12-1-T
C. Schmidt & Sons
Philadelphia, PA
225

26 ❑ 12-1-T
C. Schmidt & Sons
Philadelphia, PA
225

27 ❑ 12-1
C. Schmidt & Sons
Philadelphia, PA
90

28 ❑ 12-1-T
C. Schmidt & Sons
Philadelphia, PA
90

29 ❑ 12-1-T
C. Schmidt & Sons
Philadelphia, PA
90

30 ❑ 12-1-T
C. Schmidt & Sons
Philadelphia, PA
309

31 ❑ 12-1-T
C. Schmidt & Sons
Philadelphia, PA
DULL GRAY **250**

32 ❑ 12-1-W
C. Schmidt & Sons
Philadelphia, PA
Dull OD **350**

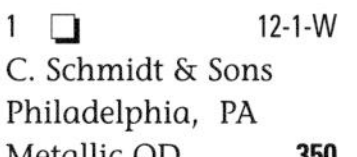

1 ❑ 12-1-W
C. Schmidt & Sons
Philadelphia, PA
Metallic OD **350**

2 ❑ 12-1-T
C. Schmidt & Sons
Philadelphia, PA
DULL GRAY **80**

3 ❑ 12-1-T
C. Schmidt & Sons
Philadelphia, PA
85

4 ❑ 12-2-T
C. Schmidt & Sons
Philadelphia, PA
Flat Bottom **85**

5 ❑ 12-1
C. Schmidt & Sons
Philadelphia, PA
Flat Bottom **110**

6 ❑ 12-1
C. Schmidt & Sons
Philadelphia, PA
Flat Bottom **85**

7 ❑ 12-1
C. Schmidt & Sons
Philadelphia, PA
Flat Bottom **85**

8 ❑ 12-1-T
New Philadelphia
New Philadelphia, OH
900

9 ❑ 12-2
Altes
San Diego, CA
400

10 ❑ 12-1-T
Terre Haute
Terre Haute, IN
750

11 ❑ 12-1
Terre Haute
Terre Haute, IN
300

12 ❑ 12-1-T
Sheridan
Sheridan, WY
1000+

13 ❑ 12-1-T
Reno
Reno, NV
110

14 ❑ 12-1-T
Reno
Reno, NV
90

15 ❑ 12-1
Reno
Reno, NV
Yellow **125**

16 ❑ 12-1
Reno
Reno, NV
Orange **125**

17 ❑ 12-1-T
Rainier
San Francisco, CA
N.O. 4% **1000+**

18 ❑ 12-1
Menominee-Marinette
Menominee, MI
135

19 ❑ 12-1-T
Chester
Chester, PA
1000+

20 ❑ 12-1-T
Fox Deluxe
Marion, IN
325

21 ❑ 12-1-T
Peter Fox of OK
Oklahoma City, OK
500

22 ❑ 12-1-T
Ph. Schneider
Trinidad, CO
1000+

23 ❑ 12-1-T
R. H. Graupner
Harrisburg, PA
1000+

24 ❑ 12-2-T
Galveston-Houston
Galveston, TX
1000+

25 ❑ 12-1
Geo. J. Renner
Akron, OH
175

26 ❑ 12-1-T
Spearman
Pensacola, FL
850

27 ❑ 12-1
Spearman
Pensacola, FL
850

28 ❑ 12-1
Spearman
Pensacola, FL
300

29 ❑ 12-1-T
Spearman
Pensacola, FL
900

30 ❑ 12-1
Spearman
Pensacola, FL
900

31 ❑ 12-2
Spearman
Pensacola, FL
500

32 ❑ 12-1
Spearman
Pensacola, FL
1000+

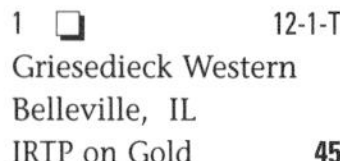

1 ❑ 12-1-T
Griesedieck Western
Belleville, IL
IRTP on Gold **45**

2 ❑ 12-1-T
Griesedieck Western
Belleville, IL
IRTP Below Gold **45**

3 ❑ 12-1-T
Griesedieck Western
Belleville, IL
4% **45**

4 ❑ 12-1
Griesedieck Western
St. Louis, MO
40

5 ❑ 12-1-T
Standard
Rochester, NY
250

6 ❑ 12-1-T
Star
Boston, MA
1000+

7 ❑ 12-1-T
Star
Boston, MA
1000+

8 ❑ 12-1-T
Stegmaier
Wilkes-Barre, PA
1000+

9 ❑ 12-1
Atlantic
Atlanta, GA
500

10 ❑ 12-1
Jones
Smithton, PA
95

11 ❑ 12-1
East Idaho
Pocatello, ID
3.2% **250**

12 ❑ 12-1
East Idaho
Pocatello, ID
DNCMT 4% **250**

13 ❑ 12-1-T
Barbey's
Reading, PA
450

14 ❑ 12-1
Sunshine
Reading, PA
125

15 ❑ 12-1
Sunshine
Reading, PA
135

16 ❑ 12-1-T
South Bethlehem
Bethlehem, PA
450

17 ❑ 12-1-T
Rainier
San Francisco, CA
N.O. 4% **800**

18 ❑ 12-1-T
Rainier
San Francisco, CA
N.O. 4% **800**

19 ❑ 12-1-T
Rainier
San Francisco, CA
700

20 ❑ 12-1-T
Carson
Carson City, NV
Red **700**

21 ❑ 12-1-T
Carson
Carson City, NV
Maroon **700**

22 ❑ 12-1-T
City
New York, NY
1000+

23 ❑ 12-1-T
City
New York, NY
1000+

24 ❑ 12-1-W
City
New York, NY
1000+

25 ❑ 12-1-T
City
New York, NY
1000+

26 ❑ 12-1-T
American
Rochester, NY
135

27 ❑ 12-1-T
American
Rochester, NY
850

28 ❑ 12-1-T
American
Rochester, NY
600

29 ❑ 12-1-T
American
Rochester, NY
1000+

30 ❑ 12-1-T
LaFayette
LaFayette, IN
125

31 ❑ 12-1
LaFayette
LaFayette, IN
125

32 ❑ 12-1
Harold C. Johnson
Lomira, WI
100

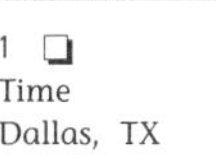

1 ❑ 12-1-T
Time
Dallas, TX
1000+

2 ❑ 12-1-T
El Rey
San Francisco, CA
N.O. 4% 1000+

3 ❑ 12-1-T
El Rey
San Francisco, CA
1000+

4 ❑ 12-1-T
El Rey
San Francisco, CA
N.O. 4% 1000+

5 ❑ 12-1-T
Sabinas
San Antonio, TX
1000+

6 ❑ 12-1-T
Sabinas
San Antonio, TX
1000+

7 ❑ 12-1-T
Peoples
Trenton, NJ
1000+

8 ❑ 12-1-T
Birk Bros.
Chicago, IL
110

9 ❑ 12-1-T
Birk Bros.
Chicago, IL
NIEO 3.2% 110

10 ❑ 12-1-T
Birk Bros.
Chicago, IL
DNCMT 4 PC 110

11 ❑ 12-1-T
Birk Bros.
Chicago, IL
STRONG 120

12 ❑ 12-1
Birk Bros.
Chicago, IL
110

13 ❑ 12-1
Birk Bros.
Chicago, IL
DNCMT 4 PC 110

14 ❑ 12-1
A.B.D. Co.
Lomira, WI
110

15 ❑ 12-1-T
Tampa Florida
Tampa, FL
300

16 ❑ 12-1
Tampa Florida
Tampa, FL
300

17 ❑ 12-1
Tampa Florida
Tampa, FL
450

18 ❑ 12-1
Tampa Florida
Tampa, FL
450

19 ❑ 12-1
Tampa Florida
Tampa, FL
450

20 ❑ 12-1-T
Tampa Florida
Tampa, FL
300

21 ❑ 12-1
Tampa Florida
Tampa, FL
300

22 ❑ 12-1
Tampa Florida
Tampa, FL
600

23 ❑ 10-1-T
Northampton
Northampton, PA
1000+

24 ❑ 10-1-T
Northampton
Northampton, PA
1000+

25 ❑ 12-1
Tube City
McKeesport, PA
175

26 ❑ 12-1-T
Red Top
Cincinnati, OH
125

27 ❑ 12-1
Red Top
Cincinnati, OH
125

28 ❑ 12-1
Red Top
Cincinnati, OH
125

29 ❑ 12-2
Uchtorff
Davenport, IA
275

30 ❑ 12-1-T
West End
Utica, NY
650

31 ❑ 12-1-T
West End
Utica, NY
700

32 ❑ 12-1-T
West End
Utica, NY
650

1 ❑ 12-1-T
West End
Utica, NY
175

2 ❑ 12-1-T
West End
Utica, NY
200

3 ❑ 12-1-T
West End
Utica, NY
100

4 ❑ 12-1-T
West End
Utica, NY
225

5 ❑ 12-1-T
West End
Utica, NY
325

6 ❑ 12-1-W
West End
Utica, NY
900

7 ❑ 12-1-T
West End
Utica, NY
100

8 ❑ 12-1-T
El Dorado
Stockton, CA
1000+

9 ❑ 12-1-T
El Dorado
Stockton, CA
450

10 ❑ 12-1-T
El Dorado
Stockton, CA
450

11 ❑ 12-1-T
El Dorado
Stockton, CA
110

12 ❑ 12-1
El Dorado
Stockton, CA
100

13 ❑ 12-1-T
Burlington
Burlington, WI
110

14 ❑ 12-1
Burlington
Burlington, WI
110

15 ❑ 12-1
Burlington
Burlington, WI
115

16 ❑ 12-2-T
Brewery Management
New York, NY
550

17 ❑ 12-1-T
Vernon
Vernon, CA
Exceeds 4%
1000+

18 ❑ 12-1
Virginia
Roanoke, VA
275

19 ❑ 12-1-T
Wacker
Lancaster, PA
1000+

20 ❑ 12-1-T
Wacker
Lancaster, PA
1000+

21 ❑ 12-1
Wacker
Lancaster, PA
600

22 ❑ 12-1-T
August Wagner
Columbus, OH
185

23 ❑ 12-1
August Wagner
Columbus, OH
185

24 ❑ 12-1
Walter
Eau Claire, WI
90

25 ❑ 12-1
Washington
Columbus, OH
500

26 ❑ 12-1-T
Webb
East Liverpool, OH
3.2-7%
175

27 ❑ 12-1
Webb
East Liverpool, OH
175

28 ❑ 12-1-T
Weber Waukesha
Waukesha, WI
150

29 ❑ 12-1
Weber Waukesha
Waukesha, WI
125

30 ❑ 12-2
Fesenmeir
Huntington, WV
275

31 ❑ 12-1-T
Largay
Waterbury, CT
1000+

32 ❑ 12-1-T
Two Rivers Bev.
Two Rivers, WI
150

CONES

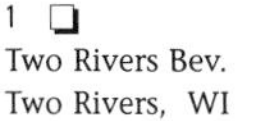
1 ❑ 12-1
Two Rivers Bev.
Two Rivers, WI
150

2 ❑ 12-2
Two Rivers Bev.
Two Rivers, WI
90

3 ❑ 12-1-T
Westminster
Chicago, IL
1000+

4 ❑ 12-1
Kiewel
Little Falls, MN
NMT 3.2%
175

5 ❑ 12-1
Kiewel
Little Falls, MN
5 PC
175

6 ❑ 12-1
Kiewel
Little Falls, MN
STRONG
175

7 ❑ 12-1-T
Geo. Wiedemann
Louisville, KY
55

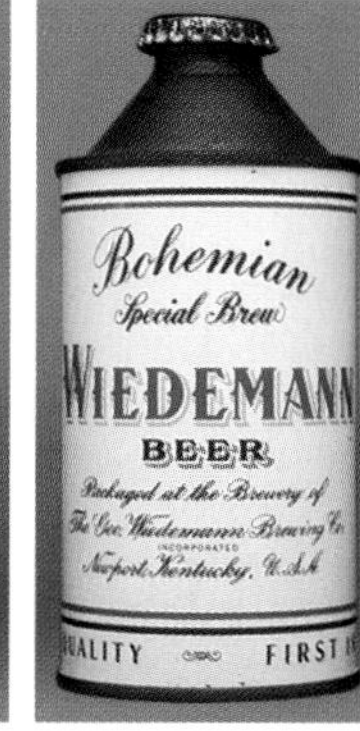

8 ❑ 12-1
Geo. Wiedemann
Louisville, KY
55

9 ❑ 12-1-T
Geo. Wiedemann
Louisville, KY
3.2%
60

10 ❑ 12-1
Geo. Wiedemann
Louisville, KY
60

11 ❑ 12-1
Geo. Wiedemann
Louisville, KY
3.2%
60

Front of 13, 14

13 ❑ 12-1-T
Pacific
San Jose, CA
150

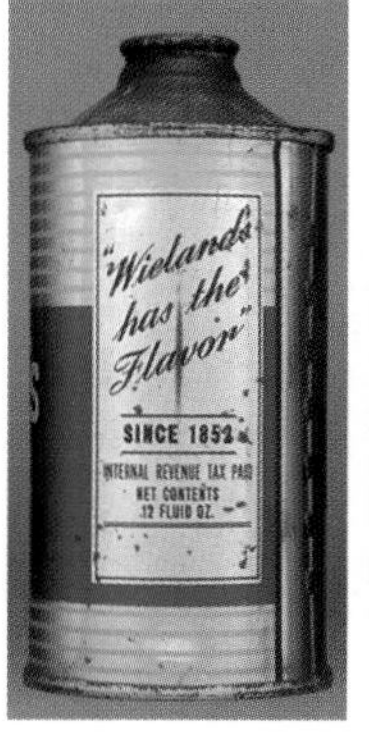

14 ❑ 12-1-T
Pacific
San Jose, CA
150

15 ❑ 12-1-T
Pacific
San Jose, CA
135

16 ❑ 12-1-T
Fernwood
Lansdowne, PA
1000+

17 ❑ 12-1-T
Wooden Shoe
Minster, OH
75

18 ❑ 12-1-T
Wooden Shoe
Minster, OH
3.2-7%
75

19 ❑ 12-1
Wooden Shoe
Minster, OH
75

20 ❑ 12-1
Wooden Shoe
Minster, OH
200

21 ❑ 12-1-T
D. G. Yuengling & Son
Pottsville, PA
900

22 ❑ 12-1-T
D. G. Yuengling & Son
Pottsville, PA
750

23 ❑ 12-1
D. G. Yuengling & Son
Pottsville, PA
750

24 ❑ 12-1-T
D. G. Yuengling & Son
Pottsville, PA
175

25 ❑ 12-1-T
D. G. Yuengling & Son
Pottsville, PA
125

26 ❑ 12-1
D. G. Yuengling & Son
Pottsville, PA
110

27 ❑ 12-1
D. G. Yuengling & Son
Pottsville, PA
110

28 ❑ 12-1
D. G. Yuengling & Son
Pottsville, PA
125

29 ❑ 12-1
Louis Ziegler
Beaver Dam, WI
400

30 ❑ 12-1
Louis Ziegler
Beaver Dam, WI
400

31 ❑ 12-1
Louis Ziegler
Beaver Dam, WI
3.2%
400

32 ❑ 12-1C
Louis Ziegler
Beaver Dam, WI
STRONG
400

When Crown Cork & Seal began to advertise its Crowntainer, the mission wasn't to convince consumers that cans — particularly crown-sealed ones — were superior to bottles as the container of choice for beer. It was to convince brewers that Crowntainers were superior to conventional cone tops. That's why ads like these ran not in popular magazines but in trade publications like *Brewer's Digest and Modern Brewery Age.*

Crowntainers Section Explanatory Notes

Twelve of the Crowntainers pictured are captioned DULL GRAY. These cans were painted with a flat gray enamel sometimes referred to as porch or battleship gray. It was used during a short period toward the end of World War II as a substitute for the Crowntainer's aluminum coating. An article in the April/May 1996 issue of *Beer Cans & Brewery Collectibles*, written by Dick Caughey, details these cans. They are not to be confused with a few other varieties of cans that came with an enamel paint such as the Hanley's 195-10.

The **Bruck's** side views 192-21, 22, 23 have the same front as 192-20.

Graupner's Old German 194-30 and **Old Graupner** 194-31 appear together under "G".

Kuebler 196-26 has the same front as 196-25 with the addition of a product list on the side.

CROWNTAINERS

12 Oz.

1 ❑ 12-2-T
Tivoli
Detroit, MI
115

2 ❑ 12-2-T
Tivoli
Detroit, MI
3.2-7% 125

3 ❑ 12-2-T
Tivoli
Detroit, MI
55

4 ❑ 12-2-T
Altes
Detroit, MI
60

5 ❑ 12-2-T
Altes
Detroit, MI
White Can 250

6 ❑ 12-2-W
Tivoli
Detroit, MI
DULL GRAY 1000+

7 ❑ 12-2-T
Beverwyck
Albany, NY
DULL GRAY 600

8 ❑ 12-2-T
Beverwyck
Albany, NY
600

9 ❑ 12-2-T
Beverwyck
Albany, NY
1000+

10 ❑ 12-2-T
Beverwyck
Albany, NY
DULL GRAY 500

11 ❑ 12-2-T
Beverwyck
Albany, NY
500

12 ❑ 12-2-T
Loewer's
New York, NY
750

13 ❑ 12-2-T
Loewer's
New York, NY
1000+

14 ❑ 12-2-T
Northampton
Northampton, PA
500

15 ❑ 12-1-T
Boston Beer
Boston, MA
1000+

16 ❑ 12-1-T
Boston Beer
Boston, MA
650

17 ❑ 12-1-T
Boston Beer
Boston, MA
1000+

18 ❑ 12-2-T
Bruckmann
Cincinnati, OH
"86 Years" 500

19 ❑ 12-1-T
Hershel Condon
Cincinnati, OH
550

20 ❑ 12-2-T
Bruckmann
Cincinnati, OH
"85 Years" 70

21 ❑ Front Like 20 12-2-T
Bruckmann
Cincinnati, OH
"85 Years" 3.2% 70

22 ❑ Front Like 20 12-2-T
Bruckmann
Cincinnati, OH
"86 Years" 70

23 ❑ Front Like 20 12-2-T
Bruckmann
Cincinnati, OH
"86 Years" 3.2% 70

24 ❑ 12-1-T
Chester
Chester, PA
1000+

25 ❑ 12-2-T
Oshkosh
Oshkosh, WI
900

26 ❑ 12-2
Oshkosh
Oshkosh, WI
1000+

27 ❑ 12-2-T
Hershel Condon
Cincinnati, OH
650

28 ❑ 12-1-T
Liebert & Obert
Philadelphia, PA
1000+

29 ❑ 12-2-T
Liebert & Obert
Philadelphia, PA
1000+

30 ❑ 12-2-T
Liebert & Obert
Philadelphia, PA
1000+

31 ❑ 12-2-T
Cremo
New Britain, CT
1000+

32 ❑ 12-2-T
Cremo
New Britain, CT
300

33 ❑ 12-2-T
Cremo
New Britain, CT
250

34 ❑ 12-2-T
Cremo
New Britain, CT
350

35 ❑ 12-1
Du Bois
Du Bois, PA
1000+

1 ❑ 12-1
Du Bois
Du Bois, PA
1000+

2 ❑ 12-2-T
Du Bois
Du Bois, PA
325

3 ❑ 12-2-T
Du Bois
Du Bois, PA
175

4 ❑ 12-2-T
Du Bois
Du Bois, PA
250

5 ❑ 12-2
Du Bois
Du Bois, PA
225

6 ❑ 12-1
Duquesne Plant #2
Stowe TWP., PA
1000+

7 ❑ 12-2-T
Ebling
New York, NY
600

8 ❑ 12-2-T
Ebling
New York, NY
200

9 ❑ 12-2-W
Ebling
New York, NY
225

10 ❑ 12-2-T
Ebling
New York, NY
200

11 ❑ 12-2-T
Ebling
New York, NY
500

12 ❑ 12-2-T
Ebling
New York, NY
50

13 ❑ 12-2-T
Ebling
New York, NY
3.2% **55**

14 ❑ 12-2-T
Ebling
New York, NY
4% **55**

15 ❑ 12-2-T
Ebling
New York, NY
STRONG **55**

16 ❑ 12-2-T
Ebling
New York, NY
250

17 ❑ 12-2-T
Ebling
New York, NY
750

18 ❑ 12-2-T
Esslinger's
Philadelphia, PA
175

19 ❑ 12-2-T
Esslinger's
Philadelphia, PA
250

20 ❑ 12-2-T
Esslinger's
Philadelphia, PA
600

21 ❑ 12-2-T
Esslinger's
Philadelphia, PA
1000+

22 ❑ 12-1-T
Falstaff
St. Louis, MO
1000+

23 ❑ 12-2-T
Frank Fehr
Louisville, KY
90

24 ❑ 12-2-T
Frank Fehr
Louisville, KY
45

25 ❑ 12-2
Frank Fehr
Louisville, KY
45

26 ❑ 12-2-T
Fitzgerald Bros.
Troy, NY
200

27 ❑ 12-2-T
Fitzgerald Bros.
Troy, NY
175

28 ❑ 12-2-T
Fitzgerald Bros.
Troy, NY
150

29 ❑ 12-2-T
Fitzgerald Bros.
Troy, NY
150

30 ❑ 12-2-T
Fitzgerald Bros.
Troy, NY
DULL GRAY **135**

31 ❑ 12-2-T
Fitzgerald Bros.
Troy, NY
75

32 ❑ 12-2-T
Fitzgerald Bros.
Troy, NY
60

33 ❑ 12-2
Fitzgerald Bros.
Troy, NY
60

34 ❑ 12-2-W
Fitzgerald Bros.
Troy, NY
500

35 ❑ 12-2-T
Fitzgerald Bros.
Troy, NY
300

1 ❑ 12-2-T
Fitzgerald Bros.
Troy, NY
50

2 ❑ 12-2-T
Fitzgerald Bros.
Troy, NY
STRONG 55

3 ❑ 12-2-T
Fitzgerald Bros.
Troy, NY
50

4 ❑ 12-2-T
Fitzgerald Bros.
Troy, NY
STRONG 55

5 ❑ 12-2
Fitzgerald Bros.
Troy, NY
50

6 ❑ 12-2
Fitzgerald Bros.
Troy, NY
STRONG 50

7 ❑ 12-2-T
Fort Pitt
Pittsburgh, PA
300

8 ❑ 12-2-W
Fort Pitt
Pittsburgh, PA
300

9 ❑ 12-2-T
Fort Pitt
Pittsburgh, PA
135

10 ❑ 12-2-T
Fort Pitt
Pittsburgh, PA
100

11 ❑ 12-2-T
Fort Pitt
Pittsburgh, PA
3.2-7% 100

12 ❑ 12-2
Fort Pitt
Pittsburgh, PA
100

13 ❑ 12-2
William Gerst
Nashville, TN
250

14 ❑ 12-1-T
Gluek
Minneapolis, MN
50

15 ❑ 12-1-T
Gluek
Minneapolis, MN
CNMT 4% 50

16 ❑ 12-1-T
Gluek
Minneapolis, MN
4% 50

17 ❑ 12-1-T
Gluek
Minneapolis, MN
CMT 4% 50

18 ❑ 12-1
Gluek
Minneapolis, MN
5% 50

19 ❑ 12-1-T
Gluek
Minneapolis, MN
DNCMT 3.2% 50

20 ❑ 12-1-T
Gluek
Minneapolis, MN
3.2% 50

21 ❑ 12-1-T
Gluek
Minneapolis, MN
CNMT 4% 50

22 ❑ 12-1
Gluek
Minneapolis, MN
DNCMT 3.2% 50

23 ❑ 12-1
Gluek
Minneapolis, MN
CNMT 4% 75

24 ❑ 12-1
Gluek
Minneapolis, MN
5% 75

25 ❑ 12-1
Gluek
Minneapolis, MN
STRONG 75

26 ❑ 12-1
Gluek
Minneapolis, MN
DNCMT 3.2% 75

27 ❑ 12-1
Gluek
Minneapolis, MN
DNCMT 4% 75

28 ❑ 12-1
Gluek
Minneapolis, MN
Paper Label 150

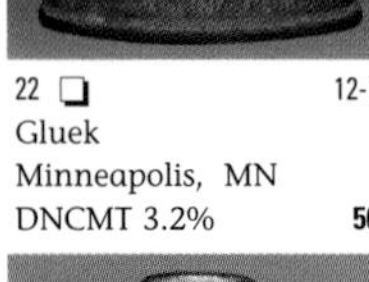

29 ❑ 12-1
Gluek
Minneapolis, MN
CMT 3.2% Paper Label 150

30 ❑ 12-1-T
R. H. Graupner
Harrisburg, PA
1000+

31 ❑ 12-1-T
R. H. Graupner
Harrisburg, PA
1000+

32 ❑ 12-1-T
Wm. Gretz
Philadelphia, PA
1000+

33 ❑ 12-1-T
Wm. Gretz
Philadelphia, PA
750

Back Of 33

35 ❑ 12-1-T
Wm. Gretz
Philadelphia, PA
Blue 350

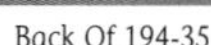

Back Of 194-35

2 ❑ 12-1-T
Wm. Gretz
Philadelphia, PA
Black
450

Back Of 2

4 ❑ 12-2-T
Griesedieck Bros.
St. Louis, MO
125

5 ❑ 12-2-T
Griesedieck Bros.
St. Louis, MO
4%
125

6 ❑ 12-1-T
Geo. J. Renner
Akron, OH
55

7 ❑ 12-1-T
Geo. J. Renner
Akron, OH
3.2-7%
55

8 ❑ 12-1
Geo. J. Renner
Akron, OH
55

9 ❑ 12-1
Geo. J. Renner
Akron, OH
3.2-7%
55

10 ❑ 12-1-T
James Hanley
Providence, RI
Enamel
75

11 ❑ 12-1-T
James Hanley
Providence, RI
Metallic
75

12 ❑ 12-1-T
James Hanley
Providence, RI
200

13 ❑ 12-1-T
James Hanley
Providence, RI
200

14 ❑ 12-1-T
James Hanley
Providence, RI
225

15 ❑ 12-1-T
James Hanley
Providence, RI
350

16 ❑ 12-2-T
Hoff-Brau
Ft. Wayne, IN
DULL GRAY
500

17 ❑ 12-2-T
Hoff-Brau
Ft. Wayne, IN
300

18 ❑ 12-2-T
Hoff-Brau
Ft. Wayne, IN
DULL GRAY
650

19 ❑ 12-2-T
Hoff-Brau
Ft. Wayne, IN
250

20 ❑ 12-2-T
John Hohenadel
Philadelphia, PA
275

21 ❑ 12-2-T
John Hohenadel
Philadelphia, PA
375

22 ❑ 12-1-T
Hudepohl
Cincinnati, OH
500

23 ❑ 12-1-W
Hudepohl
Cincinnati, OH
700

24 ❑ 12-1-T
Hudepohl
Cincinnati, OH
125

25 ❑ 12-1-T
Hudepohl
Cincinnati, OH
NMT 3.2%
750

26 ❑ 12-2-T
Hull
New Haven, CT
200

27 ❑ 12-2-T
Hull
New Haven, CT
300

28 ❑ 12-2
Iroquois
Buffalo, NY
175

29 ❑ 12-2-T
Iroquois
Buffalo, NY
150

30 ❑ 12-2
Iroquois
Buffalo, NY
125

31 ❑ 12-1-T
R. H. Graupner
Harrisburg, PA
1000+

32 ❑ 12-1
Mankato
Mankato, MN
Paper Label
300

33 ❑ 12-2-T
Wm. G. Jung
Random Lake, WI
175

34 ❑ 12-2
Wm. G. Jung
Random Lake, WI
NMT 4%
175

35 ❑ 12-2
Wm. G. Jung
Random Lake, WI
175

1 ❑ 12-2
Wm. G. Jung
Random Lake, WI
3.2% **175**

2 ❑ 12-2-T
Kamm & Schellinger
Mishawaka, IN
135

3 ❑ 12-2-T
Kamm & Schellinger
Mishawaka, IN
60

4 ❑ 12-2-T
Kamm & Schellinger
Mishawaka, IN
60

5 ❑ 12-2
Kamm & Schellinger
Mishawaka, IN
60

6 ❑ 12-2-T
Kingsbury
Sheboygan, WI
DULL GRAY **350**

7 ❑ 12-2-T
Kingsbury
Sheboygan, WI
250

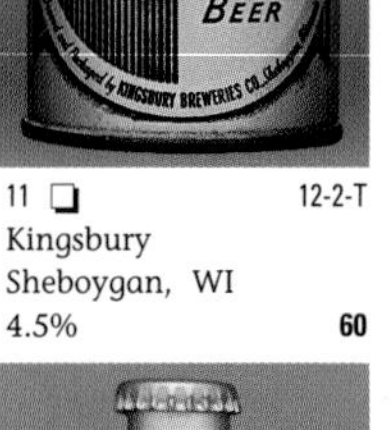

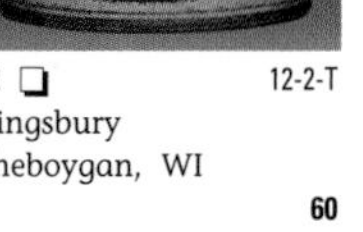

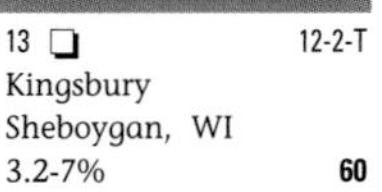

8 ❑ 12-2-T
Kingsbury
Sheboygan, WI
3.2-7% **50**

9 ❑ 12-2-T
Kingsbury
Sheboygan, WI
4% **50**

10 ❑ 12-2-T
Kingsbury
Sheboygan, WI
60

11 ❑ 12-2-T
Kingsbury
Sheboygan, WI
4.5% **60**

12 ❑ 12-2-T
Kingsbury
Sheboygan, WI
60

13 ❑ 12-2-T
Kingsbury
Sheboygan, WI
3.2-7% **60**

14 ❑ 12-2
Kingsbury
Sheboygan, WI
4% **60**

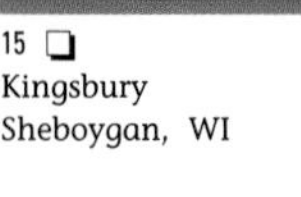

15 ❑ 12-2
Kingsbury
Sheboygan, WI
60

16 ❑ 12-2-T
Koller
Chicago, IL
90

17 ❑ 12-2-T
Koller
Chicago, IL
NMT 4% **90**

18 ❑ 12-2
Koller
Chicago, IL
90

19 ❑ 12-2-T
G. Krueger
Wilmington, DE
125

20 ❑ 12-2-T
G. Krueger
Wilmington, DE
125

21 ❑ 12-2-T
G. Krueger
Wilmington, DE
100

22 ❑ 12-2-T
Kuebler
Easton, PA
DULL GRAY **1000+**

23 ❑ 12-2-T
Kuebler
Easton, PA
DULL GRAY **600**

24 ❑ 12-2-T
Kuebler
Easton, PA
350

25 ❑ 12-2-T
Kuebler
Easton, PA
DNCMT 3.2% **350**

26 ❑ Front Like 25 12-2-T
Kuebler
Easton, PA
Product List **350**

27 ❑ 12-2
J. Leinenkugel
Chippewa Falls, WI
250

28 ❑ 12-2
J. Leinenkugel
Chippewa Falls, WI
125

29 ❑ 12-2
J. Leinenkugel
Chippewa Falls, WI
STRONG **125**

30 ❑ 12-2-T
Pilser
New York, NY
450

31 ❑ 12-2-T
Miami Valley
Dayton, OH
175

32 ❑ 12-2-T
Miami Valley
Dayton, OH
3.2-7% **150**

33 ❑ 12-2-T
Miami Valley
Dayton, OH
1000+

34 ❑ 12-2-T
Miami Valley
Dayton, OH
1000+

35 ❑ 12-2
Ebling
New York, NY
600

1 ❑ 12-2-T
Ebling
New York, NY
1000+

2 ❑ 12-1-T
Cooper
Philadelphia, PA
275

3 ❑ 12-2-T
National
Baltimore, MD
DULL GRAY **375**

4 ❑ 12-2-T
National
Baltimore, MD
375

5 ❑ 12-2-T
National
Baltimore, MD
375

6 ❑ 12-1-T
L. F. Neuweiler's Sons
Allentown, PA
PILSENER in Black **40**

7 ❑ 12-1-T
L. F. Neuweiler's Sons
Allentown, PA
PILSENER in Brown **45**

8 ❑ 12-2-T
Miami Valley
Dayton, OH
3.2-7% **1000+**

9 ❑ 12-1-T
Washington
Columbus, OH
DULL GRAY 3.2-7% **350**

10 ❑ 12-1-T
Washington
Columbus, OH
3.2-7% **300**

11 ❑ 12-3-T
Oertel
Louisville, KY
DULL GRAY **275**

12 ❑ 12-3-T
Oertel
Louisville, KY
175

13 ❑ 12-2-T
Oertel
Louisville, KY
40

14 ❑ 12-2-T
Oertel
Louisville, KY
45

15 ❑ 12-2
Oertel
Louisville, KY
40

16 ❑ 12-2
Oertel
Louisville, KY
100

17 ❑ 12-2
Oertel
Louisville, KY
750

18 ❑ 12-2-T
Eagle
Catasauqua, PA
1000+

19 ❑ 12-2-T
Eagle
Catasauqua, PA
1000+

20 ❑ 12-2-T
Northampton
Northampton, PA
135

21 ❑ 12-2
Rahr-Green Bay
Green Bay, WI
85

22 ❑ 12-1-T
Standard
Rochester, NY
300

23 ❑ 12-2-T
Old Reading
Reading, PA
750

24 ❑ 12-2-T
Old Reading
Reading, PA
125

25 ❑ 12-2
Old Reading
Reading, PA
100

26 ❑ 12-2-T
Fort Pitt
Jeanette, PA
100

27 ❑ 12-2-T
Fort Pitt
Jeanette, PA
3.2-7% **100**

28 ❑ 12-2-T
Fort Pitt
Pittsburgh, PA
100

29 ❑ 12-2-T
Rochester
Rochester, NY
Black Oval & Sign **60**

30 ❑ 12-2-T
Rochester
Rochester, NY
Brown Oval & Sign **60**

31 ❑ 12-2-T
Rochester
Rochester, NY
85

32 ❑ 12-2-T
Rochester
Rochester, NY
250

33 ❑ 12-2
Rochester
Rochester, NY
250

34 ❑ 12-2
Rochester
Rochester, NY
400

35 ❑ 12-2-W
Rochester
Rochester, NY
750

No.	Code	Brewery	Location	Notes	Value
1 ❑	12-2-T	Rochester	Rochester, NY		90
2 ❑	12-2-T	Rochester	Rochester, NY		300
3 ❑	12-2-T	Rochester	Rochester, NY		300
4 ❑	12-2-T	Rochester	Rochester, NY		350
5 ❑	12-2-T	Rochester	Rochester, NY		1000+
6 ❑	12-1-T	Henry F. Ortlieb	Philadelphia, PA		1000+
7 ❑	12-1-T	Henry F. Ortlieb	Philadelphia, PA		175
8 ❑	12-1-T	Henry F. Ortlieb	Philadelphia, PA		175
9 ❑	12-1	Henry F. Ortlieb	Philadelphia, PA		500
10 ❑	12-2-T	Pilser	New York, NY		1000+
11 ❑	12-2-T	Pilser	New York, NY		275
12 ❑	12-2-T	Metropolis	New York, NY		275
13 ❑	12-2-T	Metropolis	New York, NY		500
14 ❑	12-2-T	Potosi	Potosi, WI		135
15 ❑	12-2-T	Potosi	Potosi, WI	4%	135
16 ❑	12-2	Rahr-Green Bay	Green Bay, WI		85
17 ❑	12-2-T	Reisch	Springfield, IL		250
18 ❑	12-2	Reisch	Springfield, IL		250
19 ❑	12-2	Home	Richmond, VA		60
20 ❑	12-2	Home	Richmond, VA		75
21 ❑	12-2-T	Geo. Wiedemann	Louisville, KY		200
22 ❑	12-2-T	Geo. Wiedemann	Louisville, KY	3.2-7%	200
23 ❑	12-2-T	Koller	Chicago, IL		175
24 ❑	12-1-T	August Schell	New Ulm, MN	3.2%	175
25 ❑	12-1-T	August Schell	New Ulm, MN	STRONG	175
26 ❑	12-1-T	August Schell	New Ulm, MN	STRONG	175
27 ❑	12-1	August Schell	New Ulm, MN	3.2%	175
28 ❑	12-1	August Schell	New Ulm, MN	STRONG	175
29 ❑	12-1	Jos. Schlitz	Milwaukee, WI		1000+
30 ❑	12-1-T	C. Schmidt & Sons	Philadelphia, PA		90
31 ❑	12-1-T	C. Schmidt & Sons	Philadelphia, PA		80
32 ❑	12-1-T	C. Schmidt & Sons	Philadelphia, PA		80
33 ❑	12-1-T	C. Schmidt & Sons	Philadelphia, PA		1000+
34 ❑	12-1-T	R.H. Graupner	Harrisburg, PA		1000+
35 ❑	12-2-T	Galveston-Houston	Galveston, TX		150

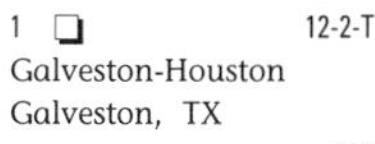

1 ❑ 12-2-T
Galveston-Houston
Galveston, TX
100

2 ❑ 12-2-T
Galveston-Houston
Galveston, TX
175

3 ❑ 12-2-T
Geo. J. Renner
Akron, OH
100

4 ❑ 12-2
Geo. J. Renner
Akron, OH
100

5 ❑ 12-1-T
Standard
Rochester, NY
175

6 ❑ 12-1-T
Standard
Rochester, NY
175

7 ❑ 12-1-T
Barbey's
Reading, PA
350

8 ❑ 12-1-T
Barbey's
Reading, PA
250

9 ❑ 12-1
Barbey's
Reading, PA
250

10 ❑ 12-1
Sunshine
Reading, PA
250

11 ❑ 12-1-W
Peoples
Trenton, NJ
375

12 ❑ 12-1-T
Peoples
Trenton, NJ
325

13 ❑ 12-2-T
Peoples
Trenton, NJ
325

14 ❑ 12-2-T
Northampton
Northampton, PA
350

15 ❑ 12-2-W
Northampton
Northampton, PA
225

16 ❑ 12-2-T
Northampton
Northampton, PA
200

17 ❑ 12-2-T
Northampton
Northampton, PA
200

18 ❑ 12-2-T
Miami Valley
Dayton, OH
200

19 ❑ 12-2-T
Brewery Management
New York, NY
600

20 ❑ 12-1-T
Wacker
Lancaster, PA
800

21 ❑ 12-1-T
Wacker
Lancaster, PA
1000+

22 ❑ 12-1-T
Washington
Columbus, OH
175

23 ❑ 12-1-T
Washington
Columbus, OH
3.2-7% **175**

24 ❑ 12-1-T
Geo. Wiedemann
Louisville, KY
3.2% **90**

25 ❑ 12-1-T
Geo. Wiedemann
Louisville, KY
110

26 ❑ 12-2-T
Geo. Wiedemann
Louisville, KY
100

27 ❑ 12-2-T
Geo. Wiedemann
Louisville, KY
3.2-7% **100**

28 ❑ 12-2
Yoerg
St. Paul, MN
3.2% **90**

29 ❑ 12-2
Yoerg
St. Paul, MN
STRONG **110**

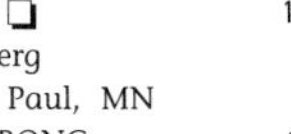

What's the best thing to do with a big idea? See if you can make it bigger. That's precisely what happened with beer cans. By December 1937, Continental Can Company had introduced the world to the quart cone top. A year and a half later, they were proudly able to report that their quarts were used for 41 brands. Quart cones never really made a comeback after World War II.

Quarts Section Explanatory Notes

The short flat-top quarts have been listed in the Quarts section while the tall flat-top quarts are shown in the Big Can section. The ages of the flat quarts in this section have been difficult to determine so they have been sequenced as best as possible by the label within the particular run of cans. As an example, the Croft 206-1 has been listed in front of the 4 and 3 Product cans as it is assumed the 6 Product cans were earlier.

Stock Ale by Croft 206-5 is listed with Croft cans.

Stegmaier's Gold Medal 210-8 has been listed with Gold Medal.

Graupner's Old German 211-8 and **Old Graupner** 211-9 appear together under G.

Indian Queen 212-2 is listed with Hohenadel Indian Queen 212-3 for brand continuity.

Unclear as to where **Neuweiler's** 215-10 and 215-13 should fall in sequence because they are IRTP but not the first label designs that were used.

Little Dutch 220-15 is listed with Wacker.

QUARTS

32 Oz. Cones/Flats

Front of 2, 3, 4

3 Line Mandatory

2 ❑ 32-1-T
P. Ballantine & Sons
Newark, NJ
135

3 ❑ 32-1-T
P. Ballantine & Sons
Newark, NJ
135

4 ❑ 32-1-T
P. Ballantine & Sons
Newark, NJ
135

Front of 6, 7, 8

4 Line Mandatory

6 ❑ 32-1-T
P. Ballantine & Sons
Newark, NJ
135

7 ❑ 32-1-T
P. Ballantine & Sons
Newark, NJ
135

8 ❑ 32-1-T
P. Ballantine & Sons
Newark, NJ
135

Front of 10

2 Line Mandatory

10 ❑ 32-1-T
P. Ballantine & Sons
Newark, NJ
135

Front of 12, 13, 14, 15

3 Line Mandatory

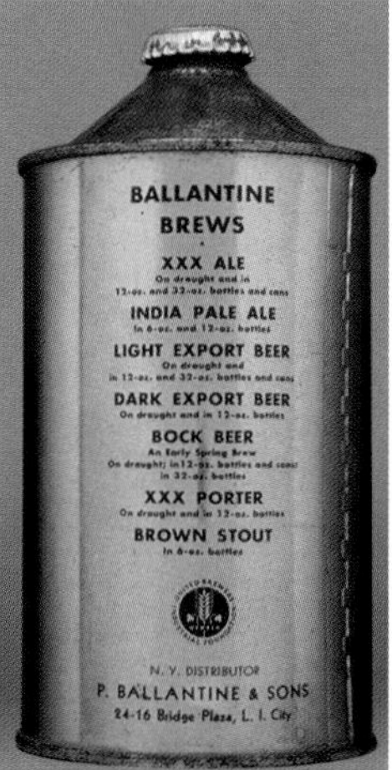

12 ❑ 32-1-T
P. Ballantine & Sons
Newark, NJ
135

13 ❑ 32-1-T
P. Ballantine & Sons
Newark, NJ
135

14 ❑ 32-1-T
P. Ballantine & Sons
Newark, NJ
135

15 ❑ 32-1-T
P. Ballantine & Sons
Newark, NJ
135

16 ❑ 32-1-T
Mount Carbon
Pottsville, PA
275

17 ❑ 32-1
Mount Carbon
Pottsville, PA
275

18 ❑ 32-1
Mount Carbon
Pottsville, PA
275

1 ❑ 32-1-T
Best
Chicago, IL
DULL GRAY **200**

2 ❑ 32-1-T
Best
Chicago, IL
175

3 ❑ 32-2
Best
Chicago, IL
300

4 ❑ 32-1-T
Beverwyck
Albany, NY
250

5 ❑ 32-1-T
Beverwyck
Albany, NY
250

6 ❑ 32-1
Beverwyck
Albany, NY
350

7 ❑ 32-1-T
Beverwyck
Albany, NY
400

8 ❑ 32-1-T
Blackhawk
Davenport, IA
Enamel **750**

9 ❑ 32-1-T
Blackhawk
Davenport, IA
Metallic DNCMT 4% **750**

10 ❑ 32-1-T
Blackhawk
Davenport, IA
750

11 ❑ 32-1-T
Blackhawk
Davenport, IA
DNCMT 4% **750**

12 ❑ 32-1-T
Loewer's
New York, NY
1000+

13 ❑ 32-1-T
Loewer's
New York, NY
1000+

14 ❑ 32-1-T
Northampton
Northampton, PA
1000+

15 ❑ 32-1-T
Enterprise
Fall River, MA
800

16 ❑ 32-1-T
Boston Beer
Boston, MA
850

17 ❑ 32-1-T
Boston Beer
Boston, MA
850

18 ❑ 32-1-T
Boston Beer
Boston, MA
350

1 ❑ 32-1-T
Peter Breidt
Elizabeth, NJ
850

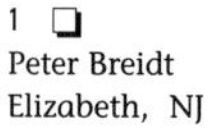

2 ❑ 32-1-T
Peter Breidt
Elizabeth, NJ
1000+

3 ❑ 32-1-T
Brockert
Worcester, MA
1000+

4 ❑ 32-2-T
Salem Assn.
Salem, OR
800

5 ❑ 32-1-T
Bruckmann
Cincinnati, OH
1000+

6 ❑ 32-1-T
Bruckmann
Cincinnati, OH
1000+

7 ❑ 32-1-T
Bruckmann
Cincinnati, OH
450

Front of 9, 10

9 ❑ 32-1-T
Bruckmann
Cincinnati, OH
85 YEARS
550

10 ❑ 32-1-T
Bruckmann
Cincinnati, OH
86 YEARS
550

11 ❑ 32-1-T
Buffalo
Sacramento, CA
400

12 ❑ 32-1-T
Burger
Cincinnati, OH
400

13 ❑ 32-1-T
Burger
Cincinnati, OH
400

14 ❑ 32-1-T
Burger
Cincinnati, OH
1000+

15 ❑ 32-1
Burger
Cincinnati, OH
800

16 ❑ 32-1-T
San Francisco
San Francisco, CA
6%
300

17 ❑ 32-1-T
San Francisco
San Francisco, CA
6%
300

18 ❑ 32-1-T
San Francisco
San Francisco, CA
4%
800

1 ❑ 32-1-T
San Francisco
San Francisco, CA
4% **800**

2 ❑ 32-2
San Francisco
San Francisco, CA
150

3 ❑ 32-1-T
Bushkill Products
Easton, PA
1000+

4 ❑ 32-1
Canadian Ace
Chicago, IL
125

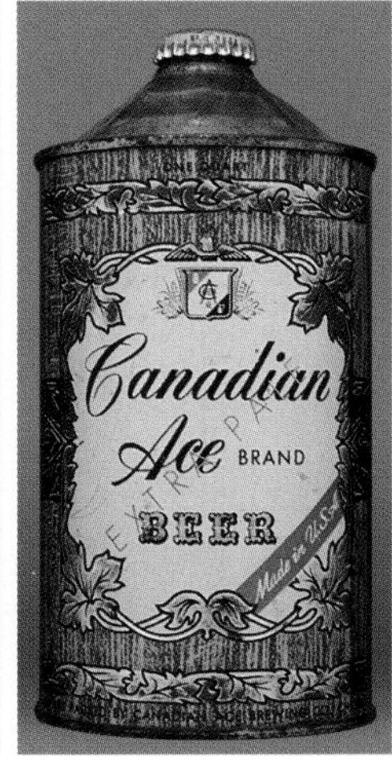

5 ❑ 32-1
Canadian Ace
Chicago, IL
100

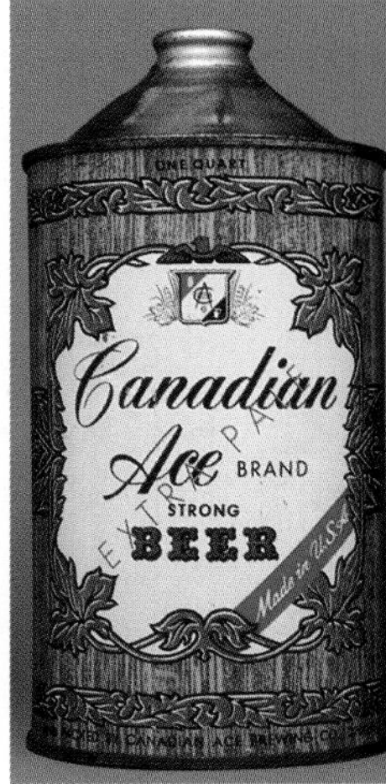

6 ❑ 32-1
Canadian Ace
Chicago, IL
STRONG **100**

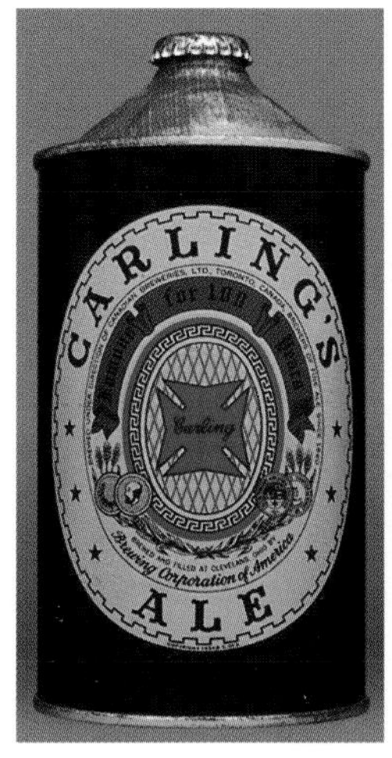

7 ❑ 32-1-T
Brewing Corp. of America
Cleveland, OH
750

8 ❑ 32-1-T
Brewing Corp. of America
Cleveland, OH
600

9 ❑ 32-1-T
Duquesne Plant #2
Stowe TWP, PA
1000+

10 ❑ 32-2-T
Terre Haute
Terre Haute, IN
700

11 ❑ 32-1-T
Chester
Chester, PA
1000+

12 ❑ 32-1-T
Enterprise
Fall River, MA
1000+

13 ❑ 32-2-T
Liebert & Obert
Philadelphia, PA
1000+

14 ❑ 32-1-T
Liebert & Obert
Philadelphia, PA
600

15 ❑ 32-1-T
Liebert & Obert
Philadelphia, PA
700

16 ❑ 32-1-T
Cremo
New Britain, CT
Matches Cone 158-16 **1000+**

17 ❑ 32-1-T
Cremo
New Britain, CT
1000+

18 ❑ 32-1-T
Croft
Boston, MA
600

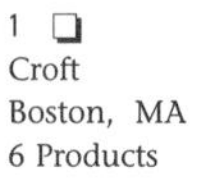

1 ❑ 32-1-T
Croft
Boston, MA
6 Products **900**

Front of 3, 4

3 ❑ 32-1-T
Croft
Boston, MA
4 Products **550**

4 ❑ 32-1-T
Croft
Boston, MA
3 Products **550**

5 ❑ 32-1-T
Croft
Boston, MA
1000+

6 ❑ 32-1-T
Croft
Boston, MA
350

QUARTS

7 ❑ 32-1-T
Croft
Boston, MA
350

8 ❑ 32-1
Croft
Boston, MA
350

9 ❑ 32-2-T
Dawson's
New Bedford, MA
400

10 ❑ 32-2-T
Dawson's
New Bedford, MA
600

11 ❑ 32-2-T
Dawson's
New Bedford, MA
700

12 ❑ 32-2-T
Dawson's
New Bedford, MA
1000+

13 ❑ 32-2-T
Dawson's
New Bedford, MA
400

14 ❑ 32-2-T
Dawson's
New Bedford, MA
700

15 ❑ 32-2
Dawson's
New Bedford, MA
800

16 ❑ 32-1
Dawson's
New Bedford, MA
800

17 ❑ 32-1
Diamond State
Wilmington, DE
1000+

18 ❑ 32-2-T
Du Bois
Du Bois, PA
800

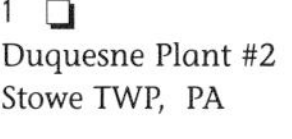

1 ❑ 32-1-T
Duquesne Plant #2
Stowe TWP, PA
400

2 ❑ 32-1-T
Ebling
New York, NY
1000+

3 ❑ 32-1-T
Ebling
New York, NY
600

4 ❑ 32-1-T
Ebling
New York, NY
1000+

5 ❑ 32-1-T
Ebling
New York, NY
275

6 ❑ 32-1
Ebling
New York, NY
DNCMT 4%
275

7 ❑ 32-1-T
Ebling
New York, NY
STRONG
275

8 ❑ 32-1-T
Edelbrew
New York, NY
1000+

9 ❑ 32-1-T
Schoenhofen-Edelweiss
Chicago, IL
125

Front of 11, 12

11 ❑ 32-1-T
Schoenhofen-Edelweiss
Chicago, IL
125

12 ❑ 32-1-T
Schoenhofen-Edelweiss
Chicago, IL
DNCMT 4%
125

13 ❑ 32-1-T
Schoenhofen-Edelweiss
Chicago, IL
75

14 ❑ 32-1-T
Schoenhofen-Edelweiss
Chicago, IL
3.2%
100

15 ❑ 32-1-T
Schoenhofen-Edelweiss
Chicago, IL
4%
100

16 ❑ 32-1
Schoenhofen-Edelweiss
Chicago, IL
100

17 ❑ 32-2
Best
Chicago, IL
700

18 ❑ 32-1-T
Westminster
Chicago, IL
1000+

QUARTS

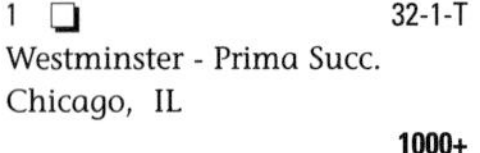

1 ❑ 32-1-T
Westminster - Prima Succ.
Chicago, IL
1000+

2 ❑ 32-1-T
Prima
Chicago, IL
1000+

3 ❑ 32-1-T
Westminster - Prima Succ.
Chicago, IL
4.0% **1000+**

4 ❑ 32-1-T
Westminster
Chicago, IL
1000+

5 ❑ 32-1-T
Otto Erlanger
Philadelphia, PA
1000+

6 ❑ 32-1-T
Otto Erlanger
Philadelphia, PA
1000+

7 ❑ 32-1-T
Otto Erlanger
Philadelphia, PA
1000+

8 ❑ 32-1-T
Esslinger's
Philadelphia, PA
225

9 ❑ 32-1-T
Esslinger's
Philadelphia, PA
200

10 ❑ 32-1-T
Esslinger's
Philadelphia, PA
275

11 ❑ 32-1-OT
Esslinger's
Philadelphia, PA
600

12 ❑ 32-1
Esslinger's
Philadelphia, PA
800

13 ❑ 32-1-T
Esslinger's
Philadelphia, PA
300

14 ❑ 32-1-T
Esslinger's
Philadelphia, PA
275

15 ❑ 32-1-OT
Esslinger's
Philadelphia, PA
650

16 ❑ 32-1-T
Esslinger's
Philadelphia, PA
Long Text **700**

17 ❑ 32-1-T
Esslinger's
Philadelphia, PA
Short Text **700**

18 ❑ 32-1
Esslinger's
Philadelphia, PA
700

1 ❏ 32-1
Esslinger's
Philadelphia, PA
400

2 ❏ 32-1-T
Esslinger's
Philadelphia, PA
1000+

3 ❏ 32-1
Fuhrmann & Schmidt
Shamokin, PA
1000+

4 ❏ 32-1-T
Fuhrmann & Schmidt
Shamokin, PA
1000+

5 ❏ 32-1-T
Fuhrmann & Schmidt
Shamokin, PA
1000+

6 ❏ 32-1-T
Fuhrmann & Schmidt
Shamokin, PA
1000+

7 ❏ 32-1-T
Fuhrmann & Schmidt
Shamokin, PA
600

8 ❏ 32-1-T
Fuhrmann & Schmidt
Shamokin, PA
300

9 ❏ 32-1
Fuhrmann & Schmidt
Shamokin, PA
300

10 ❏ 32-1-T
Chr. Feigenspan
Newark, NJ
300

11 ❏ 32-1-T
Chr. Feigenspan
Newark, NJ
400

12 ❏ 32-1-T
Fitzgerald Bros.
Troy, NY
275

13 ❏ 32-1-T
Fitzgerald Bros.
Troy, NY
800

14 ❏ 32-1-T
Fitzgerald Bros.
Troy, NY
1000+

15 ❏ 32-1-T
Fox De Luxe
Marion, IN
300

16 ❏ 32-1-T
Fox De Luxe
Grand Rapids, MI
300

17 ❏ 32-1-T
Genesee
Rochester, NY
1000+

18 ❏ 32-1-T
Genesee
Rochester, NY
250

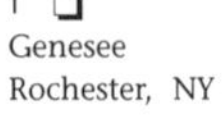

1 ❑ 32-1-T
Genesee
Rochester, NY
1000+

2 ❑ 32-1-T
Lion
Wilkes-Barre, PA
275

3 ❑ 32-1-T
Lion
Wilkes-Barre, PA
225

4 ❑ 32-1
Lion
Wilkes-Barre, PA
225

5 ❑ 32-1
Lion
Wilkes-Barre, PA
250

6 ❑ 32-1-T
Stegmaier
Wilkes-Barre, PA
250

7 ❑ 32-1-T
Stegmaier
Wilkes-Barre, PA
225

8 ❑ 32-1
Stegmaier
Wilkes-Barre, PA
500

9 ❑ 32-2-T
Mutual
Ellensburg, WA
DNCMT 4% **550**

10 ❑ 32-2-T
Mutual
Ellensburg, WA
DNCMT 4% **550**

11 ❑ 32-2-T
Silver Springs
Port Orchard, WA
DNCMT 4% **700**

12 ❑ 32-1
Tennessee
Memphis, TN
900

13 ❑ 32-2
Tennessee
Memphis, TN
900

14 ❑ 32-1-T
Fernwood
Lansdowne, PA
1000+

15 ❑ 32-1-T
Golden Age
Spokane, WA
NMT 4% **750**

16 ❑ 32-1-T
Golden Age
Spokane, WA
NMT 4% **600**

17 ❑ 32-1-T
Golden Age
Spokane, WA
NMT 4% **600**

18 ❑ 32-1-T
Golden West
Oakland, CA
N.O. 4% **350**

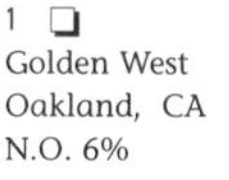
1 ❑ 32-1-T
Golden West
Oakland, CA
N.O. 6% 350

2 ❑ 32-1-T
Golden West
Oakland, CA
N.O. 6% 350

3 ❑ 32-1-T
Golden West
Oakland, CA
375

4 ❑ 32-1-T
Golden West
Oakland, CA
N.O. 4% 375

5 ❑ 32-1-T
Golden West
Oakland, CA
N.O. 4% 375

6 ❑ 32-1-T
Golden West
Oakland, CA
N.O. 4% 375

7 ❑ 32-1-T
San Francisco
San Francisco, CA
1000+

8 ❑ 32-1-T
R. H. Graupner
Harrisburg, PA
750

9 ❑ 32-1-T
R. H. Graupner
Harrisburg, PA
1000+

10 ❑ 32-1-T
Wm. Gretz
Philadelphia, PA
1000+

11 ❑ 32-1-T
Wm. Gretz
Philadelphia, PA
1000+

12 ❑ 32-1-T
Haberle Congress
Syracuse, NY
900

13 ❑ 32-1-T
Haberle Congress
Syracuse, NY
750

14 ❑ 32-1-T
Hampden
Willimansett, MA
500

15 ❑ 32-1-T
James Hanley
Providence, RI
DULL GRAY 275

16 ❑ 32-1-T
James Hanley
Providence, RI
250

17 ❑ 32-1-T
Harvard
Lowell, MA
DULL GRAY 550

18 ❑ 32-1-T
Harvard
Lowell, MA
550

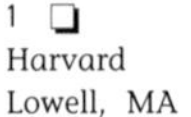
1 ❑ 32-1-T
Harvard
Lowell, MA
550

2 ❑ 32-1-T
John Hohenadel
Philadelphia, PA
1000+

3 ❑ 32-1-T
John Hohenadel
Philadelphia, PA
1000+

4 ❑ 32-1-T
John Hohenadel
Philadelphia, PA
450

5 ❑ 32-1-T
John Hohenadel
Philadelphia, PA
600

6 ❑ 32-1
John Hohenadel
Philadelphia, PA
600

7 ❑ 32-1-T
Jacob Hornung
Philadelphia, PA
1000+

8 ❑ 32-1-T
City
New York, NY
275

9 ❑ 32-1-T
City
New York, NY
275

10 ❑ 32-1-T
City
New York, NY
450

11 ❑ 32-1-T
Greater New York
New York, NY
450

12 ❑ 32-1-T
Horton Pilsener
New York, NY
200

13 ❑ 32-1-T
Horton Pilsener
New York, NY
275

14 ❑ 32-1-T
Horton Pilsener
New York, NY
450

15 ❑ 32-1-T
Metropolis
New York, NY
500

16 ❑ 32-1-T
R. H. Graupner
Harrisburg, PA
1000+

17 ❑ 32-1-T
Chas. D. Kaier
Mahanoy City, PA
1000+

18 ❑ 32-1-T
Keeley
Chicago, IL
450

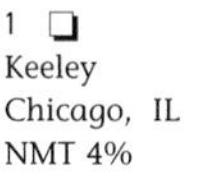
1 ❑ 32-1-T
Keeley
Chicago, IL
NMT 4% **450**

2 ❑ 32-1-T
Prima-Bismarck
Chicago, IL
Black Hops 4% **175**

3 ❑ 32-1-T
Prima-Bismarck
Chicago, IL
Green Hops **200**

4 ❑ 32-1-T
Prima-Bismarck
Chicago, IL
Green Hops 4% **200**

5 ❑ 32-1-T
Prima-Bismarck
Chicago, IL
200

6 ❑ 32-1-T
Prima-Bismarck
Chicago, IL
4% **200**

7 ❑ 32-1
Prima-Bismarck
Chicago, IL
200

8 ❑ 32-1
Prima-Bismarck
Chicago, IL
NMT 4% **200**

9 ❑ 32-2
G. Krueger
Wilmington, DE
450

10 ❑ 32-1-T
G. Krueger
Newark, NJ
150

11 ❑ 32-1-T
G. Krueger
Newark, NJ
Gold **150**

12 ❑ 32-1-T
G. Krueger
Newark, NJ
Silver **150**

13 ❑ 32-1-T
G. Krueger
Newark, NJ
Glass & Can on Tray **225**

14 ❑ 32-1-T
G. Krueger
Newark, NJ
Bottle & Can on Tray **225**

15 ❑ 32-1
G. Krueger
Newark, NJ
500

16 ❑ 32-1-T
G. Krueger
Newark, NJ
650

17 ❑ 32-1-T
G. Krueger
Newark, NJ
650

18 ❑ 32-1-T
G. Krueger
Newark, NJ
Glass & Can on Tray **250**

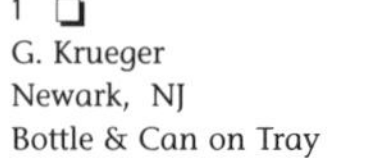

1 ❑ 32-1-T
G. Krueger
Newark, NJ
Bottle & Can on Tray
250

2 ❑ 32-1
G. Krueger
Newark, NJ
350

3 ❑ 32-1-T
Kuebler
Easton, PA
1000+

4 ❑ 32-1-T
Kuebler
Easton, PA
1000+

5 ❑ 32-1-T
Kuebler
Easton, PA
1000+

6 ❑ 32-1-T
Kuebler
Easton, PA
500

7 ❑ 32-1
Kuebler
Easton, PA
500

8 ❑ 32-1-T
Lebanon Valley
Lebanon, PA
350

9 ❑ 32-1-T
Liebmann
New York, NY
1000+

10 ❑ 32-1-T
Greater New York
New York, NY
1000+

11 ❑ 32-1-T
Loewer's
New York, NY
1000+

12 ❑ 32-2
Lucky Lager
San Francisco, CA
200

13 ❑ 32-2
Lucky Lager
San Francisco, CA
175

14 ❑ 32-1-T
Maier
Los Angeles, CA
1000+

15 ❑ 32-1-T
Manhattan
Chicago, IL
400

16 ❑ 32-1-T
Manhattan
Chicago, IL
DNCMT 4%
400

17 ❑ 32-1-T
Prima
Chicago, IL
400

18 ❑ 32-1
Mount Carbon
Pottsville, PA
1000+

1 ❑ 32-1-T
Wehle
West Haven, CT
1000+

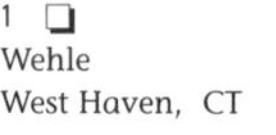

2 ❑ 32-1-T
Cooper
Philadelphia, PA
500

3 ❑ 32-1-T
Narragansett
Cranston, RI
1000+

4 ❑ 32-1-T
National
Baltimore, MD
225

5 ❑ 32-1-T
National
Baltimore, MD
225

6 ❑ 32-1
National
Baltimore, MD
225

7 ❑ 32-2
National
Baltimore, MD
350

8 ❑ 32-1-T
Louis F. Neuweiler's Sons
Allentown, PA
DULL GRAY
175

9 ❑ 32-1-T
Louis F. Neuweiler's Sons
Allentown, PA
175

10 ❑ 32-1-T
Louis F. Neuweiler's Sons
Allentown, PA
750

11 ❑ 32-1
Louis F. Neuweiler's Sons
Allentown, PA
175

12 ❑ 32-1-T
Louis F. Neuweiler's Sons
Allentown, PA
450

13 ❑ 32-1-T
Louis F. Neuweiler's Sons
Allentown, PA
750

14 ❑ 32-1
Louis F. Neuweiler's Sons
Allentown, PA
450

15 ❑ 32-1-T
Metropolis
New York, NY
1000+

16 ❑ 32-1-T
Metropolis
New York, NY
800

17 ❑ 32-1-T
Old Dutch
Brooklyn, NY
1000+

18 ❑ 32-1-T
Old Dutch
Brooklyn, NY
1000+

1 ❑ 32-1-T
Eagle
Catasauqua, PA
900

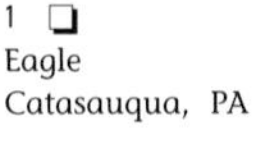

2 ❑ 32-2
Queen City
Cumberland, MD
250

3 ❑ 32-1-T
Old Missouri Sales
Chicago, IL
1000+

4 ❑ 32-1-T
Prima
Chicago, IL
1000+

5 ❑ 32-1-T
Old Reading
Reading, PA
1000+

6 ❑ 32-1-T
Old Reading
Reading, PA
1000+

7 ❑ 32-1-T
Silver Springs
Port Orchard, WA
4%
1000+

8 ❑ 32-1-T
Enterprise
Fall River, MA
1000+

9 ❑ 32-1-T
Enterprise
Fall River, MA
1000+

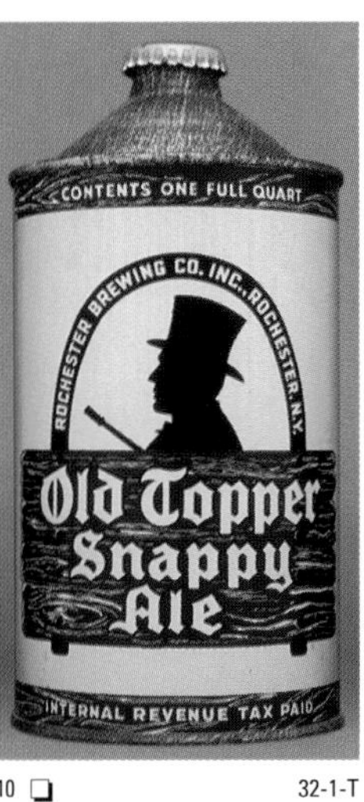

10 ❑ 32-1-T
Rochester
Rochester, NY
200

11 ❑ 32-1-T
Rochester
Rochester, NY
900

12 ❑ 32-1-T
Henry F. Ortlieb
Philadelphia, PA
500

13 ❑ 32-1-T
Henry F. Ortlieb
Philadelphia, PA
500

14 ❑ 32-1
Henry F. Ortlieb
Philadelphia, PA
500

15 ❑ 32-1
Henry F. Ortlieb
Philadelphia, PA
500

16 ❑ 32-2
Pabst
Los Angeles, CA
65

17 ❑ 32-2
Pabst
Los Angeles, CA
100

18 ❑ 32-2
Pabst
Newark, NJ
65

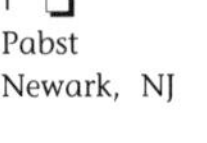

1 ☐ 32-2
Pabst
Newark, NJ
85

2 ☐ 32-2
Pabst
Newark, NJ
100

3 ☐ 32-2
Pabst
Milwaukee, WI
50

4 ☐ 32-2
Pabst
Milwaukee, WI
100

5 ☐ 32-2
Pabst
Milwaukee, WI
85

6 ☐ 32-2
Pabst
Milwaukee, WI
100

7 ☐ 32-1-T
Otto Erlanger
Philadelphia, PA
1000+

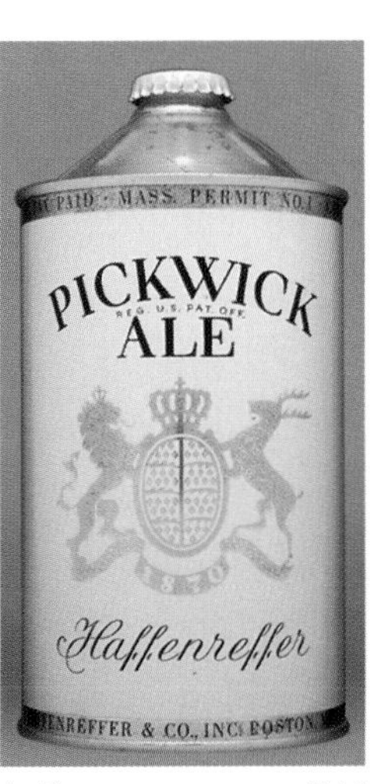

8 ☐ 32-1-T
Haffenreffer
Boston, MA
400

9 ☐ 32-1-T
Metropolis
New York, NY
250

10 ☐ 32-1-T
Pilser
New York, NY
1000+

11 ☐ 32-1-T
Pilser
New York, NY
1000+

12 ☐ 32-1-T
Pilser
New York, NY
1000+

13 ☐ 32-1-T
Prima
Chicago, IL
500

14 ☐ 32-1-T
Prima
Chicago, IL
500

15 ☐ 32-1-T
Adam Scheidt
Norristown, PA
250

16 ☐ 32-1-OT
Adam Scheidt
Norristown, PA
400

17 ☐ 32-1-OT
Adam Scheidt
Norristown, PA
400

18 ☐ 32-1-T
Largay
Waterbury, CT
1000+

QUARTS

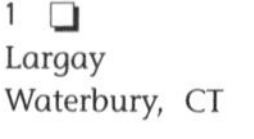

1 ❑ 32-1-T
Largay
Waterbury, CT
1000+

2 ❑ 32-1
Century
Norfolk, VA
800

3 ❑ 32-1-T
Rubsam & Horrmann
New York, NY
700

4 ❑ 32-1-T
Liebmann
New York, NY
225

5 ❑ 32-1-T
Liebmann
New York, NY
900

6 ❑ 32-1-T
Liebmann
New York, NY
900

7 ❑ 32-1-T
Liebmann
New York, NY
800

8 ❑ 32-1-T
Liebmann
New York, NY
450

9 ❑ 32-1-T
Liebmann
New York, NY
1000+

10 ❑ 32-2
Home
Richmond, VA
400

11 ❑ 32-2
Home
Richmond, VA
400

12 ❑ 32-2
Home
Richmond, VA
500

13 ❑ 32-1-T
Schmidt
Detroit, MI
650

14 ❑ 32-1-T
C. Schmidt & Sons
Philadelphia, PA
125

15 ❑ 32-1-T
C. Schmidt & Sons
Philadelphia, PA
125

16 ❑ 32-1-T
C. Schmidt & Sons
Philadelphia, PA
125

17 ❑ 32-1-T
C. Schmidt & Sons
Philadelphia, PA
DULL GRAY
150

18 ❑ 32-1-T
C. Schmidt & Sons
Philadelphia, PA
125

1 ❑ 32-1-T
C. Schmidt & Sons
Philadelphia, PA
125

2 ❑ 32-1-T
C. Schmidt & Sons
Philadelphia, PA
DULL GRAY **175**

3 ❑ 32-1
C. Schmidt & Sons
Philadelphia, PA
175

4 ❑ 32-1-T
C. Schmidt & Sons
Philadelphia, PA
175

5 ❑ 32-2-T
C. Schmidt & Sons
Philadelphia, PA
100

6 ❑ 32-2
C. Schmidt & Sons
Philadelphia, PA
100

7 ❑ 32-2
C. Schmidt & Sons
Philadelphia, PA
100

8 ❑ 32-1-T
C. Schmidt & Sons
Philadelphia, PA
1000+

9 ❑ 32-1-T
C. Schmidt & Sons
Philadelphia, PA
1000+

10 ❑ 32-1-T
Fox Deluxe
Marion, IN
850

11 ❑ 32-1-T
Stegmaier
Wilkes-Barre, PA
1000+

12 ❑ 32-1-T
Barbey's
Reading, PA
300

13 ❑ 32-1-T
Barbey's
Reading, PA
300

14 ❑ 32-1-T
Barbey's
Reading, PA
Extra Light **400**

15 ❑ 32-1
Sunshine
Reading, PA
650

16 ❑ 32-1-T
South Bethlehem
Bethlehem, PA
1000+

17 ❑ 32-1-T
Peoples
Trenton, NJ
1000+

18 ❑ 32-1-T
Peoples
Trenton, NJ
1000+

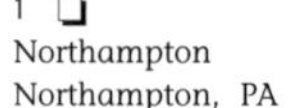

1 ❑ 32-1-T
Northampton
Northampton, PA
1000+

2 ❑ 32-2-T
Northampton
Northampton, PA
1000+

3 ❑ 32-1
Tube City
McKeesport, PA
850

4 ❑ 32-1-T
Metropolis
New York, NY
800

5 ❑ 32-1-T
Metropolis
New York, NY
800

6 ❑ 32-1-T
West End
Utica, NY
300

7 ❑ 32-1-T
West End
Utica, NY
350

8 ❑ 32-1-T
West End
Utica, NY
750

9 ❑ 32-1-T
West End
Utica, NY
400

10 ❑ 32-1-T
West End
Utica, NY
450

11 ❑ 32-1-T
Adam Scheidt
Norristown, PA
350

12 ❑ 32-1-OT
Adam Scheidt
Norristown, PA
325

13 ❑ 32-1-OT
Adam Scheidt
Norristown, PA
300

14 ❑ 32-1-T
Wacker
Lancaster, PA
1000+

15 ❑ 32-1-T
Wacker
Lancaster, PA
1000+

16 ❑ 32-1-T
Wehle
West Haven, CT
1000+

17 ❑ 32-1-T
Largay
Waterbury, CT
1000+

18 ❑ 32-1-T
Westminster
Chicago, IL
1000+

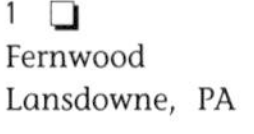

1 ❑ 32-1-T
Fernwood
Lansdowne, PA
1000+

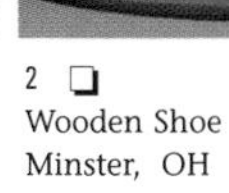

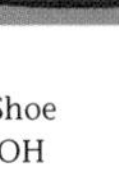

2 ❑ 32-1-T
Wooden Shoe
Minster, OH
1000+

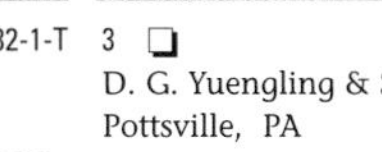

3 ❑ 32-1-T
D. G. Yuengling & Son
Pottsville, PA
1000+

4 ❑ 32-1-T
D. G. Yuengling & Son
Pottsville, PA
300

5 ❑ 32-1
D. G. Yuengling & Son
Pottsville, PA
350

6 ❑ 32-1
D. G. Yuengling & Son
Pottsville, PA
500

QUARTS

Budweiser®

BUY THE BIG SIZE

Be sure you have enough. Buy BUD® in big sizes:
• HALF QUART cans
• QUART bottles
• Full CASE of 24

HALF QUART

Budweiser

ANHEUSER-BUSCH, INC. • ST. LOUIS
NEWARK • LOS ANGELES • MIAMI • TAMPA

1954 was a year of firsts. The U.S. launched its first atomic submarine, Dr. Jonas Salk launched the first polio vaccine, Bert Parks launched the first national telecast of the Miss America Pageant and Schlitz launched the first 16-ounce flat-top beer can. As the ads shown here suggest, other brewers, both regional and national, were quick to follow.

PINTS

14, 15 & 16 Oz. Flats

Pints Section Explanatory Notes

Most of the varieties pictured in this section are differences in the size statement such as "16 Fluid Ounces" vs. "One Full Pint" and 15 vs. 16 Ounces.

A-1 224-10 has "One Full Pint" in the lower part of the gold oval.

A-1 224-11 has "16 Fluid Ounces" in the lower part of the gold oval.

Krueger 231-22 does not have Permit information between the legs of the K-man.

Miller 232-24 has "Miller" in the white background instead of "High Life."

1 ❑ 16-2
Maier
Los Angeles, CA
100

2 ❑ 16-2
Maier
Los Angeles, CA
25

3 ❑ 16-2
Maier
Los Angeles, CA
25

4 ❑ 16-1
Ace
Chicago, IL
35

5 ❑ 16-1
Canadian Ace
Chicago, IL
40

6 ❑ 16-1
Atlas
Chicago, IL
75

7 ❑ 16-2
Drewrys Ltd.
South Bend, IN
65

8 ❑ 16-2
Five Star
New York, NY
75

9 ❑ 16-2
Five Star
New York, NY
75

10 ❑ 16-2
Arizona
Phoenix, AZ
"One Full Pint"
150

11 ❑ 16-2
Arizona
Phoenix, AZ
"16 Fluid Ounces"
150

12 ❑ 16-2
Arizona
Phoenix, AZ
150

13 ❑ 16-1
Arizona
Phoenix, AZ
100

14 ❑ 16-2
Arizona
Phoenix, AZ
900

15 ❑ 16-2
Maier
Los Angeles, CA
85

16 ❑ 16-2
Atlas
Chicago, IL
50

17 ❑ 16-2
Atlas
Chicago, IL
50

18 ❑ 16-2
P. Ballantine & Sons
Newark, NJ
30

19 ❑ 16-2
P. Ballantine & Sons
Newark, NJ
15

20 ❑ 16-1
P. Ballantine & Sons
Newark, NJ
50

21 ❑ 16-2
P. Ballantine & Sons
Newark, NJ
30

22 ❑ 16-2
P. Ballantine & Sons
Newark, NJ
30

23 ❑ 16-2
P. Ballantine & Sons
Newark, NJ
30

24 ❑ 16-2
P. Ballantine & Sons
Newark, NJ
30

25 ❑ 16-2
P. Ballantine & Sons
Newark, NJ
75

26 ❑ 16-2
P. Ballantine & Sons
Newark, NJ
50

27 ❑ 16-2
P. Ballantine & Sons
Newark, NJ
50

28 ❑ 16-2
P. Ballantine & Sons
Newark, NJ
25

29 ❑ 16-2
P. Ballantine & Sons
Newark, NJ
15

30 ❑ 16-2
P. Ballantine & Sons
Newark, NJ
15

31 ❑ 16-2
P. Ballantine & Sons
Newark, NJ
15

32 ❑ 16-2
P. Ballantine & Sons
Newark, NJ
15

PINTS

1 ❑ 16-2
Maier
Los Angeles, CA
35

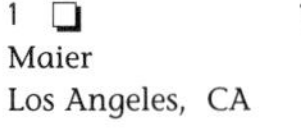

2 ❑ 16-2
Maier
Los Angeles, CA
35

3 ❑ 16-2
Best
Chicago, IL
110

4 ❑ 14-2
Carling
Atlanta, GA
25

5 ❑ 16-1
Carling
Atlanta, GA
10

6 ❑ 16-1
Carling
Atlanta, GA
10

7 ❑ 16-1
Carling
Atlanta, GA
10

8 ❑ 14-1
Carling
Atlanta, GA
35

Back of 8

10 ❑ 16-1
Carling
Natick, MA
10

11 ❑ 16-1
Carling
Natick, MA
10

Back of 11

13 ❑ 15-1
Carling
Tacoma, WA
12

14 ❑ 16-1
Carling
Tacoma, WA
12

15 ❑ 16-1
Carling
Tacoma, WA
12

16 ❑ 15-1
Carling
Tacoma, WA
12

17 ❑ 15-2
Pabst
Los Angeles, CA
10

18 ❑ 16-2
Pabst
Los Angeles, CA
10

19 ❑ 16-2
Pabst
Peoria Heights, IL
10

20 ❑ 16-1
Blatz
Milwaukee, WI
75

21 ❑ 16-1
Blatz
Milwaukee, WI
10

22 ❑ 16-1
Blatz
Milwaukee, WI
10

23 ❑ 16-2
Pabst
MIlwaukee, WI
10

24 ❑ 16-1
Blitz-Weinhard
Portland, OR
200

25 ❑ 15-2
Blitz-Weinhard
Portland, OR
40

26 ❑ 16-2
Blitz-Weinhard
Portland, OR
40

27 ❑ 15-2
Blitz-Weinhard
Portland, OR
30

28 ❑ 16-2
Blitz-Weinhard
Portland, OR
30

29 ❑ 15-2
Blitz-Weinhard
Portland, OR
50

30 ❑ 16-2
Grace Bros.
Santa Rosa, CA
300

31 ❑ 16-2
Maier
Los Angeles, CA
750

32 ❑ 16-2
Maier
Los Angeles, CA
10

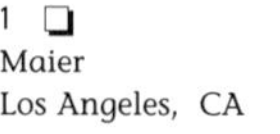
1 ❑ 16-2
Maier
Los Angeles, CA
10

2 ❑ 16-2
Maier
Los Angeles, CA
10

3 ❑ 16-2
Maier
Los Angeles, CA
12

4 ❑ 16-2
Maier
Los Angeles, CA
12

5 ❑ 16-1
Maier
Los Angeles, CA
20

6 ❑ 16-1
Grace Bros.
Santa Rosa, CA
20

7 ❑ 16-1
Grace Bros.
Santa Rosa, CA
20

8 ❑ 16-1
Grace Bros.
Santa Rosa, CA
20

9 ❑ 16-2
Maier
Los Angeles, CA
15

10 ❑ 16-2
Grace Bros.
Santa Rosa, CA
15

11 ❑ 16-2
Grace Bros.
Santa Rosa, CA
15

12 ❑ 16-2
Grace Bros.
Santa Rosa, CA
15

13 ❑ 15-2
Buckhorn
San Francisco, CA
18

14 ❑ 15-2
Theo. Hamm
San Francisco, CA
18

15 ❑ 16-2
Anheuser-Busch
Los Angeles, CA
3 City 20

16 ❑ 16-2
Anheuser-Busch
Los Angeles, CA
3 City 20

17 ❑ 16-2
Anheuser-Busch
Los Angeles, CA
3 City 10

18 ❑ 16-2
Anheuser-Busch
Los Angeles, CA
5 City 10

19 ❑ 16-2
Anheuser-Busch
Los Angeles, CA
4 City 10

20 ❑ 16-2
Anheuser-Busch
Los Angeles, CA
4 City 10

21 ❑ 16-2
Anheuser-Busch
Tampa, FL
5 City 12

22 ❑ 16-2
Anheuser-Busch
Tampa, FL
4 City 12

23 ❑ 16-2
Anheuser-Busch
St. Louis, MO
3 City 20

24 ❑ 16-2
Anheuser-Busch
St. Louis, MO
3 City 20

25 ❑ 16-2
Anheuser-Busch
St. Louis, MO
3 City 10

26 ❑ 16-2
Anheuser-Busch
St. Louis, MO
5 City 10

27 ❑ 16-2
Anheuser-Busch
St. Louis, MO
4 City 10

28 ❑ 16-2
Anheuser-Busch
St. Louis, MO
4 City 10

29 ❑ 16-2
Anheuser-Busch
St. Louis, MO
5 City 10

30 ❑ 16-2
Anheuser-Busch
St. Louis, MO
5 City 10

31 ❑ 16-2
Anheuser-Busch
St. Louis, MO
6 City 10

32 ❑ 16-2
Grace Bros.
Santa Rosa, CA
400

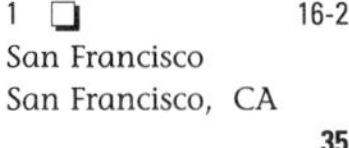

1 ❑ 16-2
San Francisco
San Francisco, CA
35

2 ❑ 16-2
San Francisco
San Francisco, CA
35

3 ❑ 16-2
San Francisco
San Francisco, CA
35

4 ❑ 16-2
Burgermeister
San Francisco, CA
35

5 ❑ 15-2
Burgermeister
San Francisco, CA
15

6 ❑ 16-2
Burgermeister
San Francisco, CA
15

7 ❑ 15-2
Burgermeister
San Francisco, CA
12

8 ❑ 15-2
Burgermeister
San Francisco, CA
10

9 ❑ 15-2
Burgermeister
San Francisco, CA
10

10 ❑ 15-2
Jos. Schlitz
San Francisco, CA
10

11 ❑ 15-2
Jos. Schlitz
San Francisco, CA
10

12 ❑ 16-2
Anheuser-Busch
Los Angeles, CA
20

13 ❑ 16-2
Anheuser-Busch
Tampa, FL
60

14 ❑ 16-2
Anheuser-Busch
St. Louis, MO
30

15 ❑ 16-2
Anheuser-Busch
St. Louis, MO
25

16 ❑ 16-2
Anheuser-Busch
St. Louis, MO
50

17 ❑ 16-2
Anheuser-Busch
St. Louis, MO
25

18 ❑ 16-2
Canadian Ace
Chicago, IL
30

19 ❑ 16-2
Canadian Ace
Chicago, IL
25

20 ❑ 16-2
Canadian Ace
Chicago, IL
25

21 ❑ 16-2
Canadian Ace
Chicago, IL
25

22 ❑ 16-2
Canadian Ace
Chicago, IL
12

23 ❑ 16-2
Canadian Ace
Chicago, IL
12

24 ❑ 16-2
Canadian Ace
Chicago, IL
12

25 ❑ 16-2
Canadian Ace
Chicago, IL
12

26 ❑ 15-2
Blitz-Weinhard
Portland, OR
35

27 ❑ 15-2
Blitz-Weinhard
Portland, OR
15

28 ❑ 16-2
Atlantic
Chicago, IL
60

29 ❑ 16-2
Atlantic
Chicago, IL
20

30 ❑ 16-2
Atlantic
Chicago, IL
20

31 ❑ 15-2
Brewing Corp. of Oregon
Portland, OR
85

32 ❑ 15-2
Brewing Corp. of Oregon
Portland, OR
55

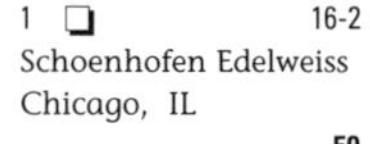

1 ❑ 16-2
Schoenhofen Edelweiss
Chicago, IL
50

2 ❑ 16-2
National
Miami, FL
15

3 ❑ 16-2
National
Miami, FL
15

4 ❑ 16-2
National
Baltimore, MD
15

5 ❑ 15-2
Adolf Coors
Golden, CO
15

6 ❑ 16-2
Croft
Cranston, RI
40

7 ❑ 16-2
Dawson's
New Bedford, MA
125

8 ❑ 16-2
Dawson's
New Bedford, MA
100

9 ❑ 16-1
Drewrys Ltd.
South Bend, IN
135

10 ❑ 16-1
Drewrys Ltd.
South Bend, IN
135

11 ❑ 16-1
Drewrys Ltd.
South Bend, IN
135

12 ❑ 16-1
Drewrys Ltd.
South Bend, IN
135

13 ❑ 16-1
Drewrys Ltd.
South Bend, IN
135

14 ❑ 16-1
Drewrys Ltd.
South Bend, IN
135

15 ❑ 16-1
Drewrys Ltd.
South Bend, IN
30

16 ❑ 16-1
Drewrys Ltd.
South Bend, IN
25

17 ❑ 16-2
Grace Bros.
Santa Rosa, CA
400

18 ❑ 16-2
Pabst
Los Angeles, CA
75

19 ❑ 16-2
Pabst
Los Angeles, CA
50

20 ❑ 16-2
Pabst
Los Angeles, CA
50

21 ❑ 16-2
Pabst
Los Angeles, CA
40

22 ❑ 16-2
Pabst
Los Angeles, CA
10

23 ❑ 15-2
Pabst
Los Angeles, CA
10

24 ❑ 16-2
Pabst
Los Angeles, CA
10

25 ❑ 16-2
Pabst
Los Angeles, CA
8

26 ❑ 16-2
Pabst
Los Angeles, CA
10

27 ❑ 16-2
Schoenhofen Edelweiss
Chicago, IL
12

28 ❑ 16-2
Drewrys Ltd.
South Bend, IN
10

29 ❑ 16-2
Drewrys Ltd
South Bend, IN
10

30 ❑ 16-1
Arizona
Phoenix, AZ
125

31 ❑ 16-1
Grace Bros.
Santa Rosa, CA
85

32 ❑ 16-1
Grace Bros.
Santa Rosa, CA
85

PINTS

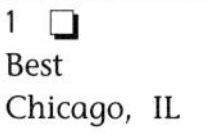

1 ❑ 16-1
Best
Chicago, IL
125

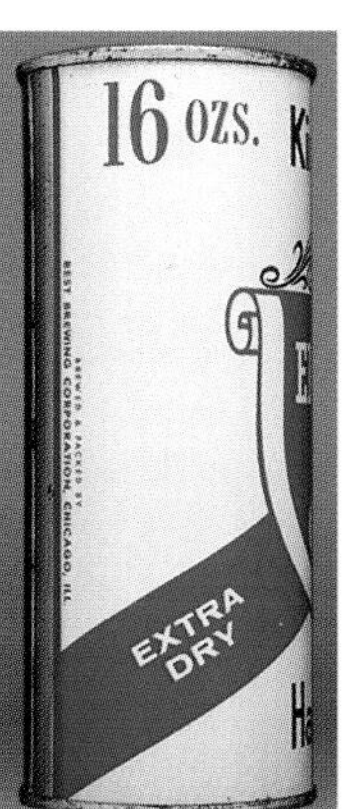

Side of 1

3 ❑ 16-1
Best
Chicago, IL
125

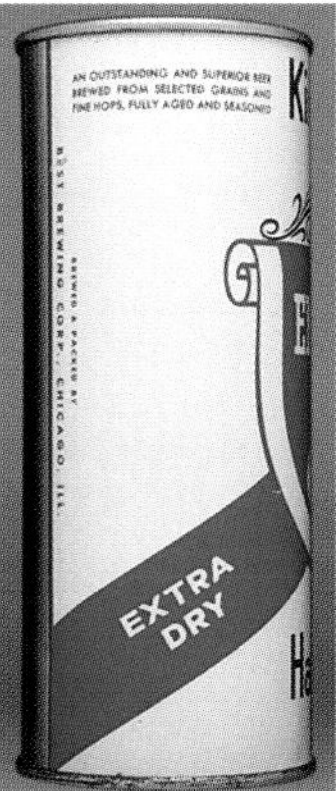

Side of 3

5 ❑ 16-1
Atlas
Chicago, IL
85

6 ❑ 16-1
Falstaff
San Jose, CA
15

7 ❑ 16-1
Falstaff
San Jose, CA
15

8 ❑ 15-2
Falstaff
San Jose, CA
10

9 ❑ 15-2
Falstaff
San Jose, CA
10

10 ❑ 15-2
Falstaff
San Jose, CA
10

11 ❑ 16-1
Falstaff
Fort Wayne, IN
15

12 ❑ 16-2
Falstaff
Fort Wayne, IN
10

13 ❑ 16-2
Falstaff
Fort Wayne, IN
10

14 ❑ 16-1
Falstaff
New Orleans, LA
12

15 ❑ 16-2
Falstaff
St. Louis, MO
10

16 ❑ 16-2
Falstaff
Omaha, NE
10

17 ❑ 16-2
General Corp.
Azusa, CA
6

18 ❑ 16-2
General
San Francisco, CA
6

19 ❑ 16-2
Lucky Lager
San Francisco, CA
6

20 ❑ 16-2
Lucky
San Francisco, CA
6

21 ❑ 15-2
Fisher
Salt Lake City, UT
10

22 ❑ 16-2
Fisher
Salt Lake City, UT
10

23 ❑ 16-2
Lucky Lager
Salt Lake City, UT
10

24 ❑ 16-1
Goebel
Detroit, MI
125

25 ❑ 16-1
Goebel
Detroit, MI
Metallic **125**

26 ❑ 16-1
Goebel
Detroit, MI
Enamel **125**

27 ❑ 16-1
Goebel
Detroit, MI
200

28 ❑ 16-2
905
Chicago, IL
50

29 ❑ 16-2
Drewrys Ltd.
South Bend, IN
50

30 ❑ 16-2
Grace Bros.
Santa Rosa, CA
55

31 ❑ 16-2
Grace Bros.
Santa Rosa, CA
50

32 ❑ 16-2
Grace Bros.
Santa Rosa, CA
50

PINTS

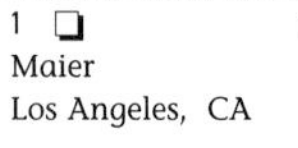

1 ❑ 16-2
Maier
Los Angeles, CA
200

2 ❑ 16-2
Regal Pale
San Francisco, CA
200

3 ❑ 16-2
Maier
Los Angeles, CA
30

4 ❑ 16-2
Grace Bros.
Santa Rosa, CA
35

5 ❑ 16-2
Grace Bros.
Santa Rosa, CA
35

6 ❑ 16-2
Grace Bros.
Santa Rosa, CA
35

7 ❑ 16-2
Schoenhofen Edelweiss
Chicago, IL
65

8 ❑ 16-2
Drewrys Ltd.
South Bend, IN
65

9 ❑ 16-2
Gunther
Baltimore, MD
50

10 ❑ 15-2
Theo. Hamm
Los Angeles, CA
15

11 ❑ 15-2
Theo. Hamm
Los Angeles, CA
12

12 ❑ 15-2
Theo. Hamm
Los Angeles, CA
12

13 ❑ 16-2
Theo. Hamm
San Francisco, CA
25

14 ❑ 15-2
Theo. Hamm
San Francisco, CA
10

15 ❑ 16-2
Theo. Hamm
San Francisco, CA
10

16 ❑ 15-2
Theo. Hamm
San Francisco, CA
8

17 ❑ 16-2
Theo. Hamm
Baltimore, MD
18

18 ❑ 16-2
Theo. Hamm
Baltimore, MD
15

19 ❑ 16-2
Theo. Hamm
Baltimore, MD
15

20 ❑ 16-2
Theo. Hamm
St. Paul, MN
10

21 ❑ 16-2
Theo. Hamm
St. Paul, MN
8

22 ❑ 16-2
Theo. Hamm
St. Paul, MN
7

23 ❑ 16-2
Theo. Hamm
St. Paul, MN
7

24 ❑ 16-2
Heidelberg
Tacoma, WA
90

25 ❑ 16-2
Heidelberg
Tacoma, WA
90

26 ❑ 16-2
Heidelberg
Tacoma, WA
60

27 ❑ 16-2
Heidelberg
Tacoma, WA
60

28 ❑ 16-2
Heidelberg
Tacoma, WA
60

29 ❑ 16-2
Heidelberg
Tacoma, WA
15

30 ❑ 15-1
Carling
Tacoma, WA
15

31 ❑ 16-1
Carling
Tacoma, WA
15

32 ❑ 15-1
Carling
Tacoma, WA
15

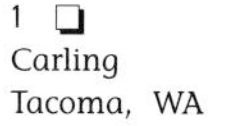

1 ❑ 16-1
Carling
Tacoma, WA
15

2 ❑ 16-2
Atlas
Chicago, IL
75

3 ❑ 16-2
Atlas
Chicago, IL
75

4 ❑ 16-2
Drewrys Ltd
South Bend, IN
60

5 ❑ 16-2
Drewrys Ltd.
South Bend, IN
60

6 ❑ 16-1
Jackson
New Orleans, LA
65 Yrs., Enam. Red
45

7 ❑ 16-1
Jackson
New Orleans, LA
65 Yrs., Met. Red
45

8 ❑ 16-1
Jackson
New Orleans, LA
25

9 ❑ 16-2
Jackson
New Orleans, LA
15

10 ❑ 16-2
Drewrys Ltd
South Bend, IN
175

11 ❑ 16-2
Maier
Los Angeles, CA
8

12 ❑ 16-2
Maier
Los Angeles, CA
8

13 ❑ 16-2
Jacob Ruppert
New York, NY
50

14 ❑ 16-2
Jacob Ruppert
New York, NY
55

15 ❑ 16-2
Jacob Ruppert
New York, NY
55

16 ❑ 16-2
Jacob Ruppert
New York, NY
35

17 ❑ 16-2
Jacob Ruppert
New York, NY
35

18 ❑ 16-2
Jacob Ruppert
New York, NY
20

19 ❑ 16-1-T
G. Krueger
Newark, NJ
600

20 ❑ 16-1
G. Krueger
Newark, NJ
40

21 ❑ 16-1-T
G. Krueger
Newark, NJ
1000+

22 ❑ 16-1-T
G. Krueger
Newark, NJ
1000+

23 ❑ 16-1
G. Krueger
Newark, NJ
25

24 ❑ 16-1
G. Krueger
Newark, NJ
65

25 ❑ 16-1
G. Krueger
Cranston, RI
65

26 ❑ 16-1
G. Krueger
Cranston, RI
20

27 ❑ 16-2
Maier
Los Angeles, CA
175

28 ❑ 16-2
Arizona
Phoenix, AZ
60

29 ❑ 16-1
Canadian Ace
Chicago, IL
55

30 ❑ 16-1
Canadian Ace
Chicago, IL
50

31 ❑ 16-1
Canadian Ace
Chicago, IL
50

32 ❑ 16-1
Pilsen
Chicago, IL
55

1 ❑ 16-1
Pilsen
Chicago, IL
55

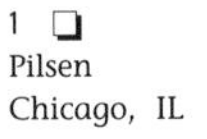

2 ❑ 16-1
Pilsen
Chicago, IL
50

3 ❑ 16-1
Lucky Lager
Azusa, CA
20

4 ❑ 16-2
Lucky Lager
Azusa, CA
10

5 ❑ 15-2
Lucky Lager
Azusa, CA
10

6 ❑ 16-2
Lucky Lager
Azusa, CA
10

7 ❑ 16-2
Lucky Lager
Azusa, CA
8

8 ❑ 16-2
Lucky Lager
Azusa, CA
8

9 ❑ 15-2
Lucky Lager
San Francisco, CA
225

10 ❑ 15-2
Lucky Lager
San Francisco, CA
225

11 ❑ 16-1
Lucky Lager
San Francisco, CA
18

12 ❑ 16-2
Lucky Lager
San Francisco, CA
8

13 ❑ 16-2
Lucky Lager
San Francisco, CA
8

14 ❑ 15-2
Lucky Lager
San Francisco, CA
8

15 ❑ 15-2
Lucky Lager
San Francisco, CA
8

16 ❑ 15-2
Lucky Lager
San Francisco, CA
8

17 ❑ 16-2
Lucky Lager
San Francisco, CA
8

18 ❑ 16-2
Lucky Lager
San Francisco, CA
8

19 ❑ 16-2
General Corp.
Salt Lake City, UT
8

20 ❑ 16-2
Lucky Lager
Salt Lake City, UT
8

21 ❑ 16-1
Miller
Milwaukee, WI
10

22 ❑ 16-1
Miller
Milwaukee, WI
8

23 ❑ 16-1
Miller
Milwaukee, WI
8

24 ❑ 16-1
Miller
Milwaukee, WI
15

25 ❑ 16-1
Miller
Milwaukee, WI
8

26 ❑ 16-1
Narragansett
Cranston, RI
20

27 ❑ 16-1
Narragansett
Cranston, RI
20

28 ❑ 16-2
National
Miami, FL
18

29 ❑ 16-2
National
Baltimore, MD
50

30 ❑ 16-2
National
Baltimore, MD
35

31 ❑ 16-2
National
Baltimore, MD
35

32 ❑ 16-2
National
Baltimore, MD
12

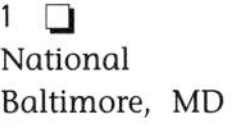

1 ❑ 16-2
National
Baltimore, MD
12

2 ❑ 16-2
905
Chicago, IL
25

3 ❑ 16-2
Drewrys Ltd
South Bend, IN
55

4 ❑ 16-2
Jos. Schlitz
Tampa, FL
1960 75

5 ❑ 14-2
Jos. Schlitz
Tampa, FL
1962 25

6 ❑ 16-2
Jos. Schlitz
Tampa, FL
1962 10

7 ❑ 14-2
Jos. Schlitz
Tampa, FL
1962 25

8 ❑ 14-2
Jos. Schlitz
Tampa, FL
1962 25

9 ❑ 16-2
Jos. Schlitz
Tampa, FL
1962 15

10 ❑ 16-2
Jos. Schlitz
Milwaukee, WI
1960 65

11 ❑ 16-2
Jos. Schlitz
Milwaukee, WI
1962 8

12 ❑ 16-2
Jos. Schlitz
Milwaukee, WI
1962 8

13 ❑ 16-2
Jos. Schlitz
Milwaukee, WI
1962 8

14 ❑ 16-2
Old Vienna
Chicago, IL
70

15 ❑ 16-2
Olympia
Olympia, WA
15

16 ❑ 16-2
Olympia
Olympia, WA
20

17 ❑ 15-2
Olympia
Olympia, WA
15

18 ❑ 16-2
Olympia
Olympia, WA
15

19 ❑ 16-2
Pabst
Los Angeles, CA
30

20 ❑ 16-2
Pabst
Los Angeles, CA
10

21 ❑ 16-2
Pabst
Peoria Heights, IL
20

22 ❑ 16-2
Pabst
Peoria Heights, IL
30

23 ❑ 16-2
Pabst
Peoria Heights, IL
10

24 ❑ 16-2
Pabst
Milwaukee, WI
15

25 ❑ 16-2
Pabst
Milwaukee, WI
25

26 ❑ 16-2
Pabst
Milwaukee, WI
25

27 ❑ 16-2
Pabst
Milwaukee, WI
6

28 ❑ 16-2
Pabst
Milwaukee, WI
6

29 ❑ 16-1
Piel Bros.
Brooklyn, NY
125

30 ❑ 16-1
Piel Bros.
Brooklyn, NY
75

31 ❑ 16-1
Piel Bros.
New York, NY
75

32 ❑ 16-2
Piel Bros.
New York, NY
30

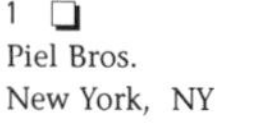

1 ❑ 16-2
Piel Bros.
New York, NY
30

2 ❑ 16-2
Piel Bros.
New York, NY
30

3 ❑ 16-2
Maier
Los Angeles, CA
65

4 ❑ 16-2
Prima
Chicago, IL
350

5 ❑ 16-2
Schoenhofen Edelweiss
Chicago, IL
50

6 ❑ 16-2
Drewrys Ltd.
South Bend, IN
40

7 ❑ 16-2
Sicks' Seattle
Seattle, WA
40

8 ❑ 16-2
Sicks' Rainier
Seattle, WA
45

9 ❑ 15-2
Sicks' Rainier
Seattle, WA
12

10 ❑ 16-2
Sicks' Rainier
Seattle, WA
12

11 ❑ 15-2
Sicks' Rainier
Seattle, WA
12

12 ❑ 16-2
Sicks' Rainier
Seattle, WA
12

13 ❑ 15-1
Sick's Rainier
Seattle, WA
100

14 ❑ 15-1
Sick's Rainier
Seattle, WA
100

15 ❑ 16-1
Sick's Rainier
Seattle, WA
100

16 ❑ 16-2
Reading
Reading, PA
5

17 ❑ 16-2
Carling
Cleveland, OH
800

18 ❑ 16-2
Regal Pale
San Francisco, CA
30

19 ❑ 16-2
Regal Pale
San Francisco, CA
30

20 ❑ 16-2
Regal Pale
San Francisco, CA
30

21 ❑ 16-2
Regal Pale
San Francisco, CA
30

22 ❑ 16-2
Maier
Los Angeles, CA
12

23 ❑ 16-2
Maier
Los Angeles, CA
30

24 ❑ 16-2
Maier
Los Angeles, CA
30

25 ❑ 16-2
Maier
Los Angeles, CA
30

26 ❑ 16-2
Regal Pale
San Francisco, CA
10

27 ❑ 16-1
Century
Norfolk, VA
75

28 ❑ 16-2
Rheingold
Los Angeles, CA
8

29 ❑ 16-2
Liebmann
New York, NY
10

30 ❑ 16-2
Rheingold
New York, NY
10

31 ❑ 16-2
Rheingold
New York, NY
8

32 ❑ 16-2
Rheingold
New York, NY
8

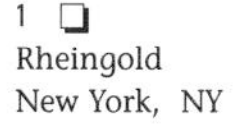

1 ❑ 16-2
Rheingold
New York, NY
5

2 ❑ 16-2
Maier
Los Angeles, CA
135

3 ❑ 16-1
Arizona
Phoenix, AZ
450

4 ❑ 16-1
F. & M. Schaefer
New York, NY
25

5 ❑ 16-1
F. & M. Schaefer
New York, NY
25

6 ❑ 16-1
F. & M. Schaefer
New York, NY
25

7 ❑ 16-1
F. & M. Schaefer
New York, NY
20

8 ❑ 16-2
F. & M. Schaefer
New York, NY
15

9 ❑ 16-1
F. & M. Schaefer
New York, NY
8

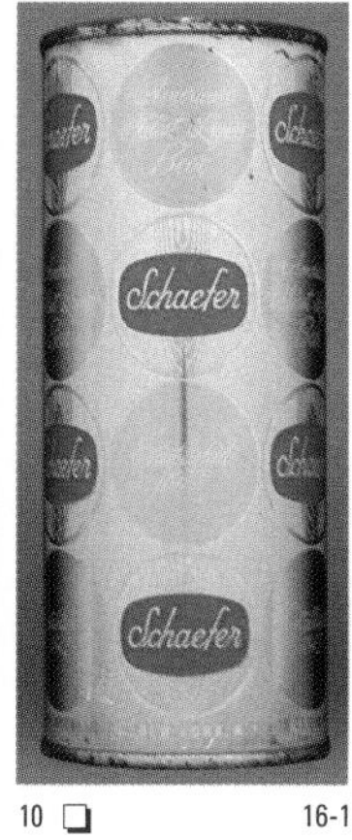

10 ❑ 16-1
F. & M. Schaefer
New York, NY
8

11 ❑ 16-1
F. & M. Schaefer
New York, NY
World's Fair
10

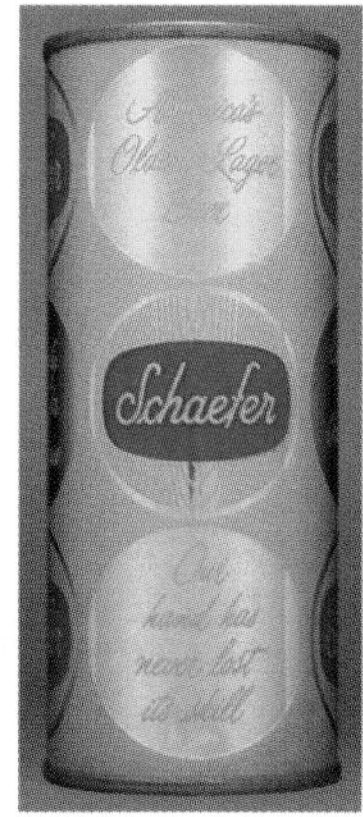

12 ❑ 16-1
F. & M. Schaefer
New York, NY
6

13 ❑ 16-2
Jos. Schlitz
Los Angeles, CA
1954
15

14 ❑ 16-2
Jos. Schlitz
Los Angeles, CA
1954
15

15 ❑ 16-2
Jos. Schlitz
Los Angeles, CA
1957
15

16 ❑ 16-2
Jos. Schlitz
Los Angeles, CA
1960
12

17 ❑ 14-2
Jos. Schlitz
Tampa, FL
1962
25

18 ❑ 16-2
Jos. Schlitz
Kansas City, MO
1957
12

19 ❑ 16-2
Jos. Schlitz
Kansas City, MO
1960
10

20 ❑ 16-2
Jos. Schlitz
Kansas City, MO
1962
10

21 ❑ 16-2
Jos. Schlitz
Kansas City, MO
1962
10

22 ❑ 16-2
Jos. Schlitz
Brooklyn, NY
1954
12

23 ❑ 16-2
Jos. Schlitz
Brooklyn, NY
1954
12

24 ❑ 16-2
Jos. Schlitz
Brooklyn, NY
1957
12

25 ❑ 16-2
Jos. Schlitz
Brooklyn, NY
1960
10

26 ❑ 16-2
Jos. Schlitz
Milwaukee, WI
1954
10

27 ❑ 16-2
Jos. Schlitz
Milwaukee, WI
1954
10

28 ❑ 16-2
Jos. Schlitz
Milwaukee, WI
1957
10

29 ❑ 16-2
Jos. Schlitz
Milwaukee, WI
1960
7

30 ❑ 16-2
Jos. Schlitz
Milwaukee, WI
1962
7

31 ❑ 16-2
Jos. Schlitz
Milwaukee, WI
1962
6

32 ❑ 16-2
Atlas
Chicago, IL
50

PINTS

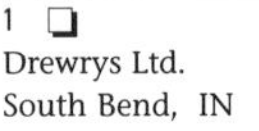

1 ❏ 16-2
Drewrys Ltd.
South Bend, IN
40

2 ❏ 16-2
Drewrys Ltd.
South Bend, IN
40

3 ❏ 16-2
Arizona
Phoenix, AZ
250

4 ❏ 16-2
Grace Bros.
Santa Rosa, CA
200

5 ❏ 16-2
Grace Bros.
Santa Rosa, CA
200

6 ❏ 16-2
Maier
Los Angeles, CA
225

7 ❏ 16-2
Atlantic
Chicago, IL
85

8 ❏ 16-1
Century
Norfolk, VA
175

9 ❏ 16-1
Century
Norfolk, VA
175

10 ❏ 16-2
Century
Norfolk, VA
350

11 ❏ 16-2
West End
Utica, NY
25

12 ❏ 16-2
Grace Bros.
Santa Rosa, CA
50

13 ❏ 16-2
Atlas
Chicago, IL
60

14 ❏ 16-2
Grace Bros.
Santa Rosa, CA
125

15 ❏ 16-2
Grace Bros.
Santa Rosa, CA
125

16 ❏ 16-2-F
Jacob Ruppert
New York, NY
70

PINTS

BIG CANS

24 & Tall 32 Oz. Flats

1 ❑ 32-2
P. Ballantine & Sons.
Newark, NJ
50

2 ❑ 32-2
P. Ballantine & Sons.
Newark, NJ
40

3 ❑ 32-2
P. Ballantine & Sons.
Newark, NJ
50

4 ❑ 32-2
P. Ballantine & Sons.
Newark, NJ
85

5 ❑ 32-2
P. Ballantine & Sons.
Newark, NJ
75

6 ❑ 24-2
Jos. Schlitz
Milwaukee, WI
75

7 ❑ 24-2
Jos. Schlitz
Milwaukee, WI
1956 **25**

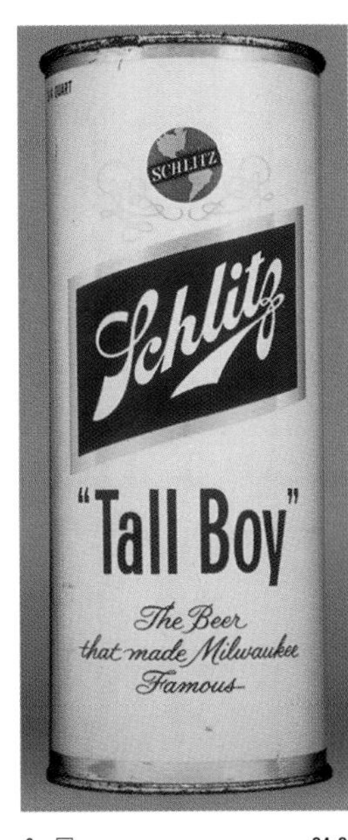

8 ❑ 24-2
Jos. Schlitz
Milwaukee, WI
1960 **7**

9 ❑ 24-2
Jos. Schlitz
Milwaukee, WI
1969 **5**

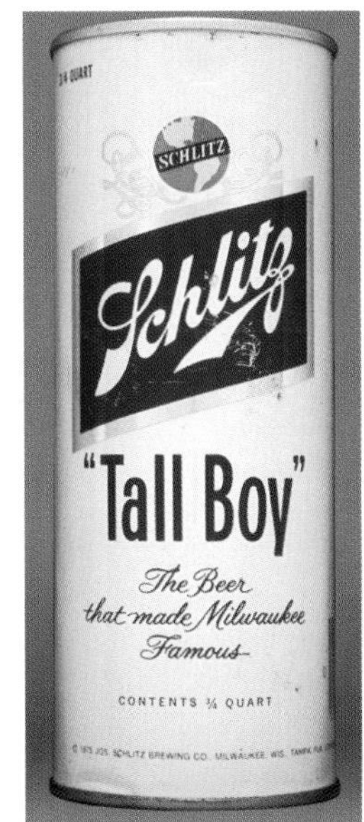

10 ❑ 24-2
Jos. Schlitz
Milwaukee, WI
1975 **5**

11 ❑ 24-2
Jos. Schlitz
Milwaukee, WI
1975 **5**

Little cans never made a very big splash in the United States. Crown Cork & Seal's efforts to introduce an 8-ounce can in 1937 went nowhere fast. And although a handful of brews (many of them malt liquors like the Country Club offering in this ad) came out in stubby 8-ounce versions in the 1950s, the beer drinker's reaction to half pints was only halfhearted at best.

SMALL CANS

$6^{1}/_{2}$, 7 & 8 Oz. Flats

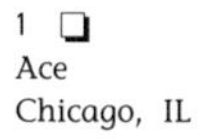
1 ❑ 7-1
Ace
Chicago, IL
75

2 ❑ 7-1
Ace
Chicago, IL
75

3 ❑ 8-1
Ace
Chicago, IL
35

4 ❑ 7-1
Ace
Chicago, IL
175

5 ❑ 7-1
Ace
Chicago, IL
100

6 ❑ 8-1
Ace
Chicago, IL
110

7 ❑ 8-1
Ace
Chicago, IL
110

8 ❑ 8-2
Pacific
Oakland, CA
750

9 ❑ 8-2
Grace Bros.
Santa Rosa, CA
18

10 ❑ 8-2
Atlas
Chicago, IL
70

11 ❑ 8-2
Atlas
Chicago, IL
60

12 ❑ 7-2
Atlas
Chicago, IL
70

13 ❑ 8-2
Atlas
Chicago, IL
18

14 ❑ 8-2
Drewrys Ltd.
South Bend, IN
18

15 ❑ 7-2
Canadian Ace
Chicago, IL
90

16 ❑ 7-2
Canadian Ace
Chicago, IL
90

17 ❑ 8-2
Canadian Ace
Chicago, IL
125

18 ❑ 8-2
Canadian Ace
Chicago, IL
100

19 ❑ 7-2
Canadian Ace
Chicago, IL
250

20 ❑ 7-2
Canadian Ace
Chicago, IL
250

21 ❑ 7-2
Adolph Coors
Golden, CO
Convex Bot., Sm. Type 5

22 ❑ 7-2
Adolph Coors
Golden, CO
Convex Bot., Lg. Type 5

23 ❑ 7-2
Adolph Coors
Golden, CO
Convex Bot., 3.2% bf 5

24 ❑ 7-2
Adolph Coors
Golden, CO
Kansas Tax, bf 5

25 ❑ 7-2
Adolph Coors
Golden, CO
Sm. Type 5

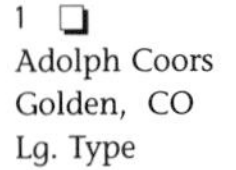
1 ❑ 7-2
Adolph Coors
Golden, CO
Lg. Type 5

2 ❑ 7-2
Adolph Coors
Golden, CO 5

3 ❑ 7-2
Adolph Coors
Golden, CO
Banquet Cream, 3.2% bf 5

4 ❑ 7-2
Adolph Coors
Golden, CO
Banquet Silver, 3.2% bf 5

5 ❑ 8-1
M. K. Goetz
St. Joseph, MO 20

Frt. of 7, 8

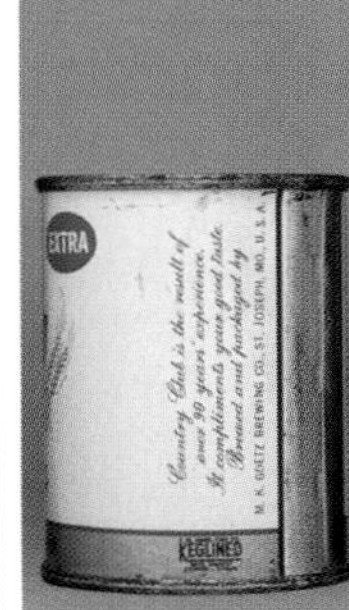
7 ❑ 8-1
M. K. Goetz
St. Joseph, MO
Over 90 Yrs. 20

8 ❑ 8-1
M. K. Goetz
St. Joseph, MO
Over 96 Yrs. 20

9 ❑ 8-1
M. K. Goetz
St. Joseph, MO 20

Frt. of 11-14

11 ❑ 8-1
M. K. Goetz
St. Joseph, MO 12

12 ❑ 8-1
M. K. Goetz
St. Joseph, MO 12

13 ❑ 8-1
M. K. Goetz
St. Joseph, MO
100 Yrs. 12

14 ❑ 8-1
M. K. Goetz
St. Joseph, MO
Over 100 Yrs. 12

15 ❑ 7-1
M. K. Goetz
St. Joseph, MO 25

Frt. of 17-19

17 ❑ 8-1
M. K. Goetz
St. Joseph, MO
Over 90 yrs. 18

18 ❑ 8-1
M. K. Goetz
St. Joseph, MO
Over 95 Yrs. 18

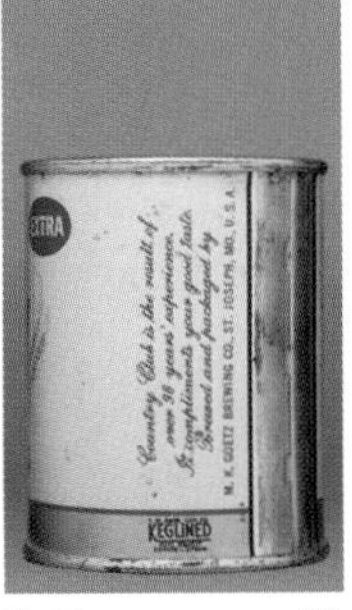
19 ❑ 8-1
M. K. Goetz
St. Joseph, MO
Over 96 Yrs. 18

20 ❑ 8-1
M. K. Goetz
St. Joseph, MO 12

Frt. of 22-26

22 ❑ 8-1
M. K. Goetz
St. Joseph, MO 12

23 ❑ 8-1
M. K. Goetz
St. Joseph, MO 12

24 ❑ 8-1
M. K. Goetz
St. Joseph, MO
100 yrs. 12

25 ❑ 8-1
M. K. Goetz
St. Joseph, MO
Over 100 Yrs. 12

26 ❑ 8-1
M. K. Goetz
St. Joseph, MO 12

Frt. of 28-30

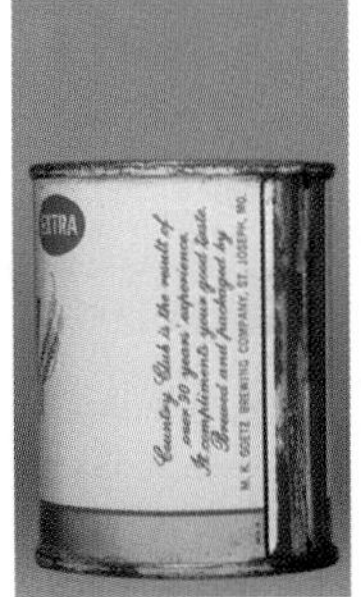
28 ❑ 8-1
M. K. Goetz
St. Joseph, MO
Over 90 Yrs. 25

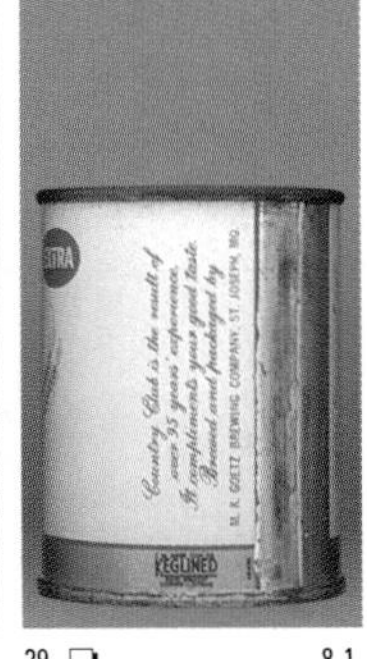
29 ❑ 8-1
M. K. Goetz
St. Joseph, MO
Over 95 yrs. 25

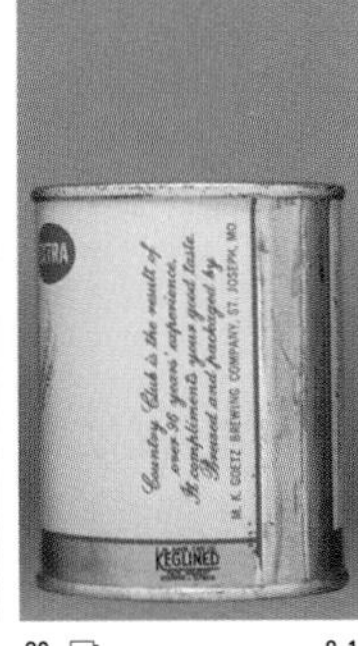
30 ❑ 8-1
M. K. Goetz
St. Joseph, MO
Over 96 Yrs. 25

31 ❑ 8-1
M. K. Goetz
St. Joseph, MO 25

Frt. of 33-37

33 ❑ 8-1
M. K. Goetz
St. Joseph, MO 10

34 ❑ 8-1
M. K. Goetz
St. Joseph, MO 10

35 ❑ 8-1
M. K. Goetz
St. Joseph, MO
100 Yrs. 10

36 ❑ 8-1
M. K. Goetz
St. Joseph, MO
Over 100 Yrs. 10

37 ❑ 8-1
M. K. Goetz
St. Joseph, MO 10

38 ❑ 8-2
Pearl
St. Joseph, MO 12

39 ❑ 8-2
Pearl
St. Joseph, MO 12

40 ❑ 8-2
Pearl
St. Joseph, MO 8

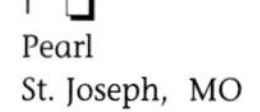

1 ❑ 8-2
Pearl
St. Joseph, MO
8

2 ❑ 8-2
Pearl
St. Joseph, MO
15

3 ❑ 8-2
Pearl
St. Joseph, MO
15

4 ❑ 8-1-OT
Peter Fox
Chicago, IL
750

5 ❑ 8-2
Gluek
Minneapolis, MN
"Malt Lager"
80

6 ❑ 8-2
Gluek
Minneapolis, MN
25

7 ❑ 8-2
Gluek
Minneapolis, MN
10

8 ❑ 8-1
Gluek
Minneapolis, MN
12

9 ❑ 8-1
Gluek
Minneapolis, MN
12

10 ❑ 8-1
Gluek
Minneapolis, MN
12

11 ❑ 8-2
G. Heileman
La Crosse, WI
8

12 ❑ 8-2
G. Heileman
La Crosse, WI
8

13 ❑ 8-1
Goebel
Oakland, CA
25

14 ❑ 8-1
Goebel
Detroit, MI
45

15 ❑ 8-1
Goebel
Detroit, MI
250

Frt. of 17, 18

17 ❑ 8-1
Goebel
Detroit, MI
25

18 ❑ 8-1
Goebel
Detroit, MI
25

19 ❑ 8-1
Goebel
Detroit, MI
45

20 ❑ 8-1
Goebel
Detroit, MI
45

21 ❑ 8-1
Goebel
Detroit, MI
Metallic Red
45

22 ❑ 8-1
Goebel
Detroit, MI
Enamel Red
45

23 ❑ 8-1
Goebel
Detroit, MI
30

24 ❑ 8-1
Goebel
Detroit, MI
30

25 ❑ 8-1
Goebel
Detroit, MI
50

26 ❑ 8-1
Goebel
Detroit, MI
40

27 ❑ 8-1
Walter
Pueblo, CO
35

28 ❑ 8-2
Pacific
Oakland, CA
200

29 ❑ 8-2
Grace Bros.
Santa Rosa, CA
50

30 ❑ 7-2
Gunther Brw. of Baltimore
Baltimore, MD
15

31 ❑ 8-2
G. Heileman
La Crosse, WI
30

32 ❑ 8-2
G. Heileman
La Crosse, WI
25

33 ❑ 8-2
G. Heileman
La Crosse, WI
30

34 ❑ 8-2
G. Heileman
La Crosse, WI
90

35 ❑ 7-2
General Corp.
San Francisco, CA
5

36 ❑ 7-2
General Corp.
San Francisco, CA
5

37 ❑ 7-2
General Corp.
San Francisco, CA
5

38 ❑ 7-2
Lucky Lager
San Francisco, CA
5

39 ❑ 7-2
Lucky Lager
San Francisco, CA
5

40 ❑ 7-2
Lucky Lager
San Francisco, CA
5

No.		Code	Brewery	Location	Note	Value
1	❑	7-2	Lucky Lager	San Francisco, CA		5
2	❑	8-2	Sterling	Evansville, IN		20
3	❑	7-1	National	Baltimore, MD		12
4	❑	8-2	L. F. Neuweiler's Sons	Allentown, PA		20
5	❑	7-2	Olympia	Olympia, WA		4
6	❑	8-2	Walter	Pueblo, CO		30
7	❑	8-2	Walter	Pueblo, CO		35
8	❑	8-2	Walter	Pueblo, CO		75
9	❑	7-2	Jacob Ruppert	New York, NY		10
10	❑	8-1	Jos. Schlitz	Milwaukee, WI	Paper Label	12
11	❑	8-1	Jos. Schlitz	Milwaukee, WI		12
12	❑	8-1	Jos. Schlitz	Milwaukee, WI	Paper Label	30
13	❑	8-1	Jos. Schlitz	Milwaukee, WI		10
14	❑	8-1	Jos. Schlitz	Milwaukee, WI	Paper Label	10
15	❑	8-1	Jos. Schlitz	Milwaukee, WI	Paper Label	15
16	❑	8-1	Pittsburgh	Pittsburgh, PA		200
17	❑	8-1	Storz	Omaha, NE		20
18	❑	8-1	Pittsburgh	Pittsburgh, PA		75
19	❑	8-1	Pittsburgh	Pittsburgh, PA		75
20	❑	8-1	Pittsburgh	Pittsburgh, PA		30
21	❑	8-1	Pittsburgh	Pittsburgh, PA		45
22	❑	8-1	Gettelman Div. of Miller	Milwaukee, WI		20
23	❑	8-1	Gettelman Div. of Miller	Milwaukee, WI		20
24	❑	8-1	Gettelman Div. of Miller	Milwaukee, WI		20

Gallons Section Explanatory Notes

Short/Typical Gallon Can - by National Can Company
Diameter with rim = 6 3/16
Height = 8 7/8

Tall Gallon - by Southern Can Company
Diameter with rim = 6 3/16
Height = 9 1/8

The height difference is all in the rims, 1/8 inch at both top and bottom.

The size of the actual can volume is identical.

GALLONS

64 & 128 Oz. (pre-1970s)

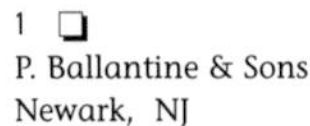

1 ❑ 128-2
P. Ballantine & Sons
Newark, NJ
125

2 ❑ 128-2
P. Ballantine & Sons
Newark, NJ
Metallic 75

3 ❑ 128-2
P. Ballantine & Sons
Newark, NJ
Enamel 75

4 ❑ 128-2
Blitz-Weinhard
Portland, OR
15

5 ❑ 128-2
Maier
Los Angeles, CA
225

6 ❑ 128-2
Maier
Los Angeles, CA
225

7 ❑ 128-2
Eastern Brewing Corp.
Hammonton, NJ
55

8 ❑ 128-2
National
Baltimore, MD
90

9 ❑ 128-2
National
Detroit, MI
100

10 ❑ 128-2
Gettelman Division of Miller
Milwaukee, WI
25

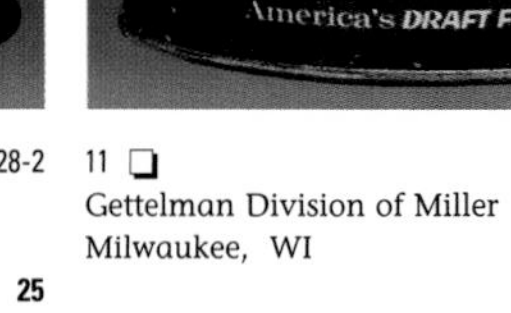

11 ❑ 128-2
Gettelman Division of Miller
Milwaukee, WI
600

12 ❑ 128-2
Maier
Los Angeles, CA
600

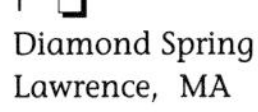
1 ❑ 128-2
Diamond Spring
Lawrence, MA
125

2 ❑ 128-2
Hudepohl
Cincinnati, OH
Short Can
30

3 ❑ 128-2
Hudepohl
Cincinnati, OH
Short Can
30

4 ❑ 128-2
Hudepohl
Cincinnati, OH
Tall Can
30

5 ❑ 128-2
Cold Spring
Cold Spring, MN
175

6 ❑ 128-2
G. Heileman
La Crosse, WI
40

7 ❑ 128-2
Fred Koch
Dunkirk, NY
600

8 ❑ 128-2
Fred Koch
Dunkirk, NY
Short Can
20

9 ❑ 128-2
Fred Koch
Dunkirk, NY
Tall Can
20

10 ❑ 64-2
Atlantic
Chicago, IL
200

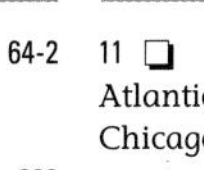
11 ❑ 128-2
Atlantic
Chicago, IL
65

12 ❑ 128-2
Associated
Evansville, IN
75

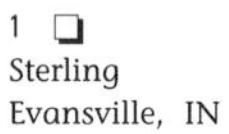

1 ❑ 128-2
Sterling
Evansville, IN
90

2 ❑ 128-2
Eastern Brewing Corp.
Hammonton, NJ
20

3 ❑ 128-2
Associated
Evansville, IN
125

4 ❑ 128-2
Maier
Los Angeles, CA
175

5 ❑ 128-2
Maier
Los Angeles, CA
175

6 ❑ 128-2
Jacob Ruppert
New York, NY
100

7 ❑ 128-2
Standard
Rochester, NY
25

8 ❑ 128-2
Standard
Rochester, NY
25

9 ❑ 128-2
Sterling
Evansville, IN
175

10 ❑ 128-2
Sterling
Evansville, IN
125

11 ❑ 128-2
Standard
Rochester, NY
25

12 ❑ 128-2
Standard
Rochester, NY
25

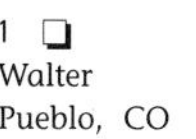
1 ❑ 128-2
Walter
Pueblo, CO
350

2 ❑ 128-2
Atlantic
Chicago, IL
1000+

3 ❑ 252 (1/16 BBL.) -1
Burger
Cincinnati, OH
1000+